CAMBRIDGE MONOGRAPHS
ON MECHANICS AND APPLIED MATHEMATICS

GENERAL EDITORS

G. K. BATCHELOR, PH.D., F.R.S.
Professor of Fluid Dynamics at the University of Cambridge
J. W. MILES, PH.D.
Professor of Applied Mathematics, University of California, La Jolla

THE ROTATION OF THE EARTH
A GEOPHYSICAL DISCUSSION

*Recipient of the Monograph Prize of the American
Academy of Arts and Sciences for the year 1959 in the
field of physical and biological sciences*

T0296970

Tidal friction (see Chapter 11). From Neugebauer (1957), Plate 6: Warka fragments.

THE
ROTATION OF THE
EARTH
A GEOPHYSICAL DISCUSSION

BY

WALTER H. MUNK
University of California

AND

GORDON J. F. MACDONALD

CAMBRIDGE UNIVERSITY PRESS
CAMBRIDGE
LONDON · NEW YORK · MELBOURNE

CAMBRIDGE UNIVERSITY PRESS
Cambridge, New York, Melbourne, Madrid, Cape Town, Singapore, São Paulo, Delhi

Cambridge University Press
The Edinburgh Building, Cambridge CB2 8RU, UK

Published in the United States of America by Cambridge University Press, New York

www.cambridge.org
Information on this title: www.cambridge.org/9780521104067

First published 1960
Reprinted with corrections 1975
This digitally printed version 2009

A catalogue record for this publication is available from the British Library

ISBN 978-0-521-20778-2 hardback
ISBN 978-0-521-10406-7 paperback

TO
SIR HAROLD JEFFREYS

CONTENTS

10 CHANDLER WOBBLE

11 HISTORICAL VARIATIONS

12 GEOLOGICAL VARIATIONS

PREFACE

'Disturbed by Newcomb's suspicions of the Earth's irregularities as a time-keeper, I could think of nothing but precession and nutation, and tides and monsoons, and settlements of the equatorial regions, and meltings of the polar ice' (Thomson, 1876). Lord Kelvin had just returned from a visit abroad to deliver the Presidential Address before the British Association. He was scheduled to speak on recent scientific progress in America. Instead he devoted his talk entirely to the rotation of the Earth. For us, too, the subject has been irresistible.

This book is an account of certain irregularities in the rotation of the Earth which are not ordinarily included in the gravitational theory. Such irregularities are a nuisance to the astronomer. They complicate his time-scale and limit the accuracy with which he can predict eclipses and other events. He has now circumvented the problem by legislation: by redefining time in terms of the length of year rather than the length of day.

It has become increasingly clear that irregularities in rotation are largely caused by events on, and in, the Earth; conversely, that such events can be effectively studied by means of the measured irregularities. It is the purpose of this book to make this method of study readily accessible to the geophysicist, and so make a geophysical asset out of an astronomical nuisance.

The astronomers were the first to attempt geophysical interpretations of the irregularities they had discovered. The earliest discussions usually appear in the form of brief geophysical supplements to the astronomical papers, and they reflect the faith that the relative simplicity of celestial mechanics can be carried over into the interpretation. Irregularities are fitted by a few sine waves, one speaks of the 'proper motion' of observatories, and the Himalayan complex is suddenly raised a foot. Unfortunately, the terrestrial mechanics is more involved than the celestial mechanics: 'Es sind . . . nicht mehr die einfachen Verhältnisse der Himmelsmechanik massgebend, sondern wir befinden uns hier bereits auf dem verschlungenen Gebiete der Geophysik' (Klein and Sommerfeld, 1903).

The subject-matter appears to have been fashionable during the late nineteenth century and was treated at length by Routh, George Darwin, Kelvin, and other great Victorians. During that era physicists were still aware of their external surroundings and had not yet become obsessed with the atom. The last part of Thomson and Tait's *Treatise on Natural Philosophy* contains the first systematic account. The subject was reopened in the light of modern geophysical knowledge by Jeffreys. His contributions dominate the subject.

At the moment the astronomical evidence poses a dozen related problems; these are summarized in ch. 1. Seven have been examined with some degree of success only during the last ten years, and new evidence has turned up for two others. Several problems considered as solved in the 1920's are now wide open. The entire subject needs to be tidied. In doing so we do not imply that the rate of progress in this field has leveled off. To the contrary, we expect new problems to be posed and solved problems to be unsolved at an accelerating rate. The introduction of the cesium frequency standard into the Time Service late in 1955 has already left its mark; moon cameras, satellites and computers add a new dimension.

The diversity of the subject is appalling. It touches on every branch of geophysics. By the time it is covered, information will have been gained concerning wind and air masses, atmospheric, oceanic and bodily tides, sea level, rigidity and anelasticity of the Earth's mantle, and motion in its fluid core. In each case the information is limited to certain integral quantities taken over the entire globe. This is the weakness of the method—and its strength. In principle such integrated quantities can be evaluated by appropriate summations of individual station values. For competitive accuracy one usually finds that the stations are too unevenly distributed, and too few. This is true now; we doubt whether it will ever be any different.

Here we have attempted to give only as much information about astronomical instruments and methods as is required for an intelligent use of the data. With regard to geodynamics, we have stated assumptions, sketched the derivations, and given the formulae for the actual calculations. Cited references may be helpful in further developments of the theory. The geophysical discussion is intended for a reader without special training in the various branches of this science. Errors and omissions will undoubtedly be found. We would appreciate hearing about them.

It is a pleasure to record our gratitude to those who have made helpful comments; among them F. Birch, H. Bondi, G. Clemence, C. Eckart, W. Elsasser, F. Gilbert, R. Haubrich, W. Markowitz, R. Revelle, L. Slichter and H. Urey. We are grateful to Gretchen Chambers and Janice Von Herzen for the preparation of the manuscript. Elizabeth Strong has taken a most active part in all phases of the investigations leading to this book. Our research (without which there would have been no incentive to prepare this book) has been generously supported by the Office of Naval Research, the National Science Foundation, the Guggenheim Foundation, and the Institute of Geophysics, University of California. Contrast this with the earliest American work pertinent to the subject: transit observations in an observatory surreptitiously smuggled into the Naval Depot of Charts and Instruments (Dupree, 1957, p. 62).

NOTATION

Only the symbols that appear throughout the book are here identified; a notation that is used in one section only is defined locally. Several symbols have different meanings in different chapters. This is impossible to avoid in a subject covering dynamics, geophysics, and astronomy. When a choice had to be made between an established convention and some degree of ambiguity, our decision was with convention.

Cartesian tensors are employed; subscripts i, j, k, l, m, are reserved for this purpose. Thus m_i designated m_1, m_2, m_3. But U_n is a potential of degree n, and *not* a tensor; nor is the load deformation ψ_L, nor the cross-correlation R_{uv}. Bold type indicates complex numbers, $\mathbf{m} = m_1 + im_2$. Some parameters appear in dimensional, dimensionless, complex, and operational forms: e.g., $\tilde{\mu}$ is the (dimensional) rigidity, μ a dimensionless rigidity, $\boldsymbol{\mu} = \mu_{\text{real}} + i\mu_{\text{imaginary}}$ a complex rigidity, and $\hat{\mu}$ a rigidity operator.

A, B, C	the principal moments of inertia; $C = 8 \cdot 068 \times 10^{44}$ g cm^2, $C - A = C - B = 2 \cdot 6 \times 10^{42}$ g cm^2
BIH	Bureau International de L'Heure (§ 9.1)
C_{ij}	inertia tensor (3.1.4)
$\hat{D}$	the operator d/dt
E	energy
ET	Ephemeris Time (§ 8.2, 11.2)
G	gravitational constant, $6 \cdot 670 \times 10^{-8}$ cm^3 g^{-1} sec^{-2}
GET	Newcomb's Great Empirical Term (§ 11.2)
H	precessional constant, $3 \cdot 273 \times 10^{-3}$ (2.3.1); also magnetic field strength
H_i	absolute angular momentum (3.1.3)
$H(x)$	Heaviside step function
IRM	isothermal remnant magnetization (§ 12.2)
ILS	International Latitude Service
K	kinetic energy
$K_{\mathbb{C}}$	general lunar coefficient (7.4.3)
$K\text{–}V$	Kelvin–Voigt body (§ 5.11)
L_i	torque

$L_\odot, L_\mathbb{C}$ L_3 due to Sun and Moon, respectively

$\Delta L_\odot, \Delta L_\mathbb{C}, \Delta L_\yen$ longitude discrepancies of Sun, Moon, and Mercury (11.2.1, 11.2.3)

$M, M_\oplus$ mass of Earth, $5{\cdot}976 \times 10^{27}$ g

M abbreviation for Maxwell body (§ 5.11)

NRM   natural remnant magnetization (§ 12.2)

O   order symbol; $y = 0(x)$ implies $\lim\limits_{x \to 0} |\,y/x\,| < \infty$

PZT   photographic zenith tube

Q (§ 4.3)

R autocorrelation (§ A.2)

S surface; dS surface element; also the power spectrum (§ A.2)

S_n surface spherical harmonic of degree n (5.10.1)

T length of record; also $ET_\odot$, the Ephemeris Time determined by the Sun

TRM   thermal remnant magnetization (§ 12.2)

U_i velocity vector relating to non-rotating coordinates

UT   Universal Time (§ 8.2, 11.2)

U_n, V_n, W_n spherical solid harmonic of degree n

U, V, W spherical solid harmonic of degree 2

V volume; dV volume element

WWD   weighted discrepancy difference (11.2.4)

X_i non-rotating coordinate

a radius of Earth, $6{\cdot}371 \times 10^8$ cm

a_{ij} anelastic strain (§ 4.1)

a_n^m, b_n^m spherical harmonic coefficients of degree n and order m

$a_\mathbb{C}, b_\mathbb{C}, c_\mathbb{C}$ terms in the longitude of the Moon; similarly for Sun and Mercury (11.2.3, 11.2.6)

b tidal amplitude factor (7.4.2)

$\mathbf{c}_n^m$ complex spherical harmonic coefficient $a_n^m + ib_n^m$

c_{ij} perturbations in inertia tensor (6.1.1)

c/s, c/year   cycles per second, cycles per year

d_{ij} total rate of strain (§ 4.1)

f frequency in cycles per unit time; as subscript, referring to 'fluid' (§. 5.4)

f_c	Coriolis frequency, $2\Omega \cos\theta$
f_0	resonance (or Chandler) frequency (1/14) cycles per month
f_N	Nyquist frequency (A.2.15)
$f_{\mathbb{C}}, f_{\odot}, f_{\mathbb{Q}}$	discrepancies in the longitudes of the Moon, Sun, and Mercury (11.2.3, 11.2.4)
g	acceleration of gravity, 980 g cm^{-2}
g_{ij}	metric strain tensor of material in the relaxed state (§ 4·1)
h, h'	Love numbers
h_i	relative angular momentum
i	imaginary unit
i, j, k, l, m	tensor subscript indices
k, l, k', l'	Love numbers
l.o.d.	length of day
m	a mass
m_i	direction cosines of rotation-axis
$\binom{m}{n}$	indicating order m and degree n of special harmonic
$n_{\mathbb{C}}, n_{\odot}, n_{\mathbb{Q}}$	orbital velocities of Moon, Sun, and Mercury
n_i	outwardly directed normal to a surface element, dS
p_n^m	spherical harmonic of order m and degree n (A.1.1)
p	hydrostatic pressure
p_{ij}	stress tensor
q	surface load, g cm^{-2}; also the angular velocity of the Earth relative to the 'mean Sun', 15° per mean solar hour
r	distance from center of Earth
$r(\tau)$	lag window (§ A.2)
s	d$s = \sin\theta\, d\theta\, d\lambda$ is surface element on unit sphere; also frequency divided by resonance frequency
s_{ij}	frictional stress
t	time
Δt	time interval between tabulated (or observed) values; also ET$_\odot$ − UT (11.2.2)
u_i	velocity vector relative to rotating coordinates
x_i	rotating coordinates
Γ	length of the mean solar day, 86,400 sec

Ψ'_i	modified excitation function (6.3.3)
Ω	mean diurnal rotation, 7.292×10^{-5} radians sec^{-1}
α, β, γ	damping factors
$\beta(\lambda, t)$	tidal phase (7.4.2)
δ_{ij}	Kroenecker delta or substitution tensor (§ 3.1)
$\delta(x)$	Dirac delta function
ε	ellipticity
ε_{ij}	elastic strain
ε_{ijk}	alternating tensor (3.1.2)
η	dynamic viscosity
θ	colatitude
ι	isostatic factor
κ	transfer function (§ 6.4); also electrical conductivity
λ	east longitude; also Lamé's constant
μ	rigidity
ν	kinematic viscosity; also degress of freedom (A.2.8)
ξ	elevation of sea surface above mean sea level
ρ	density; $\bar{\rho} = 5.53$ g cm^{-3} is mean density of Earth, $\rho_w = 1.025$ g cm^{-3} density of sea water
σ	frequency in radians per unit time
σ_0	resonance (or Chandler) frequency, $(2\pi/14)$ radians per month
σ_r	resonance (or Eulerian) frequency, $(2\pi/10)$ radians per month
τ	a time constant; also the amount by which the Earth is slow
τ_0	finite strength
τ_{ij}	elastic stress
υ	surface tension
ϕ_i	excitation function (§ 6.1, 6.3)
$\psi_i, \psi_{i(L)}, \psi_{i(D)}$	excitation functions for rigid Earth, load and rotational deformations (§ 6.3)
ω_i	angular velocity
$\mathscr{C}$	'global' functions defined in (A.1)
$\leftmoon \; \odot \; \mercury \; \oplus$	subscripts referring to Moon, Sun, Mercury, and Earth

☾   in cos ☾, sin ☾: the longitude of the 'mean Moon', (table 7.4.1)

☉   in cos ☉, sin ☉: the longitude of the 'mean Sun' measured from the beginning of the year (not from 21 March)

$\langle x \rangle$ time average of x

CHAPTER 1

PREVIEW

A preview of the treatment might be helpful. Following a qualitative discussion of the irregularities in rotation (ch. 2), the following four chapters are devoted to fundamentals. The solution to any
problem must satisfy (1) the equations governing the dynamics of
rotating bodies and (2) the equations governing the appropriate
stress–strain relations. In ch. 3 the dynamic equations are presented
in a form sufficiently general to impose no restrictions on deformation. The stress–strain relations are discussed in ch. 4. In most
problems the stress–strain relations can be introduced in the form of
dimensionless parameters, the Love numbers (ch. 5). Perturbation
methods are given in ch. 6. For a reading of the subsequent part of
this book, dealing with the observations and their interpretation,
the chapters on fundamentals may not be prerequisite.

The remaining chapters are devoted to a discussion of irregularities
in the rotation of the Earth. The irregularities fall into two cate-
gories: (1) a wobble of the Earth and (2) changes in the rate of
rotation or, more simply, changes in the length of day (l.o.d.). The
evidence with regard to wobble is outlined in the upper half of
fig. 1.1, with regard to l.o.d. in the lower half. It has been customary
to discuss problems involving wobble separately from those involving the l.o.d., possibly because the methods of observation are
so different. But the two subjects are closely related, and much is
gained when they are treated side by side.

The methods of observation are dealt with in chs. 7 and 8. The
remaining chapters present a discussion of the irregularities organized
according to frequency. Ch. 9 deals with irregularities with a
time-scale on the order of one year or less. Evidence for the wobble
comes from observations by the International Latitude Service
(ch. 7), for the l.o.d. from a comparison of astronomic time with
precise clock time (ch. 8). The annual wobble is largely due to
seasonal shifts in air mass. The annual variation in the l.o.d. is
caused by winds, shorter period terms by bodily tides. Ch. 10

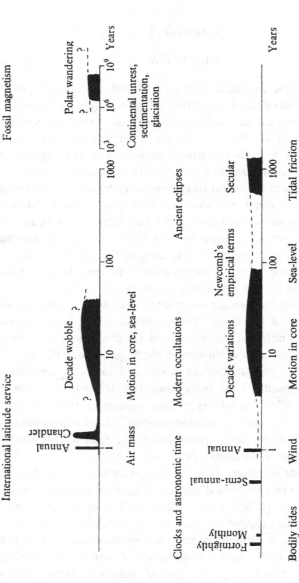

Fig. 1.1. The spectrum of rotation. The *wobble* components (*top*) and *length of day* components (*bottom*) are schematically arranged according to their time-scale in years. Vertical lines indicate discrete frequencies, shaded portions indicate a continuous, or noisy, spectrum. Principal source of the observations is shown above lines, presumable geophysical cause beneath lines.

deals with the 14-month Chandler wobble, the period of which is governed by the ellipticity and rigidity of the Earth; the wobble is generated by random impulses of unknown origin and damped by some unknown imperfections from elasticity, or by some other means. The longer periods are discussed in ch. 11. The evidence from latitude observations is inconclusive. Evidence for the l.o.d. comes largely from modern occultations and ancient eclipses. Remarkably large irregular variations in the l.o.d. with a decade time-scale may be due to electromagnetic coupling of the mantle to a turbulent, fluid core. Fluctuations with a century time-scale (Newcomb's empirical terms) may be due to changes in the Earth's moment of inertia. Changes over the last few thousand years are predominantly the result of tidal friction, but here again changes in inertia (presumably associated with a variable sea-level) must play an important part.

Ch. 12 deals with wobble on a geologic time-scale. Paleomagnetism and other indirect evidence is sometimes interpreted to indicate very large displacements of the pole during the last few hundred million years. Vertical unrest of continents and convective motion in the Earth's mantle have been suggested as causes. For a discussion of this problem the dynamics has to be extended to the case of an anelastic Earth.

The clearcut distinction made in the literature between the regular variations of short period (annual, semi-annual etc.) and the irregular variations of long period is not suitable from the geophysical point of view. The short-period spectral 'lines' must be imbedded in a continuous spectrum caused by weather; long-period spectral lines, such as the one associated with the 18.6-year tide, must be superimposed on the continuous spectrum of the variations. The concept of the 'power spectrum' is central to our discussion. A brief account of the method is given in the Appendix.

The literature on the subject is enormous, but there appears to have been no previous attempt to give a unified account of the spectrum of both wobble and rotation. The first systematic account of the wobble problem appeared in the *Treatise on Natural Philosophy* (Thomson and Tait, 1883) and in the *Theorie des Kreisels* (Klein and

Sommerfeld, 1903). Jeffreys's (1959) book *The Earth* contains a summary of short-period wobble in ch. 7, and of the effect of tidal friction on long-period changes in the l.o.d. in ch. 8. Essentially the same subject-matter is covered in ch. 16 and by ch. 6 of vol. II, *Physics of the Earth Series* (Lambert, 1931; Lambert *et al.* 1931). Changes in the l.o.d. of all periods are discussed by Spencer-Jones (1956) in the *Handbuch der Physik* and in ch. 1 of *The Earth as a Planet* (Kuiper, 1954), and by Gondolatsch (1953). Melchior (1957) has given the most up-to-date summary of the wobble problem.

CHAPTER 2

PRECESSION, NUTATION AND WOBBLE

1. Wobble and precession

Consider time-exposures of stars with a camera pointing vertically upward, that is, opposite to local gravity.* On the photographs the star trails appear as portions of concentric circles. For two positions on the Earth, the *poles of rotation*, the concentric circles would be centered on the photographs. The *celestial poles* vertically above the poles of rotation are those two points in the sky where a star would have no diurnal motion. (Polaris is near the celestial north pole.) The *rotation axis* (or *celestial axis*) extends through the poles of rotation from one celestial pole to the other.

If the positions are checked after a month, it will be found that the poles of rotation have moved a few feet from their previous positions. By the time a year has passed, the poles will have moved in a roughly elliptical path with a mean diameter of twenty feet.

A stake is driven in the ground somewhere near the center of the ellipse. It marks the *pole of reference*. The *axis of reference* extends from the center of the Earth through this pole. The pole of rotation revolves about the pole of reference. With respect to an observer on a fixed star, the rotation axis remains fixed, and the reference pole (averaged for diurnal motion) revolves about the rotation pole.

In the discussion so far the celestial pole has been considered as fixed in its position near Polaris.† But over a period of many years the celestial pole shifts appreciably. In 5000 years it will be near α Cephei; 5000 years ago it was near α Draconis and the southern cross could be seen in England. These changes in the orientation of the rotation axis *in space* are associated with the precession of the equinox. There are also shorter-period changes in the position of the celestial pole, the forced nutations. Such changes in the orientation of the Earth's rotation axis in space are altogether different from the wobble of the Earth relative to this axis.

* The small effect of tidal deflexions of the local vertical is discussed in § 7.4.
† Neglecting 'sway,' § 6.6.

It would be unfortunate if wobble and precession could be detected only by an observer at the pole. In practice the measurements of appropriate angles at any latitude will do. The angles which are to be determined are the declination of a star and the latitude of a place. Fig. 2.1 shows the situation for the idealized cases of wobble only, and precession (or forced nutation) only. The celestial pole P is near Polaris; S is some star, and its polar distance or co-declination (90°—declination) is SOP; A is some fixed place on Earth and ZA is the direction of local gravity. The colatitude (90°—latitude) of A

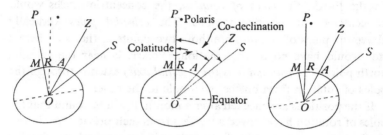

Fig. 2.1. Precession (*left*) and wobble (*right*). The undisturbed position is shown at the center. The cut is at right angles with the equatorial plane and through the Earth's center O. S is some star. A is some fixed place and Z its zenith. P is the north celestial pole. M the pole of reference and R the pole of rotation.

is defined by the angle ZOP between the celestial pole and the zenith. The left figure illustrates precession. The celestial pole has shifted away from Polaris. This changes the declination but not the latitude. The right figure illustrates wobble. The pole of reference M has shifted to the left of the pole of rotation R. The declination is unchanged, but now the latitude is altered. Hence precession is determined from declination; wobble is determined from latitude. A brief account of the instruments and methods actually used is given in ch. 7.

2. Causes of precession and forced nutation

Changes of the Earth's rotation axis in space are caused largely by the pull of the Moon and Sun on the Earth's equatorial bulge. There would be no such effect if the Earth were spherical; nor if the equatorial plane were coincident with the plane of the Sun's orbit (the ecliptic plane) and the plane of the Moon's orbit. Actually the equatorial plane is inclined by $23\frac{1}{2}°$ to the ecliptic, and the inclination to the Moon's orbit is not far different. If it were not for the Earth's rotation, the action of Sun and Moon would be to bring these planes into coincidence. But because of the gyroscopic effect of rotation the action is at right-angles to what one would otherwise expect. The obliquity of the ecliptic remains near $23\frac{1}{2}°$, and the celestial pole describes a circle about the pole of the ecliptic in 26,000 years, a motion known as the precession of the equinox. From the observed precession and the mass of the Moon one can evaluate the precessional constant

$$H = \frac{C - A}{C} = 0.00327293 \pm 0.00000075, \qquad (2.2.1)$$

where A, A, C are the moments of inertia about the Earth's principal axes, C being taken about the axis of greatest moment (Jeffreys, 1952, p. 145).

There is a slight additional precession due to other planets. The complex interplay of the Sun's and Moon's orbits is associated with oscillations of shorter periods, among them the 19-yearly lunar nutation. The shorter periods involve chiefly a motion of the celestial pole to and from the ecliptic pole, a nodding which gives the motion its name, *nutation*.*

We shall not consider precession and forced nutation any further. They are treated in many textbooks (for example, Routh, 1905, ch. 11). But we do not wish to imply that there are no remaining problems of geophysical interest. For example, a genuine discrepancy by one part in 600 between the observed amplitude of the 19-yearly nutation and that computed for a rigid Earth has been explained by allowing for the fluidity of the core together with certain modifications in the elastic properties of the mantle (Jeffreys and Vicente, 1957a, b).

3. Wobble and length of day

Rotation can be represented by a vector drawn parallel to the rotation axis and proportional in length to the rate of rotation. Relative to a reference system suitably attached to the Earth (§ 3.2),

---

* The 'forced nutations' are not to be confused with the Chandler wobble (ch. 10) which is often referred to as 'free nutation' or 'Eulerian nutation'.

the x_1, x_2-components then designate the wobble, and the x_3-component (drawn roughly parallel to the reference axis) is associated with the length of the sidereal day. Wobble can be expressed either as an angular displacement of the rotation axis relative to the axis of reference, or as a linear displacement of the pole of rotation relative to the pole of reference. For comparison,

$$0\overset{''}{\cdot}0100 \quad \text{and} \quad 1\cdot01 \text{ ft} \qquad (2.3.1)$$

are equivalent, a relationship that makes it convenient to use centiseconds of arc and feet of displacement as interchangeable units.

The sidereal day is defined as the time-interval between successive transits of a star across the celestial meridian (a great circle through the celestial poles and the zenith). Measurements of the l.o.d. require a good clock in addition to a proper telescope. Measurements of wobble require only the determination of an angle. In principle, measurements of l.o.d. and wobble are quite independent, but in practice the segregation is not so clear-cut (7.4.8, 8.3.5).

CHAPTER 3

DYNAMICS*

1. Fundamental equations

The Eulerian equations of motion in a coordinate system x_i ($i = 1, 2, 3$), rotating with angular velocity ω_i relative to coordinates X_i fixed in space and coinciding with x_i for the moment, are

$$L_i = \frac{dH_i}{dt} + \varepsilon_{ijk}\omega_j H_k. \tag{3.1.1}$$

L_i are the components of torque, H_i those of angular momentum. ε_{ijk} is the 'alternating' tensor, defined by the following properties:

$$\begin{aligned}
\varepsilon_{ijk} = &\ 0 \text{ if any two subscripts are equal,} && i = j, i = k, j = k,\\
= &\ + 1 \text{ if subscripts are in even order,} && 1, 2, 3, 1, 2 \ldots,\\
= &\ - 1 \text{ if subscripts are in odd order,} && 1, 3, 2, 1, 3 \ldots
\end{aligned}$$
$$\tag{3.1.2}$$

By the usual summation convention, in any expression containing a repeated suffix, that suffix is to be given all possible values and the results then added.

Equations (3.1.1) are quite general. They refer, for example, to a system of particles moving among themselves. It is convenient to separate angular momentum into two parts:

$$H_i = C_{ij}(t)\omega_j + h_i(t), \tag{3.1.3}$$

where

$$C_{ij} = \int_V \rho(x_k x_k \delta_{ij} - x_i x_j)\, dV \tag{3.1.4}$$

is the (variable) tensor of inertia for matter contained in a volume V, and δ_{ij} is the Kronecker delta (or substitution tensor), with the properties $\delta_{ij} = 1$ if $i = j$, $\delta_{ij} = 0$ if $i \neq j$. The second part of (3.1.3) designates a relative angular momentum

$$h_i = \int_V \rho\varepsilon_{ijk}x_j u_k\, dV \tag{3.1.5}$$

---

* For a more complete treatment the reader is referred to the classic works of v. Oppolzer (1886), Tisserand (1891), and Klein and Sommerfeld (1903). A treatment by Woolard (1953) has been found helpful.

due to motion u_i relative to the x_i-system. Substituting (3.1.3) into (3.1.1) leads to

$$L_i = \frac{\mathrm{d}}{\mathrm{d}t}(C_{ij}\omega_j + h_i) + \varepsilon_{ijk}\omega_j(C_{kl}\omega_l + h_k). \qquad (3.1.6)$$

All subsequent inquiries concerning irregularities in rotation take the form of special solutions to (3.1.6). This equation was given by Liouville in 1858 (Routh, 1905: § 22), and it will be referred to as the Liouville equation.

2. Frames of reference

Routh (1905, § 22) and other textbooks distinguish between the rotating reference-axes in the Eulerian equations (3.1.1) and the 'body axes' from which changes in the body can be described (3.1.4, 5). The two rotating axes can be combined without loss of generality, as we have done, at the expense (possibly) of not making the best use of the symmetry of the situation.

The choice of the rotating x_i-system is altogether arbitrary. It could, for example, rotate at some rate in a sense opposite to the Earth's rotation. For convenience the coordinate axes must now be attached to the Earth in some way. In most papers on the subject the coordinate system is said to rotate 'with the Earth'. If the Earth were rigid, there would be no further difficulty. Winds, ocean currents, and the fluid core introduce complications. To get around this difficulty the axes can be attached to the 'solid' Earth. However, there are tidal distortions and, for processes on a geologic time-scale, convective motion in the mantle has been postulated; in all events we are faced with relative motion of different parts of the crust. Such motion is known to take place along geologic faults and has been postulated by Wegener on a grandiose scale as a drift of continents with respect to one another.

Clearly we require a set of rigid axes that are kinematically defined so as not to impose any restraints on the deforming Earth. There are a number of possible choices.

(1) *Tisserand's mean axes of body* are defined so that $h_i = 0$. Thus if winds, ocean currents, and all other relative motion ceased, these axes would rotate with the resulting frozen body. An alternate

and equivalent definition is instructive. For a rigid body rotating with angular velocity ω_i the velocity of any particle in space is $(\mathrm{d}X_i/\mathrm{d}t) = \varepsilon_{ijk}\omega_j X_k$. For a deformable body we can select some value of ω_i, say $\bar{\omega}_i$, which makes

$$\int_V \rho \left(\frac{\mathrm{d}X_i}{\mathrm{d}t} - \varepsilon_{ijk}\omega_j X_k\right)^2 \mathrm{d}V$$

a minimum. For the $\bar{\omega}_i$-axes it can be shown that $h_i = 0$, so that $\bar{\omega}_i$ is the angular velocity of the mean axes. Jeffreys (1952, p. 206) refers his calculations to the mean axes.

(2) *The principal axes, or axes of figure*, are defined so that the products of inertia C_{ij}, $i \neq j$, vanish. Darwin (1877) chose the principal axes for his treatment of polar wandering.

The derivation in § 3.1 refers to any set of rigid rotating axes, and thus includes the mean axes and principal axes as special cases. The conditions $h_i = 0$ and $C_{ij} = 0$, $i \neq j$, respectively, lead to considerable simplifications of (3.1.6). These are the obvious choices for mathematical simplicity. But there are disadvantages to these fundamental axes. Winds and other relative motion rotate the mean axes slowly relative to the observatories, and it may be necessary to correct the observed values. The principal axes also shift relative to the observatories. Moreover, Jeffreys's choice of the mean axes makes it awkward to include the effects of relative motion, and he does ignore such effects. The variable angular momentum of the atmosphere shifts the mean axes relative to the equatorial bulge and accordingly would produce variations in C_{ij} even if the Earth were rigid. We have found it expedient to use

(3) the '*geographic*' axes which are attached in a 'prescribed way' to the observatories. There are difficulties due to relative motion of the observatories. For many problems the relative motion can be neglected. Stations of the International Latitude Service have moved rather well together, the attempt having been made to locate them away from active faults. If the relative motion is not negligible, then we choose a set of rigid axes which are attached in some prescribed way to the observatories. Geophysical observations, astronomical observation, the relative motion of the observatories, equations (3.1.1–6), everything is referred to these axes. We may wish to

contemplate a least-squares fit to stakes driven in the ground at suitable intervals over the whole Earth, a procedure that avoids a bias from the positioning of observatories and gets rid of tidal distortions to a first order.

For all three systems only the *motion* of the axes has been defined. The choice remains of determining the position. It is convenient to place the origin at the Earth's center of mass so that $\int \rho x_i \, dV$ vanishes, to align the x_3-axis reasonably well with the axis of figure (or rotation), to draw the x_1-axis through the meridian of Greenwich, and the x_2-axis towards 90° east of Greenwich.

3. Further discussion of the Liouville equation

The reader will find references to half a dozen papers where incorrect results have been obtained because of the misinterpretation or omission of terms in the fundamental relation (3.1.6). To avoid such pitfalls, it may be worth while to discuss further what the terms mean and what they do not mean.

(1) L_i is an exterior torque acting on the body contained within the volume V. The surface enclosing this volume can be chosen at will. For example, in considering the effect of winds we may exclude the atmosphere, in which case there will be an exterior torque resulting from wind stress; or we may include the entire planet Earth and set $L_i = 0$. The torque approach and the momentum approach lead, of course, to identical results (see § 6.9 for some sample calculations), the choice depending on instrumentation: is it easier to measure the wind stress or the angular momentum of the atmosphere?

(2) L_i are the components of torque along the *rotating* axes x_i. The torque exerted by steady winds blowing along a *fixed* longitude has components that do not vary in time, $L_i \neq f(t)$. A torque fixed in space, on the other hand, produces components that do vary in time with diurnal frequency Ω. For example, radiation pressure from a fixed star must differ slightly over the Earth's two hemispheres, and the resulting torque will have variable components such as $L_1 = \cos \Omega t$, $L_2 = \sin \Omega t$.

(3) The derivation has referred only to a time 0 when the rotating x_i-system and the fixed X_i-system coincided for the moment. At

this instant the components L_i, h_i, and C_{ij} in the x_i and X_i-systems are the same, and the derivation can refer to the inertial frame of reference X_i in the usual way. But for any subsequent time, t', the derivation is identical provided a fixed X_i'-system is chosen which coincides with the rotating x_i-system at time t'. With the understanding that components of torque, momentum, and inertia are always taken along the x_i-system, equations (3.1.6) are valid for all times.

(4) The interpretation of dC_{ij}/dt and dh_i/dt requires consideration. It is easier, and safer, first to perform the integrations (3.1.4, 5) inherent in the definitions of C_{ij} and h_i, and then to differentiate the time-dependent integrals. As previously stated, the integrals are taken over a volume enclosed by some material surface S. If the differentiation is to be performed first, then certain complications arise due to distortion of S with time. If the enclosing surface is fixed relative to the reference coordinates, $S = S(x_i)$, one must allow for the flux of inertia and relative angular momentum through the boundary S.

(5) The quantities h_i and C_{ij} depend on the fields of density $\rho(x_k, t)$ and relative velocity $u_i(x_k, t)$. In our equations ρ and u_i appear as independent variables. There are, of course, restraints imposed by the conservation of mass, momentum and energy and by the equations of state. For example, there can be no change in the density field $\rho(t)$ unless there is some motion $u_i(t)$. It would be possible to transform the equations so that the conservation principles are automatically satisfied. We have taken the point of view that for each geophysical application the fields of ρ and u_i are to be specified in accordance with the proper physical laws, but there is nothing to prevent the reader from working out examples which imply, for example, creation and destruction of matter.

(6) With L_i, h_i, and C_{ij} thus specified, the equations can be solved for the angular velocity $\omega_i(t)$ of the reference system x_i relative to the fixed system X_i. Let y_i designate a coordinate system rotating with the same angular velocity $\Omega = (\omega_i \omega_i)^{\frac{1}{2}}$ as the x_i-system, but with y_3 directed always along the axis of instantaneous rotation. Then

$$x_i = \frac{\omega_i}{\Omega} y_3, \qquad (3.3.1)$$

so that ω_i/Ω are the direction cosines of the rotation axes relative to the reference axes. In all problems except polar wandering we align the x_3-axis nearly parallel to the y_3-axis so that $d\omega_3/dt$ is nearly the acceleration in the diurnal rotation and ω_1, ω_2 the components of wobble.

CHAPTER 4

DEFORMATION

The real difficulty of our subject concerns the deformation of the Earth under prescribed stresses. For problems involving wobble with periods of the order of one year or shorter, it is customary to allow for a purely elastic deformation of the Earth by the introduction of the appropriate Love numbers. The next chapter is devoted to this topic. By this choice of notation various problems can be solved with considerable ease; but this ease is deceptive inasmuch as it hides the fact that the corresponding problem in elasticity had to be first solved to make the Love number available. For the damping of the Chandler wobble and problems involving a longer time-scale than one year, the importance of the deformation problem becomes explicit, and some deviation from perfect elasticity in the Earth's mantle may have to be allowed for. The choice of a meaningful anelastic stress-strain model involves much speculation, to say the least. Precise solutions are hardly possible and probably not worthwhile at the present state of the art. But some general statements concerning rotational stability can be made from energy considerations.

The problem of stress and strain plagued Kelvin and George Darwin. Kelvin assumed the Earth could be treated as an elastic body even for long-period deformations, while Darwin assumed that the Earth behaves plastically, even for small stresses. Today the situation is identical to that of 1900, and proponents of the Darwin and of the Kelvin view can be found. There are few problems in geophysics in which less progress has been made.

1. Stress, strain and deformation

A fiber is stretched by the application of an external force. On release of the external force the fiber may return at once to its original length, or it may return very slowly, or it may assume a new, greater, length. If the fiber returns at once to its initial length on the release of the external stress, then the response is termed elastic. Whether or not a given material responds elastically to a given external force depends not only on the nature of the material but also on the magnitude of the external force and possibly on its duration.

In discussing the general problem of the deformation of solids, it is essential to distinguish between the elastic (or recoverable) strain and the total deformation (Eckart, 1948). In one-dimensional problems, such as the deformation of a fiber by an external force applied parallel to its length, the elastic strain ε is defined as

$$\varepsilon = \frac{s - g}{g},$$

where s is the actual distance between two points of the fiber, and g is the distance between the same two points at an instant after removal of external forces.

The rate of change of one-dimensional elastic strain is

$$\frac{d\varepsilon}{dt} = (1 + \varepsilon) \left(\frac{1}{s}\frac{ds}{dt} - \frac{1}{g}\frac{dg}{dt} \right). \qquad (4.1.1)$$

$\frac{ds}{dt}$ is the actual rate at which two points in the fiber move apart. $\frac{dg}{dt}$ is the rate of change of the length between the two points at zero external force. $\frac{1}{s}\frac{ds}{dt}$ is termed the rate of (total) deformation while $\frac{1}{g}\frac{dg}{dt}$ is the rate of anelastic deformation. The rate of deformation nearly equals the rate of elastic strain, provided the strain and the rate of anelastic deformation are small:

$$\frac{d\varepsilon}{dt} \approx \frac{1}{s}\frac{ds}{dt} \quad \text{for} \quad \begin{cases} \varepsilon \ll 1 \\ \dfrac{1}{g}\dfrac{dg}{dt} \ll \dfrac{1}{s}\dfrac{ds}{dt}. \end{cases}$$

The classical theory of infinitesimal elasticity is based on these two assumptions.

The rate of elastic strain (4.1.1) vanishes if the rates of total deformation and of anelastic deformation both vanish, or if the rate of total deformation equals the rate of anelastic deformation:

$$\frac{1}{s}\frac{ds}{dt} = \frac{1}{g}\frac{dg}{dt}.$$

In the latter case the length of the fiber changes at a finite rate, yet the elastic strain remains constant. The rate of total deformation may vanish so that there is no motion, yet the elastic strain may change provided the rate of anelastic deformation is finite.

The notion of the anelastic rate of deformation can be carried over into three dimensions (Eckart, 1948; Knopoff and MacDonald, 1958). The relations are relatively simple if the elastic strains are small compared with unity. In Earth problems the total elastic strain is often very large because of high hydrostatic pressure. However, the large total strain can be thought of as made up of two parts, a large initial strain due to self attraction of the material elements and a small strain due to small changes in the forces acting on the system. We will be interested only in the effect of the small superimposed strains. Since the concept of total deformation plays no role in the development, there is no limit on the magnitude of the deformation. We require only that the superimposed elastic strain be small.

In three dimensions the total rate of deformation d_{ij} is defined by

$$d_{ij} = \frac{1}{2}\left(\frac{\partial u_i}{\partial x_j} + \frac{\partial u_j}{\partial x_i}\right),$$

where u_i are the velocity components and d_{ij} is a symmetric tensor. The elastic strain is defined by

$$\varepsilon_{ij} = \tfrac{1}{2}(\delta_{ij} - g_{ij}),$$

where g_{ij} is the metric tensor of the material in the relaxed state, free from external stress. The rate of anelastic deformation a_{ij} is then

$$a_{ij} = d_{ij} - \frac{d\varepsilon_{ij}}{dt}$$

provided the elastic strains are small. A material is termed anelastic if $a_{ij} \neq 0$.

The total stress p_{ij} acting on matter within the Earth consists of three parts:

$$p_{ij} = -p\delta_{ij} + \tau_{ij} + s_{ij}.$$

p is the hydrostatic pressure; τ_{ij} is the elastic stress, given by

$$\tau_{ij} = \lambda\varepsilon_{kk}\delta_{ij} + 2\mu\varepsilon_{ij},$$

where λ and μ are elastic constants; s_{ij} is the frictional stress. The form of s_{ij} depends on the frictional model. For example, in a Kelvin–Voigt body (§ 5.11) s_{ij} is assumed to depend on the rate of deformation, according to

$$s_{ij} = \lambda_{K-V} d_{kk} \delta_{ij} + 2\mu_{K-V} d_{ij}.$$

The only stress acting on an Earth in hydrostatic equilibrium is the hydrostatic pressure. The magnitude of the pressure at any point within the Earth is determined by the distribution of density within the Earth. In all problems we will assume the Earth initially in hydrostatic equilibrium, $p_{ij} = -p_0 \delta_{ij}$, and then examine small deviations from this equilibrium condition:

$$p_{ij} = -p_0 \delta_{ij} + \lambda \varepsilon_{kk} \delta_{ij} + 2\mu \varepsilon_{ij} + s_{ij}.$$

The description of the relation between stress and deformation is incomplete unless we describe the dependence of the rate of anelastic deformation on the stress. The exact nature of this dependence is not at all clear. There are, however, certain limiting conditions imposed by experimental studies. There is no anelastic deformation resulting from a hydrostatic stress, provided the material is homogeneous (Bridgman, 1949, p. 149). The rate of anelastic deformation should thus be independent of the hydrostatic pressure. If the applied stresses are non-hydrostatic, then behavior of the material depends on the magnitude of the stress differences. For stress differences below a critical value (ranging in rocks from 10^2 to 10^4 bars) crystalline solids behave elastically in short-time laboratory experiments. Stresses exceeding critical stress difference lead to fracture or plastic flow. The response of crystalline materials to small stresses imposed for long periods of time is a subject of some controversy, and the existence of a finite strength for such prolonged stresses cannot be considered as established by experiment.

In a Maxwell body (§ 5.11) the relation between the rate of anelastic deformation and the elastic stress is

$$a_{ij} = \lambda_M \tau_{kk} \delta_{ij} + 2\mu_M \tau_{ij}.$$

The experimentally imposed requirement that $a_{ii} = 0$ results in $\lambda_M = -\frac{2}{3}\mu_M$, hence

$$a_{ij} = 2\mu_M(\tau_{ij} - \tfrac{1}{3}\tau_{kk}\delta_{ij}). \tag{4.1.2}$$

Such a relation implies that permanent or plastic deformation is possible even for vanishingly small non-hydrostatic stresses. If the material shows finite strength, (4.1.2) is valid only for stress differences exceeding some critical value. We adopt the convention of labeling the three principal stresses of the tensor τ_{ij} as τ_1, τ_2, τ_3, with $\tau_1 \leqslant \tau_2 \leqslant \tau_3$. For materials with finite strength τ_0, the deformation tensor a_{ij} is given by (4.1.2) only if $|\tau_3 - \tau_1| > \tau_0$; otherwise $a_{ij} = 0$. This simple criterion will be sufficient for the problems discussed in this book. Many other criteria are possible (Nadai, 1950).

2. Energy and stability

The kinetic energy of the planet Earth is

$$K = \tfrac{1}{2} \int_V \rho U_i U_i \, dV,$$

where U_i are the components of the velocity relative to the fixed space-axes X_i and V is a volume enclosing that part of the universe which we regard as the Earth. In terms of the components of velocity u_i relative to the rotating axes x_i, the kinetic energy is

$$K = \tfrac{1}{2} \int_V \rho u_i u_i \, dV + \int_V u_i \varepsilon_{ijk} \omega_j x_k \, dV + \tfrac{1}{2} \int_V \rho[\omega_i^2 x_i^2 - (\omega_i x_i)^2] \, dV$$
$$= k + \omega_i h_i + \tfrac{1}{2} H_i H_j (C_{ij})^{-1}, \tag{4.2.1}$$

where $C_{ik}(C_{kj})^{-1} = \delta_{ij}$. In stability problems it is convenient to refer to Tisserand's axes (§ 3.2), for then $\omega_i h_i$ vanishes and all the energy of relative motion is contained in the relative kinetic energy k, a positive definite quantity. A growth of a disturbance will manifest itself in the increase of k. If the Earth can be considered free from the action of external forces, the total angular momentum is constant, and the total kinetic energy can change only through changes in k and in the moment of inertia C_{ij}.

The principle of conservation of energy applied to a rotating body free of internal heat sources and thermally and mechanically isolated from the rest of the universe leads to

$$\frac{d}{dt}(k + W) = -\int_V \tau_{ij} a_{ij} \, dV - \int_V s_{ij} d_{ij} \, dV, \tag{4.2.2}$$

where
$$W = H_i H_i (C_{ij})^{-1} + P + E$$

and P is the potential energy associated with body forces, E the elastic energy. The rate of change of elastic energy consists of two parts:

$$\frac{\mathrm{d}E}{\mathrm{d}t} = -\int_V \frac{1}{\rho} p_0 \frac{\mathrm{d}p}{\mathrm{d}t}\,\mathrm{d}V + \int_V \tau_{ij} \frac{\mathrm{d}\varepsilon_{ij}}{\mathrm{d}t}\,\mathrm{d}V.$$

The first part designates the effect of compression; the second part, the effect of the perturbation stresses τ_{ij}.

The terms on the right of (4.2.2) represent the energy dissipated by anelastic deformation and frictional resistance, respectively. By the second law of thermodynamics both dissipation terms are positive, thus

$$\frac{\mathrm{d}}{\mathrm{d}t}(k + W) \leqslant 0.$$

The stability characteristics of an anelastic rotating body are contained in this equation. We give the following results without proof.

(1) A freely rotating solid, free from frictional stresses and rotating with an angular velocity such that the maximum stress difference is less than the strength of the solid. Then $s_{ij} = 0$, $a_{ij} = 0$; hence $\frac{\mathrm{d}}{\mathrm{d}t}(k + W) = 0$ and $k + W = $ constant. At equilibrium the relative kinetic energy k vanishes and W is a constant, zero, say:

$$k = 0, \quad W = 0 \text{ at equilibrium.}$$

The motion is perturbed through the action of an internal heat source or by an external force. Then k or W or both are non-vanishing, and

$$k + W = \text{constant.}$$

W is an absolute minimum provided the rotation-axis coincides with the axis of greatest moment of inertia; k is positive definite and cannot increase indefinitely at the expense of W. A small perturbation can lead only to harmonic motion with energy periodically interchanged between kinetic, rotational, potential, and elastic energy. The system is totally stable.

(2) Friction, with W an absolute minimum. For small motions conservation of energy leads to

$$k + W = \text{constant} - \alpha t,$$

where α is a positive constant. The effect of friction is to damp out the perturbations, and the motion is said to be secularly stable (Lyttleton, 1953). Anelasticity does not alter the conditions for secular stability and contributes only to the dissipation.

(3) W not an absolute minimum (e.g., rotation about the axis of least moment of inertia). Without anelasticity, a slight perturbation will lead to oscillations (possibly damped) about the equilibrium position, and the motion is said to be ordinarily stable. With anelasticity present a perturbation will grow at a rate initially described by

$$k + W = \text{constant} - \beta t,$$

where β is a positive constant, with W decreasing until it finally reaches an absolute minimum. Such secular instability (in the sense used by Lyttleton) can arise in anelastic materials, provided a given disturbance results in non-vanishing elastic stresses and rates of anelastic deformation.

As a simple example, consider a planet in free rotation about the axis of greatest moment of inertia such that W is an absolute minimum. The external form of the planet is assumed in equilibrium with the angular velocity of rotation. A meteorite lands on the planet at some point other than the equator. The mean position of the rotation axis is slightly displaced and W is no longer an absolute minimum (the moment of inertia is maximized only if the added mass is at the greatest possible distance from the rotation-axis). The small displacement of the rotation axis creates stresses within the planet, and no further displacement of the axis will occur if the elastic stresses nowhere exceed the strength of the material, i.e., a_{ij} is everywhere zero. If a_{ij} does not vanish, even for small stresses, then the body is secularly unstable, and a large motion will result with the equatorial bulge adjusting to the changing (with respect to the mean axis) position of the rotation axis (§ 12.6, 12.8).

3. Generalized Q

In many problems we will use the 'specific dissipation function'

$$\frac{1}{Q} = \frac{1}{2\pi E} \oint \frac{dE}{dt}\, dt \qquad (4.3.1)$$

as a dimensionless measure of the rate at which energy is dissipated in a vibrating system. Here $\oint(\mathrm{d}E/\mathrm{d}t)\,\mathrm{d}t$ is the energy dissipated over a complete cycle, and E is the peak energy stored in the system during a cycle. The quantity $1/Q$ is particularly useful in discussions of observational data since it does not depend on the detailed mechanism by which energy is dissipated.

Consider the case of a linear, damped oscillator:

$$\frac{\mathrm{d}^2x}{\mathrm{d}t^2} + 2\alpha\,\frac{\mathrm{d}x}{\mathrm{d}t} + \sigma_0^2 x = \sigma_0^2 a \cos \sigma t.$$

The specific dissipation turns out to be

$$\frac{1}{Q} = \frac{2\alpha}{\sigma_0}\frac{\sigma}{\sigma_0} \qquad (4.3.2)$$

and depends therefore on frequency. For the special case of a forced oscillation at resonance frequency, $\sigma = \sigma_0$, we write

$$\frac{1}{Q_0} = \frac{2\alpha}{\sigma_0}, \qquad (4.3.3)$$

and this corresponds to the usual definition in electrical circuit theory. Accordingly, $Q_0 a$ is the amplitude at resonance. The sharpness of resonance is indicated by

$$\frac{1}{Q_0} = \frac{2\Delta\sigma}{\sigma_0}, \qquad (4.3.4)$$

where $\sigma_0 \pm \Delta\sigma$ are the circular frequencies at the half-power points (amplitude $2^{-\frac{1}{2}}Q_0 a$). The phase lag ϕ is given by

$$\tan \phi = \frac{\sigma\sigma_0}{\sigma^2 - \sigma_0^2}\frac{1}{Q_0} = \frac{\sigma_0^2}{\sigma^2 - \sigma_0^2}\frac{1}{Q}. \qquad (4.3.5)$$

The free oscillation decays according to

$$e^{-\alpha t} \cos \sigma_0 t. \qquad (4.3.6)$$

CHAPTER 5

LOVE NUMBERS AND ASSOCIATED COEFFICIENTS

If the Earth were rigid we could at once apply the Liouville equation (3.1.6) to compute changes in rotation arising from specified geophysical events. For a deformable Earth the equation still holds, but allowance must be made for secondary effects such as the yield of the Earth under superficial loads and the shift in the equatorial bulge resulting from changes in rotation. Such mass shifts surely must be taken into account together with the specified shifts which brought about the disturbance in the first place. In return for this complication, a comparison of geophysical and astronomical observations will give information concerning the elastic (and anelastic) properties of the Earth.

Deformations result from *body* forces, such as tidal forces or the centrifugal force associated with rotation, and from *surface* stresses, such as atmospheric pressure or wind drag. A sudden load applied to the surface will initiate elastic waves traveling with speeds of the order of kilometers per second. Fundamental modes of free vibrations of the Earth associated with such waves have periods of the order of one hour. If the period of the forcing function is long compared to this, then it can be assumed that the elastic deformations follow instantly and are given by static considerations. The oceans and fluid core have response times much longer than the free vibrations, and there is some question as to the frequency below which static theory applies.

1. The Love numbers* h, k and l

Consider the response of the Earth to a disturbing potential $U(r)$ of degree 2; tidal forces of the Moon and Sun, and centrifugal forces

* The numbers as written above all refer to deformations which are spherical harmonics of degree 2; hence $k = k_2$. These are by far the most important. But there will be occasional need for the numbers k_n, k'_n, h_n, l_n etc., that arise from disturbances of any degree n. The notation l was introduced by Lambert, and is implicit in the discussion by Toshi Shida.

arising from rotation can be written as gradients of such a potential U. The resulting deformation defines the Love numbers as follows: the ground is lifted by hU_{surface}/g, and the additional gravitational potential at the (displaced) surface arising solely from this redistribution of mass is kU. Thus $1 + k$ is a factor allowing for the attraction of the bulge by itself, and the response by hU/g takes this self-attraction into account. A fluid surface covering the globe would remain equipotential and be lifted by $(1 + k)U/g$ relative to the centre of the Earth and by $(1 + k - h)U/g$ relative to the sea bottom.

In addition to the vertical displacement of the solid surface by hU/g there is a horizontal displacement with components

$$\frac{l}{g}\frac{\partial U}{\partial \theta}, \quad \frac{l}{g}\frac{1}{\sin\theta}\frac{\partial U}{\partial \lambda}, \tag{5.1.1}$$

where θ is colatitude and λ east longitude.

The Love numbers are dimensionless parameters which neatly summarize some of the elastic properties of the Earth. Their evaluation belongs to the subject of elasticity. Information comes from a great variety of sources. The great advantage of writing the equation in terms of these parameters is that the equations can readily be adapted to any future improvement of our knowledge concerning the elastic behavior of the Earth.

The most detailed calculations are those by Takeuchi (1950) on the basis of certain distributions of density and elastic properties within the Earth as derived from seismic and other evidence. His results are (Jeffreys, 1952, p. 213)

$$\left.\begin{array}{lll} k_T = 0 \cdot 290, & h_T = 0 \cdot 587, & l_T = 0 \cdot 068 \\ k_T = 0 \cdot 281, & h_T = 0.610, & l_T = 0 \cdot 082 \end{array}\right\} \tag{5.1.2}$$

for two models proposed by Bullen. There are other methods of measuring Love numbers, as will be shown, and for various reasons the results are not immediately comparable (§ 10.4 and § 10.8).

2. Rotational deformation

We consider the distortion of the Earth due to any potential, U, of degree 2. The distortion gives rise to an exterior gravitational potential kU, by definition of k. But the gravitational potential near the boundary of a body departing slightly from spherical symmetry is given by MacCullagh's formula (Jeffreys, 1952: § 4.025–2 and 7.04–13). In the present case the deformation is a spherical harmonic of degree 2, and the pertinent terms in Mac-Cullagh's formula can be written $(Gm/r) + V$, with

$$V = \frac{G}{2r^5}(-3C_{ij} + C_{kk}\delta_{ij}). \tag{5.2.1}$$

Consider the special case of the centrifugal potential, which equals $\frac{1}{2}\omega^2$ times the square of the distance from the rotation-axis, or

$$\tfrac{1}{2}[\omega^2 r^2 - (\omega_i x_i)^2], \quad \omega^2 = \omega_i \omega_i, \quad r^2 = x_i x_i.$$

This can be organized into the terms $\frac{1}{3}\omega^2 r^2 + U$, where

$$U = \tfrac{1}{6} x_i x_j(-3\omega_i \omega_j + \omega_k \omega_k \delta_{ij}) \tag{5.2.2}$$

is a spherical harmonic of degree 2. The term $\frac{1}{3}\omega^2 r^2$ leads to a purely radial deformation which consists of a contraction near the center of the Earth and an extension in the outer parts (Love, 1927, p. 143). Substitution of (5.2.2) into (5.2.1) yields

$$C_{ij} = I\delta_{ij} + \frac{ka^5}{3G}(\omega_i \omega_j - \tfrac{1}{3}\omega^2 \delta_{ij}), \tag{5.2.3}$$

where $I = \frac{1}{3}(C_{11} + C_{22} + C_{33})$ is the inertia of the sphere in the absence of rotational deformation. Increased diurnal rotation increases C_{33} and diminishes C_{11} and C_{22} in accordance with $\delta C_{11} = \delta C_{22} = -\frac{1}{2}\delta C_{33}$, so that I remains unchanged. For a homogeneous ellipsoid this is readily demonstrated from the definition (3.1.4).

3. The secular Love number

The Love number k in (5.2.3) can be interpreted as a measure of the Earth's yield to centrifugal deformation in the course of its development during the last five billion years or so. Without loss of generality we can place the x_3-axis along the rotation vector. Then $\omega_1 = 0$, $\omega_2 = 0$, $\omega_3 = \Omega$, and

$$C_{11} = C_{22} = A = I - \frac{k_s a^5}{9G}\Omega^2, \quad C_{33} = C = I + \frac{2k_s a^5}{9G}\Omega^2, \tag{5.3.1}$$

so $$k_s = \frac{3GHC}{a^5\Omega^2} = \frac{\frac{3}{2}\beta H}{\varepsilon_f}, \quad \beta = \frac{C}{Ma^2}, \quad \varepsilon_f = \frac{\frac{1}{2}\Omega^2 a}{g}, \quad (5.3.2)$$

where H is the precessional constant (2.2.1). If all mass were concentrated at the center, $C = 0$ and hence $k_s = 0$. For a homogeneous sphere $C = \frac{2}{5}Ma^2$, and with $M = 5.98 \times 10^{27}$ g for the mass of the Earth, one obtains $k_s = 1.14$. The correct value lies between these limits. From the value of H and the international gravity formula one obtains (Jeffreys, 1952, p. 145)

$$C = 0.3336\,Ma^2, \quad (5.3.3)$$

as compared to $C = 0.4\,Ma^2$ for the homogeneous Earth, and thus

$$k_s = 0.96. \quad (5.3.4)$$

4. The fluid Love numbers

The foregoing calculation of k involves the observed rate of precession and the shape of the ellipsoid derived from gravity measurements. There have been no assumptions concerning the stress–strain relations within the Earth.

We next compute the 'fluid' Love number k_f based on the assumption that the Earth is in hydrostatic equilibrium, i.e., that it has the shape of a rotating fluid with a density distribution equal to that of the actual Earth. To a first order the surface ellipticity is then given by

$$\varepsilon = h_f(\tfrac{1}{2}\Omega^2 a/g). \quad (5.4.1)$$

If all the mass were concentrated at the center, $h_f = 1$ and $\varepsilon = 1/580$. For a homogeneous Earth it has been shown by Kelvin that $h_f = \frac{5}{2}$, so that $\varepsilon = 1/232$. The observed ellipticity, 1/297, lies between these values. Using the observed value of ε gives

$$h_f = (580/297) = 1.96. \quad (5.4.2)$$

But, for a fluid surface, $h_f = 1 + k_f$ (§ 5.1); thus

$$k_f = 0.96, \quad (5.4.3)$$

which is equal numerically to the secular Love number, k_s.

A more precise argument is given by Bullard (1948): from the observed precession and Bullen's distribution of density within the Earth, Bullard obtains $\varepsilon^{-1} = 297.338 \pm 0.050$, *assuming hydrostatic equilibrium.* The eccentricity can be independently derived without the assumption of hydrostatic equilibrium, either from gravity observations or from the motion of the Moon. The resultant values, $296 \cdot 17 \pm 0 \cdot 68$ and $296 \cdot 72 \pm 0 \cdot 65$, do not differ significantly from the preceding value. The most recent analysis of gravity observations (Heiskanen, 1957; Heiskanen and Vening Meinesz, 1958) gives $297 \cdot 0$ to $297 \cdot 2$, and these are in even better agreement with the hydrostatic value. According to these observations, the shape of the actual Earth does not differ measurably from that of an equivalent rotating fluid. There might be a difference of as much as one-third per cent and in fact the satellite observations indicate that this is so (§ 12.8). Such a difference, if it were real, would be a certain measure of the Earth's finite strength to resist deformation from stresses, no matter for how long a time these stresses are applied (§ 12.8). In the case of no finite strength $k_f = k_s$ exactly for sustained stresses, and there is no stability in the above sense. The question whether $k_f \approx k_s$ or $k_f = k_s$ is clearly an important one.

5. The tidal-effective Love numbers

From studies of Earth tides and the Chandler wobble one obtains $h = 0 \cdot 59$, $k = 0 \cdot 29$. The close agreement with the values (5.1.2) derived by Takeuchi from seismic evidence is to some extent misleading (§ 10.3). There is a marked contrast with the values $h_s = 1 \cdot 96$, $k_s = 0 \cdot 96$ obtained from the figure of the Earth. A number of hypotheses may explain the difference, and it is not known which of these (if any) is correct: one hypothesis is based on the relative magnitude of stresses, with the secular Love numbers referring to stress differences above a critical strength, tidal-effective Love numbers to stress differences below the critical strength. A second hypothesis is based on the relative duration of stresses, with the secular Love numbers referring to stresses much more prolonged than some critical duration, tidal-effective Love numbers to stresses much less prolonged than this critical duration. Still another possibility is that the Earth was originally molten and that it now has the figure at the time of congealment. The agreement between k_s and k_f then implies little or no changes in rotation. This is extremely unlikely (§ 11.9). But there can be no doubt that the Earth responds far differently to ordinary tide potentials and the annual wobble than it responds to diurnal rotation. We may regard h, k, as the asymptotic case of generalized Love numbers when the frequency is high and the disturbance infinitesimal; h_s, k_s, as the asymptotic values for low frequency and large amplitude. An example of generalized Love numbers is discussed in § 5.11.

6. The 'equivalent' Earth

Kelvin has shown (Jeffreys, 1952, pp. 211, 363–4) that, for an incompressible homogeneous sphere of rigidity $\tilde{\mu}$,

$$h = \frac{5/2}{1 + \mu}, \quad k = \frac{3/2}{1 + \mu}, \quad l = \frac{5/4}{1 + \mu}; \quad \mu = \frac{19}{2}\frac{\tilde{\mu}}{\rho g a}. \quad (5.6.1)$$

Kelvin's solution yields the fluid Love numbers $h_f = 2\cdot5$, $k_f = 1\cdot5$, compared to the observed values $h_f = 1\cdot96$, $k_f = 0\cdot96$ for the Earth (there is no observed value for l_f). For a better fit to actual conditions the simplest procedure is to adopt the relations

$$h = \frac{h_f}{1 + \mu}, \quad k = \frac{k_f}{1 + \mu}, \quad (5.6.2)$$

which are of the same form as Kelvin's solution but use the observed values $h_f = 1\cdot96$, $k_f = 0\cdot96$. The usefulness of this model of an equivalent Earth depends on how well the two known values h and k can be computed by an appropriate choice of the single parameter μ. The set of values

$$\mu = 2\cdot3, \quad h = 0\cdot59, \quad k = 0\cdot29 \quad (5.6.3)$$

are in excellent accord with the best estimates of tidal-effective rigidity. Accordingly, this value $\mu = 2\cdot3$ is a satisfactory measure of the tidal-effective rigidity of the Earth.* By inference we can roughly estimate the fluid Love number l_f to be $0\cdot98$ as compared to $5/4$ for homogeneous Earth; for, with this value and $\mu = 2\cdot3$, one obtains

$$l = \frac{l_f}{1 + \mu} = 0\cdot07 \quad (5.6.4)$$

in agreement with Takeuchi's calculation.

* Kelvin's famous observation that the effective rigidity of the Earth is nearly that of steel is based on a value $\mu = 4\cdot1$.

7. Love numbers of degree n

We shall have occasion to refer to Love numbers of any degree n. Their definitions are given simply by writing* U_n in place of $U(= U_2)$ in § 5.1. For a homogeneous incompressible sphere the formulae are†

$$h_n = \frac{n + \frac{1}{2}}{n - 1} \frac{1}{1 + \mu N}, \quad k_n = \frac{\frac{3}{2}}{n - 1} \frac{1}{1 + \mu N},$$

$$l_n = \frac{n + \frac{1}{2}}{n(n - 1)} \frac{1}{1 + \mu N}, \tag{5.7.1}$$

where
$$N = \frac{2(2n^2 + 4n + 3)}{19n}. \tag{5.7.2}$$

For the case $n = 2$, $N = 1$ and (5.7.1) reduces to (5.6.1).

8. Load deformation and the coefficients h', k'

Consider the effect of a variable surface load $q(t)$ g cm^{-2}, e.g., a blanket of snow. We are concerned with the term q_n of degree n in the surface spherical harmonic expansion of q. The interior potential resulting from a gravitating layer q_n is

$$U_n = \frac{4\pi G q_n a}{2n + 1} \left(\frac{r}{a}\right)^n. \tag{5.8.1}$$

The deformation is due to two opposing effects. First there is a normal stress (positive upwards) $p_n = -gq_n$ which will depress the surface. Then there is the gravitational attraction of the Earth by the snow, and this will raise the ground. By the combined effect of pressure and attraction the surface is raised by $h'_n U_n/g$ and the gravitational potential arising from the distortion is $k'_n U_n$. This defines h'_n, k'_n. The depression is found to be somewhat larger than the gravitational uplift, and h'_n, k'_n are negative, as expected.

The case of an incompressible homogeneous sphere serves as a simple example. It has been shown by Chree (1889) that the effect of any radial stress, p_n, is equivalent to that of a body potential, $U_n = p_n/\rho$. Writing $p_n = -gq_n$, $3g = 4\pi Ga\rho$, the combined potentials of the gravitating layer, of the deformation arising from the

---

* Thus U_n is not a tensor; n is not a tensor subscript. See index of notation.
† These formulae for Love numbers are not found in Love's great work on elasticity; they have to be extracted from five equations in five unknowns (Love 1927, Section 177).

'load potential' p_n/ρ, and of the deformation arising from the gravitational attraction are

$$U_n + k_n \frac{p_n}{\rho} + k_n U_u = U_n(1 + k'_n),$$

hence

$$k'_n = \left(\frac{p_n}{\rho U_n} + 1\right) k_n = \left(-\frac{2n+1}{3} + 1\right) k_n = -\tfrac{2}{3}(n-1)k_n,$$

$$(5.8.2)$$

and similarly for h'_n. Thus if one has obtained the Love numbers k_n, h_n due to a potential that does not load the Earth, such as the rotational and tidal potential, the response to one that does is given by

$$h'_n, k'_n = [-\tfrac{2}{3}(n-1)] \, h_n, k_n. \qquad (5.8.3)$$

Combining this with (5.7.1) gives

$$h'_n = -\tfrac{2}{3}(n+\tfrac{1}{2}) \frac{1}{1 + \mu N}, \quad k'_n = -\frac{1}{1 + \mu N}, \qquad (5.8.4)$$

$$h'_2 = h' = -\frac{\tfrac{5}{3}}{1 + \mu}, \quad k' = -\frac{1}{1 + \mu}. \qquad (5.8.5)$$

At first sight it appears surprising that the two opposing effects of load depression and gravitational uplift are of the same order, $-\tfrac{5}{3}$ and $+1$ for the case $n = 2$. The local distortion by loads of limited extent, corresponding to harmonics of high degree n, is determined by load depression rather than gravitational uplift in the ratio $\tfrac{2}{3}n$. This may be the reason why we are unaccustomed to the importance of gravitational uplift when $n = 2$. The two opposing effects cancel when $n = 1$, and this is closely related to the fact that an expansion, into spherical harmonics, of the gravity of a spheroid yields no term for $n = 1$ (Jeffreys, 1952, 4·026). This term is equivalent to a rigid body displacement and does not affect surface gravity. For $n = 0$, N becomes infinite, hence $h'_0 = 0$, $k'_0 = 0$: a uniform load over an incompressible Earth causes no deformation.

9. Load deformation, second order

It will be noted that for a fluid Earth $\mu = 0$, and $k_f' = -1$. Thus the potential of a surface load after yielding is $(1 + k_f')U = 0$. The

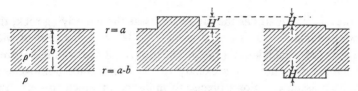

Fig. 5.1. A layer of density ρ' extends from the surface at $r = a$ to an interface at $r = a-b$ (*left*). A thickening by H' (*center*) leads to yielding of the layer by an amount H (*right*).

inertia terms are proportional to $(1 + k_f')U$ and vanish also. But potential and inertia cannot be identically zero. The inertia of an ocean covered by icebergs is larger than after the bergs have melted. It is necessary to allow for the slight difference in the distances of the icebergs and of the displaced water from the rotation axis.

The undisturbed surface is at a distance a from the Earth's center (fig. 5.1). A block formed out of material from a superficial layer leads to an increase in mass per unit area of $\rho'H'$. The center of mass of this block is at a distance $a + \frac{1}{2}H'$ from the Earth's center *before yielding*. Again we are concerned only with surface harmonics of degree 2, so that $q_2 = \rho'H_2' = \rho'H'$. The exterior potential before yielding is

$$U_I = \tfrac{4}{5}\,\pi Gr^{-3}(\rho'H')(a + \tfrac{1}{2}H')^4.$$

In the event (H'/a) approaches 0 this reduces to the first order expression (5.8.1) for the case $n = 2$. After yielding by an amount H we are left with a load (positive or negative) $\rho'(H' - H)$ at an average distance $a + \frac{1}{2}H' - \frac{1}{2}H$ and a load $(\rho' - \rho)H$ at $a - b - \frac{1}{2}H$. The corresponding potential of degree 2 is

$$U_F = \tfrac{4}{5}\pi Gr^{-3}[\rho'(H' - H)(a + \tfrac{1}{2}H' - \tfrac{1}{2}H)^4$$
$$+ (\rho' - \rho)H(a - b - \tfrac{1}{2}H)^4].$$

But $-H_2 = h'U_I/g$ and $U_F = (1 + k')U_I$ at $r = a$, by definition of h', k'. For a homogeneous Earth (except for the surface load) $3g = 4\pi Ga\rho$ and $h' = -\tfrac{5}{3}(1 + \mu)^{-1}$; furthermore b, H, H' are all small compared to a and H, H' small compared to b. One finds after some manipulation

$$k' = -\frac{1-\iota}{1+\mu}, \quad \iota = 4\frac{b}{a}\left(1-\frac{\rho'}{\rho}\right), \quad 1+\mu = \frac{\rho'H_2'}{\rho H_2}. \quad (5.9.1)$$

In the case the load has the same density as the underlying rock, the *isostatic* factor ι vanishes, and $k' = -(1+\mu)^{-1}$ in agreement with the first order calculation (5.8.5). The numerical factor in the expression for ι does not apply exactly to the Earth since homogeneity has been assumed. For a precise formulation in which conservation of mass is explicit a development along the lines of § 5.12 is indicated.

10. Fluid Earth with surface tension

Another case of interest is that of a thin elastic crust over a fluid Earth. Perhaps this case is adequately covered by considering a liquid sphere with a surface tension $\tilde{v}$ over and above that arising from sphericity; presumably the crust is not in tension except when distorted.

Let U_2 be any second degree disturbing potential. The distorted surface is at $r = a(1 + \varepsilon S_2)$, where

$$S_n = p_n^m (\cos\theta)(\cos m\lambda, \sin m\lambda), \quad (m = 0 \text{ to } n) \quad (5.10.1)$$

is the surface spherical harmonic and p_n^m are the associated Legendre functions (A.1.1). The potentials arising from the excess surface tension and from the distortion of the sphere are

$$V = -\frac{3\tilde{v}}{\rho r^2} a\varepsilon S_2, \quad W = \tfrac{4}{3}\pi Ga^3\left(\frac{1}{r} + \frac{3a^2\varepsilon S_2}{5r^3}\right)$$

respectively. For a homogeneous sphere, $3g = 4\pi Ga\rho$; $U_2 + V_2 + W_2$ = constant at the distorted surface, and the terms containing S_2 give

$$h = \frac{(U_2/g)_{\text{surf}}}{a\varepsilon S_2} = -\frac{(V_2 + W_2)_{\text{surf}}}{ga\varepsilon S_2} = \frac{5/2}{1+v}; \quad k = \frac{3/2}{1+v}, \quad (5.10.2)$$

where

$$v = \frac{15}{2}\frac{\tilde{v}}{\rho ga^2} \quad (5.10.3)$$

is a dimensionless surface tension which enters in the same way as the dimensionless rigidity μ in (5.6.1). Similarly h' and k' are obtained from (5.8.5) by replacing μ with v.

11. Love operators and complex Love numbers

In the study of polar wandering and of the damping of the Chandler wobble, we shall refer to solutions where the Earth is considered a Maxwell body or Kelvin–Voigt body (M or K–V). In a Maxwell (or elastoviscous) body the total rate of deformation is written as a sum of an elastic and a viscous term (§ 4.1):

$$\frac{1}{s}\frac{\mathrm{d}s}{\mathrm{d}t} = \frac{1}{2\tilde{\mu}}\frac{\mathrm{d}}{\mathrm{d}t}(\tau_{\text{elastic}}) + \frac{1}{2\tilde{\eta}_M}(\tau_{\text{elastic}}), \qquad (5.11.1)$$

where $\tilde{\mu}$ is the rigidity, $\tilde{\eta}$ the dynamic viscosity, and τ_{elastic} the elastic stress. In a Kelvin–Voigt (or firmoviscous) body the total stress is written as a sum of an elastic stress and a viscous stress:

$$\tau_{\text{elastic}} = 2\tilde{\mu}\varepsilon + 2\tilde{\eta}_{K-V}\frac{\mathrm{d}\varepsilon}{\mathrm{d}t}. \qquad (5.11.2)$$

The Kelvin–Voigt body is distinguished by the fact that there is no permanent strain associated with deformation. The dissipation results from an added stress proportional to the rate of strain; in the Maxwell body the dissipation results from permanent deformation. The Maxwell body is often portrayed as a spring and dashpot in series; the Kelvin–Voigt body as a spring and dashpot in parallel. Love operators can be written for any combinations of springs and dashpots.

Once the elastic problem has been solved, the appropriate solutions for the M and K–V bodies can be found by replacing the dimensionless rigidity* μ with the operators

$$\hat{\mu}_M = \frac{\mu\hat{D}}{\hat{D} + \tau^{-1}}, \quad \hat{\mu}_{K-V} = \mu(1 + \tau\hat{D}), \qquad (5.11.3)$$

respectively. Here $\tau = \tilde{\eta}/\tilde{\mu}$ is the characteristic time constant, and $\hat{D}$ the operator d/dt (Jeffreys and Jeffreys, 1950, chs. 7 and 8). The Love operators

$$\hat{k} = \frac{k_f}{1 + \hat{\mu}}, \quad \hat{k}' = -\frac{1 - \iota}{1 + \hat{\mu}} \qquad (5.11.4)$$

* The operators refer to the dimensionless rigidity μ (5.6.1) rather than to the dimensional rigidity $\tilde{\mu}$. We shall not use an operator such as $\tilde{\mu}(1 + \tau\hat{D})$.

are a convenient notation. For the case of simple harmonic motion $e^{i\sigma t}$, the operator $\hat{D}$ becomes $i\sigma$, and $\hat{\mu}$, $\hat{k}$, $\hat{k}'$ become the complex numbers $\tilde{\mu}$, $\tilde{k}$, $\tilde{k}'$

For elastic waves, the generalized "Q" (§4.3) becomes

$$\frac{1}{Q_M} = \frac{\tilde{\mu}}{\tilde{\eta}_M}\frac{1}{\sigma}, \quad \frac{1}{Q_{K-V}} = \frac{\tilde{\eta}_{K-V}}{\tilde{\mu}}\sigma, \qquad (5.11.5)$$

respectively.

12. A systematic development

Jeffreys (1952) has emphasized the need for a systematic development of the Love number corrections. Jeffreys (1916) and Rosenhead (1929) end up with a sea surface that is not isobaric and Schweydar (1916) treats an Earth in which mass is not conserved.

The customary scheme is to write

$$h_n U_n/g, \quad k_n U_n, \quad (1 + k_n)U_n \qquad (5.12.1)$$

for the amount the ground is lifted, the added potential due to the deformation and the final potential, all in terms of an initially prescribed potential U_n that does not load the Earth. For a potential, U'_n that does load, we have introduced the notation

$$h'_n U'_n/g, \quad k'_n U'_n, \quad (1 + k'_n)U'_n. \qquad (5.12.2)$$

A tidal potential, U_2, or the potential of a snow load, $U' = \sum_{n=\theta}^{\infty} U'_n$, may serve as examples for the two cases.

The difficulty is that the distortion of the sea bottom causes an ocean tide which gives rise to an additional potential and to additional load deformations, and these are linked in a complicated way to the initial disturbance. It is perhaps better to refer to the 'final' potentials, V_n, V'_n (which include the initially prescribed potentials, U_n, U'_n) and to define the Love numbers K_n, H_n, K'_n, H'_n such that

$$H_n V_n/g, \quad K_n V_n, \quad V_n; \quad H'_n V'_n/g, \quad K'_n V'_n, \quad V'_n \qquad (5.12.3)$$

have the same meanings as the quantities in (5.12.1) and (5.12.2). It follows that

$$H_n = \frac{h_n}{1 + k_n}, \quad K_n = \frac{k_n}{1 + k_n}; \quad H'_n = \frac{h'_n}{1 + k'_n}, \quad K'_n = \frac{k'_n}{1 + k'_n}.$$

$$(5.12.4)$$

Owing to any initial disturbing potential, the ground is raised by

$$g^{-1} \sum_{n=0}^{\infty} [H_n V_n + H_n' V_n'],$$

and the sea surface relative to the Earth's center of mass by

$$g^{-1} \sum_{n=0}^{\infty} [V_n + V_n'],$$

so $\xi = \sum_{n=0}^{\infty} \xi_n = g^{-1} \mathscr{C}(\theta, \lambda) \sum_{n=0}^{\infty} [(1 - H_n)V_n + (1 - H_n')V_n']$ (5.12.5)

is the elevation of the sea surface over the sea bottom; here $\mathscr{C}(\theta, \lambda)$ is the 'ocean function' (see § A.1), defined by

$$\mathscr{C}(\theta, \lambda) = 1, \text{ where there are seas;}$$
$$\mathscr{C}(\theta, \lambda) = 0, \text{ where there is land.}$$

The final potential V_n is determined by the initial disturbance and the deformation potentials:

$$V_n = U_n + K_n V_n, \qquad (5.12.6)$$

whereas V_n' depends also on the displacement of the sea surface:

$$V_n' = U_n' + K_n' V_n' + \frac{3g}{2n+1} \frac{\rho_w}{\rho} \xi_n. \qquad (5.12.7)$$

V_n' must be considered as unknown since this potential depends on the ocean tide ξ which in turn depends in a complicated way on the disturbing potentials.

Equations (5.12.6) and (5.12.7) are solved for V_n and V_n', and the resulting expressions substituted into (5.12.5):

$$\sum_{n=0}^{\infty} \xi_n = g^{-1} \mathscr{C}(\theta, \lambda) \sum_{r=0}^{\infty} (J_r U_r + J_r' U_r' + \frac{3g}{2r+1} J_r' \xi_r), \quad (5.12.8)$$

where

$$J_r = \frac{1 - H_r}{1 - K_r} = 1 + k_r - h_r, \quad J_r' = \frac{1 - H_r'}{1 - K_r'} = 1 + k_r' - h_r'.$$

Equation (5.12.8) displays the complicated dependence of the ocean tide on the initial disturbing potentials U_r and U_r'. Suppose we wish to solve for $s + 1$ harmonics, from $n = 0$ to $n = s$. Then (5.12.8) gives $s + 1$ equations for $s + 1$ unknown, ξ_0 to ξ_s. The corresponding potentials follow from (5.12.6) and (5.12.7). All $4(s + 1)$ Love numbers are presumed to be given from the known properties of the Earth.

A convenient manner of dealing with (5.12.8) is to expand $\mathscr{C}(\theta, \lambda)$ into spherical harmonics (A.1.2),

$$\mathscr{C}(\theta, \lambda) = \sum_{n=0}^{\infty} \sum_{m=0}^{n} (a_n^m p_n^m \cos m\lambda + b_n^m p_n^m \sin m\lambda), \quad (5.12.9)$$

and, similarly,

$$\xi = \sum_{a=0}^{\infty} \sum_{b=0}^{a} (y_a^b p_a^b \cos b\lambda + z_a^b p_a^b \sin b\lambda),$$

$$U_r = \sum_{m=0}^{r} (u_r^m p_r^m \cos m\lambda + v_r^m p_r^m \sin m\lambda)$$

and similarly for U' using u', v'. The evaluation of the coefficients y_a^b, z_a^b requires the calculation of triple products of the associated function p_n^m (A.1).

As an important special case, consider the initial potential

$$U_2 = u_2^0 p_2^0.$$

The coefficients of the ocean tide are given by

$$gy_2^0 = J_2 u_2^0(a_0^0 + \tfrac{1}{2}a_2^0 + \tfrac{2}{7}a_4^0) + 3gJ_2'\{\tfrac{2}{5}y_2^0(2a_0^0 + a_2^0 + \tfrac{4}{7}a_4^0)$$
$$+ (\tfrac{141}{350}y_2^1 a_2^1 + z_2^1 b_2^1) + \tfrac{1}{7}(y_2^1 a_4^1 + z_2^1 b_4^1) + \tfrac{2}{7}(y_2^2 a_4^2 + z_2^2 b_4^2)\}$$

and similar expressions for y_2^1, y_2^2, z_2^1, z_2^2. The five equations determine all second degree components of ξ. The coefficients are not independent, since conservation of water requires that

$$y_0^0 = 0.$$

For a p_2^0 disturbance, we have

$$gy_0^0 + \tfrac{1}{5}J_2 u_2^0 a_2^0 + \tfrac{3}{100}gJ_2'\{y_2^0 a_2^0 + \tfrac{3}{2}(y_2^1 a_2^1 + z_2^1 b_2^1) + 6(y_2^2 a_2^2 + z_2^2 b_2^2)\}.$$

For disturbing potentials other than p_2^0, the requirement that the mass of the Earth be conserved is automatically satisfied.

CHAPTER 6

SOLUTIONS TO THE APPROXIMATE
LIOUVILLE EQUATION

The Liouville equation is greatly simplified by a perturbation scheme which is valid for all problems to be treated except polar wandering. The deformation of the Earth is allowed for by various Love numbers.

1. Perturbations

The following scheme is convenient as long as the poles of figure and rotation do not wander too far from the reference pole:

$$\left. \begin{aligned} C_{11} &= A + c_{11}, & C_{22} &= A + c_{22}, & C_{33} &= C + c_{33}, \\ C_{12} &= c_{12}, & C_{13} &= c_{13}, & C_{23} &= c_{23}, \\ \omega_1 &= \Omega m_1, & \omega_2 &= \Omega m_2, & \omega_3 &= \Omega(1 + m_3) \end{aligned} \right\}, \quad (6.1.1)$$

where A, A, C are the moments of inertia referred to the principal axes; Ω is the mean angular velocity of the Earth, 0.729×10^{-4} radians per sidereal second; c_{ij}/C, m_i and $h_i/(\Omega C)$ are small dimensionless quantities whose products and squares can be neglected. The Liouville equation (3.1.6) then reduces to the simple form

$$\frac{\dot{m}_1}{\sigma_r} + m_2 = \phi_2, \quad \frac{\dot{m}_2}{\sigma_r} - m_1 = -\phi_1, \quad (6.1.2)$$

$$\dot{m}_3 = \dot{\phi}_3, \quad (6.1.3)$$

where σ_r and ϕ_i are defined by

$$\sigma_r = \frac{C - A}{A} \Omega \quad (6.1.4)$$

and

$$\left. \begin{aligned} \Omega^2(C - A)\phi_1 &= \Omega^2 c_{13} + \Omega \dot{c}_{23} + \Omega h_1 + \dot{h}_2 - L_2 \\ \Omega^2(C - A)\phi_2 &= \Omega^2 c_{23} - \Omega \dot{c}_{13} + \Omega h_2 - \dot{h}_1 + L_1 \\ \Omega^2 C \phi_3 &= -\Omega^2 c_{33} - \Omega h_3 + \Omega \int_0^t L_3 \, dt \end{aligned} \right\}, \quad (6.1.5)$$

with $(\cdot)$ designating d/dt. The left sides of (6.1.2, 3) are determined by astronomic observations, the right sides by geophysical observations. The dimensionless 'excitation function' ϕ_i contains all

38   THE ROTATION OF THE EARTH

possible geophysical effects on the motion of the Earth.* The
evaluation of ϕ_i is the principal task of this book.
Variations in the l.o.d. are dealt with by (6.1.3). Wobble is dealt
with by (6.1.2); m_1, m_2, 1, are the direction cosines of the rotation
axis (3.3.1). In the complex system

$$\mathbf{m} = m_1 + im_2, \quad \boldsymbol{\phi} = \phi_1 + i\phi_2 \tag{6.1.6}$$

we have simply $\qquad i\dfrac{\dot{\mathbf{m}}}{\sigma_r} + \mathbf{m} = \boldsymbol{\phi}.$ \hfill (6.1.7)

Variations in the inertia tensor C_{ij} arising from rotational deformation are given by (5.2.3, 4). These can be put into a more convenient
form by using the definition (5.3.2) of k_s and setting $k_s = k_f$:

$$C_{ij} = \text{constant} + \frac{k}{k_f}(C - A)\frac{\omega_i\omega_j - \frac{1}{3}\omega^2\delta_{ij}}{\Omega^2}.$$

In terms of the perturbation convention we then have

$$\left.\begin{array}{ll} C_{11} = C_{22} = -\dfrac{2}{3}\dfrac{k}{k_f}(C-A)m_3, & C_{33} = +\dfrac{4}{3}\dfrac{k}{k_f}(C-A)m_3, \\[2mm] C_{13} = \dfrac{k}{k_f}(C-A)m_1, & C_{23} = \dfrac{k}{k_f}(C-A)m_2. \end{array}\right\} \tag{6.1.8}$$

When these expressions are substituted in (6.1.5), one obtains

$$\boldsymbol{\phi} = \left[1 - \frac{1}{\Omega}\frac{d}{dt}\right]\left[\frac{C_{13} + iC_{23}}{C - A}\right]$$

$$= \frac{k}{k_f}\left[1 - \frac{1}{\Omega}\frac{d}{dt}\right]\mathbf{m} = \psi_D - i\Omega^{-1}\dot{\psi}_D, \tag{6.1.9}$$

$$\phi_3 = -\frac{C_{33}}{C} = -\frac{4}{3}\frac{k}{k_f}Hm_3 = \psi_{D,3} \tag{6.1.10}$$

for that part of the excitation function arising from the rotational
deformation. We call this the *deformation axis* and introduce the
notation $\psi_{D,i}$.

2. Free wobble

In the case of free nutation of a rigid Earth, $\boldsymbol{\phi} = 0$ and $\exp(i\sigma_r t)$
is a solution to (6.1.7). The period $2\pi/\sigma_r$ is about ten months. The

* Equations in essentially this form have been given by Young (1953). The excitation function has previously been discussed by Klein and Sommerfield (1903, p. 712–
715). They in turn refer to earlier work by V. Volterra in 1895, A. Wangerin in 1899
and E. Jahnke in 1899.

role of the Earth's deformation is to increase the period of free
nutation by a substantial fraction (~ 40 per cent). This effect is not
altogether obvious and in fact had not been anticipated prior to
Chandler's discovery of a 14-month period in the latitude variation
(§ 7.1). A qualitative explanation is as follows: for a rigid Earth the
frequency of free nutation is proportional to the equatorial bulge
(6.1.4); for a deformable Earth it is dependent only on that part of
the equatorial bulge which does not adjust to the instantaneous
rotation-axis.

Consider the excitation function (6.1.9) due solely to rotational
deformation. Equation (6.1.7) becomes

$$\frac{i\dot{\mathbf{m}}}{\sigma_r} + \mathbf{m} = \boldsymbol{\psi}_D - i\Omega^{-1}\dot{\boldsymbol{\psi}}_D \approx \boldsymbol{\psi}_D. \qquad (6.2.1)$$

The approximation depends on $\sigma_r/\Omega = (C - A)/A$ being a small
number. The error is $0 \cdot 1$ per cent. Equations (6.1.7) and (6.1.9) can
be written in the form

$$i\dot{\mathbf{m}} + \sigma_0\mathbf{m} = 0,$$

which differs from the corresponding equation for a rigid Earth
((6.1.7) with $\boldsymbol{\phi} = 0$)

$$i\dot{\mathbf{m}} + \sigma_r\mathbf{m} = 0$$

in that the frequency of free nutation has been reduced from σ_r to

$$\sigma_0 = \sigma_r \frac{k_f - k}{k_f}. \qquad (6.2.2)$$

For the 'equivalent' Earth (§ 5.6), the ratio is

$$\frac{\sigma_0}{\sigma_r} = \frac{\mu}{1 + \mu} = \frac{2 \cdot 3}{3 \cdot 3} = 0 \cdot 70. \qquad (6.2.3)$$

The principal axis of a body with moments and products of
inertia A, A, C; c_{12}, c_{13}, c_{23} is inclined (approximately) by $c_{13}/(C - A)$,
$c_{23}/(C - A)$ relative to the reference axis x_3. It follows from (6.1.8)
and (6.1.9) that $\boldsymbol{\psi}_D$ designates the inclination of the principal axis of
the rotationally deformed equatorial bulge, the 'deformation axis.'
The positions where the $\boldsymbol{\psi}_D$-axis pierces the surface are the deformation poles.

Table 6.1

Model	'Tidal-effective' Love number	Deformation axis	Nutation frequency
Rigid Earth	$k = 0$	$\psi_D = 0$	$\sigma_0 = \sigma_r$ = 1 cycle in 10 months
'Fluid' Earth	$k = k_f$	$\psi_D = \mathbf{m}$	$\sigma_0 = 0$
'Actual' Earth	$k = 0\cdot29$	$\psi_D = 0\cdot30\ \mathbf{m}$	$\sigma_0 = \dfrac{\sigma_r}{1\cdot4}$ = 1 cycle in 14 months

The three models in table 6.1 are instructive: in the case of the 'fluid' Earth, the equatorial bulge adjusts completely to the rotation and there is no rotational stability and no nutation: $\sigma_0 = 0$. Only that portion of the equatorial bulge that remains frozen during the wobble (about 70 per cent) provides any stability.

3. Forced wobble

Consider a wobble induced by any prescribed event. The procedure to be followed is to compute first the excitation function, $\psi(t)$, as if the Earth were rigid. The effect of rotational deformation is to produce an additional excitation,

$$\psi_D = (k/k_f)\mathbf{m},$$

which must be taken into account in addition to the prescribed excitation. If the initially prescribed event does not load the Earth (e.g., winds), then the total excitation function consists of the two parts

$$\phi = \psi + \psi_D; \tag{6.3.1a}$$

if it does load the Earth, then

$$\phi = \psi + \psi_D + \psi_L, \tag{6.3.1b}$$

where, in view of § 5.8, $\psi_L = k'\psi$ (6.3.2)

is an additional excitation arising from load deformation.*

The effect of rotational deformation is then to produce a larger excitation (and wobble) than would have obtained on a rigid Earth;

---

* The excitation axis is then a generalization of the 'principal axis', or 'axis of figure'. If the prescribed excitation, ψ, is due entirely to relative motion, then the deformation axis ψ_D is also the principal axis; if it is due entirely to a shift in matter, then the excitation axis ϕ is the principal axis.

the effect of load deformation is to reduce the overall excitation (k' is negative). One may choose to interpret rotational deformation as positive feedback, load deformation as negative feedback. It is convenient to define a 'modified excitation',

$$\Psi = \kappa_{\text{wobble}}\,\psi, \qquad (6.3.3)$$

where κ_{wobble} is a 'transfer function' equal to

$$\kappa_{\text{wobble}} = \frac{k_f}{k_f - k}, \qquad (6.3.4a)$$

and

$$\kappa_{\text{wobble}} = (1 + k')\frac{k_f}{k_f - k}, \qquad (6.3.4b)$$

depending upon whether the prescribed event does not or does load the Earth. The equation (6.1.7) can then be written in the following equivalent forms:

$$\dot{\mathbf{m}} = i\sigma_r(\mathbf{m} - \boldsymbol{\phi}), \qquad (6.3.5)$$

$$\dot{\mathbf{m}} = i\sigma_0(\mathbf{m} - \Psi). \qquad (6.3.6)$$

The two forms differ with respect to frequency and the excitation function. The total excitation, $\boldsymbol{\phi}$, includes the rotational deformation ψ_D, and when this is combined with $\mathbf{m}$ in the sense of § 6.2 then σ_r becomes σ_0, and ϕ becomes Ψ.

4. Transfer function

The transfer function κ for the equivalent Earth is obtained from the definitions of k and k' given in (5.6.2) and (5.8.5) respectively. The values are

$$\kappa_{\text{wobble}} = \frac{1 + \mu}{\mu} = 1\cdot 43, \qquad (6.4.1a)$$

$$\kappa_{\text{wobble}} = 1 \qquad (6.4.1b)$$

depending on whether the prescribed event does not or does load the Earth. In the latter case the amplification from rotational deformation cancels the diminution from load deformation, and the induced wobble is the same as if the Earth were rigid. At first glance the result might appear surprising. But the load deformation enters only insofar as it contributes to the products of inertia, and these are

THE ROTATION OF THE EARTH

Table 6.2. The transfer function κ.

	(a): no load	(b): load
wobble	1·43	1·00
l.o.d.	1·00	0·70

spherical harmonics of degree 2 and of the same form as the rotational deformation.*

For problems involving the l.o.d. the rotational deformation as well as the load deformation acts in a sense to reduce the total excitation as compared to what it would be on a rigid Earth. To maintain symmetry with the wobble notation, it is convenient to write

$$\phi_3 = \Psi'_3 = \kappa_{\text{l.o.d.}}\,\psi_3, \qquad (6.4.2)$$

where, according to (6.2.3),

$$\kappa_{\text{l.o.d.}} = \left(1 + \frac{4}{3}\frac{k}{k_f}H\right)^{-1} = 0\cdot999, \qquad (6.4.3a)$$

$$\kappa_{\text{l.o.d.}} = 1 + k' = \frac{\mu}{1+\mu} \qquad (6.4.3b)$$

depending on whether the prescribed event does not or does load the Earth.

The working rule is then to compute the excitation function, ψ_i, as if the Earth were rigid, and to multiply by a factor κ as given in table 6.2 to obtain the modified excitation Ψ'_i. The result can readily be compared to the astronomic observations by virtue of the relations

$$\dot{\mathbf{m}} = i\sigma_0(\mathbf{m} - \Psi'), \qquad (6.4.4a)$$

$$\dot{m}_3 = \dot{\Psi}'_3, \qquad (6.4.4b)$$

which follow from (6.3.6), (6.3.3) and (6.4.2).

---

* This intimate relation between rotational and load deformations has apparently been overlooked. The procedure has been to derive the amplification arising from rotational deformation, and then, by reference to Rosenhead's (1929) paper, to multiply the results by a numerical factor 0·69 to allow for the load deformation.

The numerical values in table 6.2 must be used with caution. The values are based on the tidal-effective rigidity as obtained from the Chandler wobble (§ 10.3). At high frequencies the values may fail because the response of oceans and core may be frequency-dependent (§ 10.4); at very low frequencies the anelastic deformation of the mantle may play an important role, and the transfer function is then no longer a pure number (§ 5.11, 12.6).

5. A geometric interpretation

The effects of rotational and load deformations on wobble can now be summarized geometrically (fig. 6.1). Initially the excitation and rotation poles are taken at the origin of the geographic coordinates, hence $\boldsymbol{\phi} = 0$, $\mathbf{m} = 0$. At time 0 the excitation pole is displaced suddenly (in the direction 19° E. on the figure) by some specified event.

In the upper diagram the Earth is assumed to be rigid, and the pole of rotation $\mathbf{m}$ revolves about the excitation pole ($\boldsymbol{\psi} = \boldsymbol{\Psi} = \boldsymbol{\phi}$), as shown. In the central diagram allowance is made for deformation in the case of an excitation that does not load the Earth ($\boldsymbol{\psi}_L = 0$). The new feature is that the equatorial bulge adjusts itself to the disturbed position of the pole of rotation; the plane of the bulge tends to align normal to the rotation axis, but on account of elastic constraint is only about one-third successful: $\boldsymbol{\psi}_D = (k/k_f)\mathbf{m} = 0.30\mathbf{m}$. The total excitation pole $\boldsymbol{\phi} = \boldsymbol{\psi} + \boldsymbol{\psi}_D$ consists of that part $\boldsymbol{\psi}$ which was computed under the assumption of a rigid Earth plus the additional part $\boldsymbol{\psi}_D$ arising from the deformation. The pole of rotation $\mathbf{m}$ revolves about the mean position* $\boldsymbol{\Psi}$ of the excitation pole with a radius amplified by a factor $k_f/(k_f - k) = 1.43$ as compared to the rigid case, but the separation $\mathbf{m} - \boldsymbol{\phi}$ between the rotation pole and the *instantaneous* excitation pole is the same as in the rigid case. The speed of wandering of the rotation pole is proportional to $\mathbf{m} - \boldsymbol{\phi}$ (6.3.5) and therefore unchanged by the deformation, but the period for a complete revolution is increased from

---

* The fact that the modified excitation pole $\boldsymbol{\Psi}$ is also the mean position of the instantaneous pole $\boldsymbol{\psi}$ can be proven as follows. Assuming $\boldsymbol{\Psi}$ to be at the center of the concentric circles prescribed by $\mathbf{m}$ (radius $\mathcal{R}$) and by $\boldsymbol{\phi}$, we may write $\mathbf{m} = \boldsymbol{\Psi} + \alpha\mathcal{R}$, $\boldsymbol{\phi} = \boldsymbol{\Psi} + \alpha(k/k_f)\mathcal{R}$, where α is a unit vector. Form $(k/k_f)\mathbf{m} - \boldsymbol{\phi}$ and allow for (6.2.3). The result is $\boldsymbol{\psi}_D - \boldsymbol{\phi} = \boldsymbol{\Psi}[(k/k_f) - 1]$. We now make use of (6.3.1), (6.3.2), and (6.3.4) to obtain $\boldsymbol{\Psi} = \kappa\boldsymbol{\psi}$ which in accord with the definition (6.3.3) of $\boldsymbol{\Psi}$.

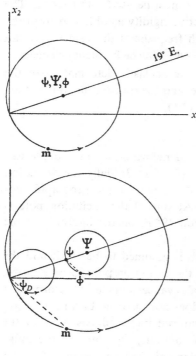

Fig. 6.1. The coordinates x_1, x_2 are drawn from the undisturbed poles in the direction of Greenwich and 90° east of Greenwich, respectively. The disturbed excitation pole ψ is displaced in the direction 19° E. The resulting paths of the pole of rotation, m, are shown for a rigid Earth (*top*); for a deformable Earth in the case of a non-loading excitation (*center*), and of a loading excitation (*bottom*). The initially specified excitation, ψ, is the same for all three cases. The positions of the poles of deformation ψ_D, load ψ_L, total excitation ϕ, and of the modified excitation Ψ are shown.

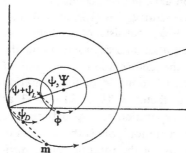

10 months to 14 months on account of the corresponding increase in radius.

The situation is more complicated if the excitation pole loads the Earth (lower diagram). The excitation $\psi + \psi_L$ after allowing for load deformation is then smaller than the excitation ψ due to the initially prescribed surface loads. This reduction is cancelled by the amplification due to rotational deformation, so that the radius of the circle described by the pole of rotation is now the same as for the rigid Earth. As always, the speed of the wandering pole is proportional $\mathbf{m} - \boldsymbol{\phi}$. This is reduced by a factor 1·4 as compared to the rigid case, and accordingly the period of revolution increased from 10 to 14 months.

6. Sway

Our emphasis has been on the motion of the rotation axis relative to the reference system. If there is no outside torque, the axis that remains truly fixed in space is that of the absolute angular momentum H_i, also called the 'invariable' axis. The rotation axis executes a slight 'sway' relative to the invariable axis, and hence the celestial pole is not absolutely fixed relative to Polaris, even in the absence of precession. Observationally, sway is undistinguishable from precession and forced nutation; these all involve variations in the polar distance (or declination) of a star. In our terminology, sway is confined to variations due entirely to events on the planet Earth, whereas precession and nutation depend on attractive forces of Sun, Moon and planets.

The position, with respect to the reference axes, of the invariable pole is

$$\frac{\mathbf{H}}{C\Omega} = \frac{H_1 + iH_2}{C\Omega}.$$

A line through the Earth's center and the invariable pole retains its direction in space. Dividing both sides of (3.1.3) by $C\Omega$ and applying the perturbation scheme (6.1) gives

$$\frac{\mathbf{H}}{C\Omega} = \mathbf{m} + H(\boldsymbol{\phi} - \mathbf{m}). \qquad (6.6.1)$$

The precessional constant H (no relation to $\mathbf{H}$) is a small quantity; it follows that the invariable and rotation poles are separated by a distance small compared to the displacement of either pole except if $|\mathbf{m}| \ll |\boldsymbol{\phi}|$. The condition $|\mathbf{m}| \ll |\boldsymbol{\phi}|$ applies to frequencies large compared to the frequency, σ_0, of free nutation.

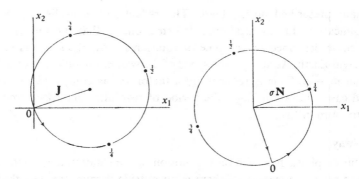

Fig. 6.2. Positions of the pole of rotation at times 0, $\tfrac{1}{4}T$, $\tfrac{1}{2}T$, $\tfrac{3}{4}T$ due to (*left*) an excitation $\mathbf{\Psi} = \mathbf{J}H(t)$, and (*right*) $\mathbf{\Psi} = \mathbf{N}\delta(t)$.

7. Solutions for various excitation functions

The solution to (6.4.4a) for the wobble due to any excitation $\mathbf{\Psi}(t)$ is

$$\mathbf{m}(t) = e^{i\sigma_0 t}\left[\mathbf{m}_0 - i\sigma_0 \int_{-\infty}^{t} \mathbf{\Psi}(\tau)e^{-i\sigma_0 \tau}\,d\tau\right], \qquad (6.7.1)$$

where $\mathbf{m}_0$ is an arbitrary complex constant.

Consider first $\mathbf{\Psi}(t) = \mathbf{J}H(t)$, where $H(t)$ is the Heaviside step function. Then $\mathbf{m}_0 = 0$, $\mathbf{m} = 0$ for $t < 0$, and

$$\mathbf{m} = \mathbf{J}(1 - e^{i\sigma_0 t}), \quad t \geqslant 0. \qquad (6.7.2)$$

The rotation pole $\mathbf{m}$ describes a circle about the modified excitation pole $\mathbf{J}$ in a period $T = 2\pi/\sigma_0$ (fig. 6.2). The direction is west-to-east, in the same sense as the diurnal rotation and positive in the adopted coordinate system.

As a second example, let $\mathbf{\Psi}(t) = \mathbf{N}\delta(t)$, where $\delta(t)$ is the Dirac function. The solution is

$$\mathbf{m} = -i\sigma_0 \mathbf{N}e^{i\sigma_0 t}, \quad t \geqslant 0. \qquad (6.7.3)$$

The rotation pole shifts abruptly at time 0, and then revolves positively and $90°$ in advance of the previous solution (fig. 6.2).

As a third example, we take the case of a harmonic excitation of any frequency, σ:

$$\mathbf{\Psi} = \mathbf{\Psi}^c \cos \sigma t + \mathbf{\Psi}^s \sin \sigma t. \qquad (6.7.4)$$

The four numbers Ψ_1^c, Ψ_2^c, Ψ_1^s, Ψ_2^s determine phase and amplitude of the excitation. It is often convenient to use the alternate forms

$$\Psi = \Psi^+ e^{i\sigma t} + \Psi^- e^{-i\sigma t} = |\Psi^+| e^{i(\sigma t + \lambda^+)} + |\Psi^-| e^{-i(\sigma t - \lambda^-)}. \quad (6.7.5)$$

The first term in either expression designates positive (west-to-east) motion of the excitation pole around a circle of radius $|\Psi^+|$; east longitude λ^+ at time 0 is given by $\tan \lambda^+ = \arg \Psi^+$. The second term represents a corresponding circular motion in the opposite sense. Linear and circular components are connected by the formulae

$$\Psi^c = \Psi^+ + \Psi^-, \quad \Psi^s = i(\Psi^+ - \Psi^-). \quad (6.7.6)$$

The motion (6.7.5) is elliptical. The semi-major and semi-minor axes have the magnitude

$$|\Psi^+| + |\Psi^-|, \quad |\Psi^+| - |\Psi^-|,$$

respectively, and the ellipticity equals

$$\frac{2|\Psi^-|}{|\Psi^+| + |\Psi^-|}.$$

The major axis lies in east longitude $\frac{1}{2}(\lambda^+ + \lambda^-)$ and is occupied at a time t so that $\sigma t = \frac{1}{2}(\lambda^- - \lambda^+)$. Analogous formulae can be written for $\mathbf{m}$.

The solution consists of two parts: the particular solution

$$\mathbf{m} = \mathbf{m}_0\, e^{i\sigma_0 t} \quad (6.7.7)$$

shows positive motion with the frequency, σ_0, of the free nutation, whereas the forced wobble is given by

$$\mathbf{m} = \frac{\sigma_0 \Psi^+}{\sigma_0 - \sigma}\, e^{i\sigma t} + \frac{\sigma_0 \Psi^-}{\sigma_0 + \sigma}\, e^{-i\sigma t}, \quad (6.7.8)$$

with the forced frequency σ. By appropriate superposition of free and forced motion it is, of course, possible to develop complex patterns. This is true particularly if free and forced frequencies are close to each other, as in fact they are for the Chandler motion and the annual wobble. Various examples have been worked out by Wanach (1919), and a few of these have been reproduced by Lambert (1931, p. 256–8). It is more instructive to consider the forced motion

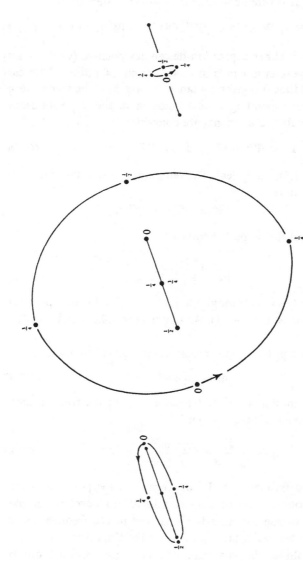

Fig. 6.3. The excitation pole oscillates with frequency $\sigma = 2\pi/T$ along a fixed longitude, occupying positions at times $0, \frac{1}{4}T, \frac{1}{2}T, \frac{3}{4}T$, as indicated on the central diagram. The elliptical path of the rotation pole is shown for $\sigma = \frac{1}{4}\sigma_0$, (left), (center); and $\sigma = 4\sigma_0$, (right), where $\sigma_0 = 2\pi/(1\cdot2$ years) is the frequency of free nutation.

by itself. The free motion can be regarded as having been destroyed by friction or separated in the record by suitable analysis.

Fig. 6.3 shows the forced wobble for the case of plane polarization, $|\Psi^+| = |\Psi^-|$. In the case of low frequencies (left) the pole of rotation describes an ellipse with the major axis along the axis of the excitation pole. In the asymptotic case of $\sigma \to 0$, the two poles oscillate together, $\mathbf{m} \to \Psi$. As the frequency approaches resonance, $\sigma \to \sigma_0$, the path of pole of rotation approaches a circle, and the radius becomes infinite (if there is no dissipation). The center diagram is for the case of the annual frequency, which lies slightly above resonance. For even higher frequencies (right) the eccentricity of the path of $\mathbf{m}$ becomes larger again, but in this case with the major axis at right angles to that of Ψ. As $\sigma \to \infty$ the wobble becomes vanishingly small. In all events the motion of the pole along its elliptical path is in a positive (west-to-east) sense.

For the case of circular polarization of the excitation pole, whether positive ($\Psi^- = 0$) or negative ($\Psi^+ = 0$), the pole of rotation revolves about a circular path in the same sense as the excitation pole. For Ψ^- the two poles are always in phase. The radius of the circle described by the rotation pole equals that of the excitation pole at vanishingly low frequencies, is half as large at resonance, and vanishes at very high frequency (6.7.8). For Ψ^+ the poles are in phase at low frequency and out of phase at high frequency, with an abrupt transition at resonance. The radius of the circle described by the rotation pole again equals that of the excitation pole at vanishingly low frequency, and is infinitesimal at vanishingly high frequency; at resonance it becomes infinite.

It is desirable to plot the amplitude of the wobble as a continuous function of frequency. The mean square polar excursions are

$$\left.\begin{array}{l}\langle |\Psi|^2 \rangle = \langle \Psi_1'^2 + \Psi_2'^2 \rangle = |\Psi^+|^2 + |\Psi^-|^2 \\ \langle |\mathbf{m}|^2 \rangle = \langle m_1^2 + m_2^2 \rangle = |\mathbf{m}^+|^2 + |\mathbf{m}^-|^2 \end{array}\right\}, \qquad (6.7.9)$$

where $\langle \ \rangle$ designates a time average. But according to (6.7.8)

$$\mathbf{m}^+ = \frac{\Psi^+}{1 - s}, \quad \mathbf{m}^- = \frac{\Psi^-}{1 + s},$$

with $s = \sigma/\sigma_0$.

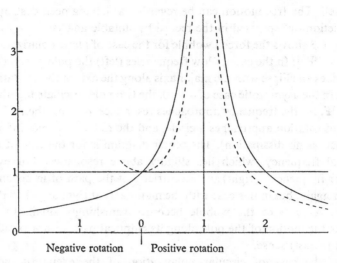

Fig. 6.4. The amplitude transmission, I, as a function of the dimensionless frequency, $s = \sigma/\sigma_0$. Solid: circular (positive and negative) polarization; dashed: plane and random polarizations.

Hence

$$\langle |\mathbf{m}|^2 \rangle = \frac{(1 + s^2)\langle |\Psi|^2 \rangle + 2s\,(|\Psi^+|^2 - |\Psi^-|^2)}{(1 - s^2)^2}.$$

The amplitude transmission ratio

$$I = [\langle |\mathbf{m}|^2 \rangle / \langle |\Psi|^2 \rangle]^{\frac{1}{2}} \qquad (6.7.10)$$

has the following values for various special cases:

circular positive, $\qquad \Psi = \Psi^+ \qquad\qquad : \dfrac{1}{1 - s}$

circular negative, $\qquad \Psi = \Psi^- \qquad\qquad : \dfrac{1}{1 + s}$

plane polarization, $\qquad |\Psi^+| = |\Psi^-| \qquad\quad : \dfrac{\sqrt{1 + s^2}}{1 - s^2}$

random polarization, $\langle |\Psi|^2 \rangle = 2\,\langle |\Psi^+|^2 \rangle = 2\,\langle |\Psi^-|^2 \rangle : \dfrac{\sqrt{1 + s^2}}{1 - s^2}$

The cases are illustrated in fig. 6.4. At vanishingly low frequency the transmission ratio is unity for all three cases. But it should be recalled that the transmission ratio has been defined with respect to the modified excitation function Ψ. The asymptotic value for the corresponding ratio with respect to the (non-modified) excitation ψ is $\sigma_r/\sigma_0 = 1 \cdot 40$.

Perhaps the most likely geophysical case is one where the plane of polarization of the excitation function and its phase are randomly distributed. For this case, I is given by the same formula as for plane polarization. This can be demonstrated as follows. The excitation and rotation poles describe wiggly paths. These can be frequency-analyzed, and for each spectral component the relationship is the one given by equation (6.7.9). Consider the record to be divided into decades. Frequency components are computed, and an ensemble average is formed from all available decades. Since there is no preference for positive or negative circular polarization, $|\Psi^+|$ must equal $|\Psi^-|$ in the ensemble average. Equation (6.7.9) then reduces to the case of plane polarization.

8. The excitation function

Equations (6.1.5) give the total excitation, including rotational and load deformations. In all practical problems we are forced to evaluate the excitation, ψ_i, as if the Earth were rigid, and to allow for the secondary deformation by way of the transfer function. With this understanding (6.1.5) can be used as written, with ϕ_i replaced by ψ_i.

Equations (6.1.5) are well suited for computing the excitation function whenever changes in relative angular momentum are well separated from changes in the products of inertia. This usually arises whenever one or the other is zero. For example, a flywheel fixed to the ground and rotating at a variable rate gives changes in the relative angular momentum, but not in inertia. In the case of a melting icecap the angular momentum of the flowing water is negligible but changes in inertia may not be.

Equation (6.1.5) is not convenient if it is desired to separate explicitly the effects arising due to changes in the distribution of matter from those due to relative motion. The reason is that $\dot{c}_{ij}$ and h_i both involve relative motion; moreover, these are of the same order. Equations in 'separated form' have been used by Munk and Groves (1952) to estimate the annual wobble induced by winds and ocean currents.

Equations (6.1.5) can be written

$$\left.\begin{array}{l}\Omega^2(C-A)\phi\\ \quad=\displaystyle\int_V \Delta\rho\ \mathbf{F}(\text{matter})\ dV + \int_V \rho\ \mathbf{F}(\text{motion})\ dV + \mathbf{F}(\text{torque})\\ \Omega^2 C\phi_3\\ \quad=\displaystyle\int_V \Delta\rho\ F_3(\text{matter})\ dV + \int_V \rho\ F_3(\text{motion})\ dV + F_3(\text{torque})\end{array}\right\},$$

where $\Delta\rho(x_i, t)$ is the density departure associated with the excitation functions $\phi_k(t)$, and where

matter $\quad F_1 = -\,\Omega^2 x_1 x_3,\quad F_2 = -\,\Omega^2 x_2 x_3,\quad F_3 = -\,\Omega^2(x_1^2 + x_2^2)$

motion $\quad\begin{cases} F_1 = -\,2\Omega x_3 u_2 + x_3\dot{u}_1 - x_1\dot{u}_3\\ F_2 = 2\Omega x_3 u_1 + x_3\dot{u}_2 - x_2\dot{u}_3\\ F_3 = \Omega(-\,x_1 u_2 + x_2 u_1)\end{cases}$

torque $\quad F_1 = -\,L_2,\quad F_2 = L_1,\quad F_3 = \Omega\displaystyle\int_0^t L_3\,dt$

are functions depending on distribution of matter, relative motion (velocity and acceleration), and on torque, respectively. Spherical coordinates are often convenient. Let u_λ, u_θ, u_r designate eastward, southward and upward components of velocity, and $dV = r^2 \sin\theta\ dr\ d\theta\ d\lambda$ the differential volume. Then

matter $\quad\begin{cases} F_1 = -\,r^2\Omega^2 \sin\theta \cos\theta \cos\lambda,\ F_2 = -\,r^2\Omega^2 \sin\theta \cos\theta \sin\lambda,\\ F_3 = -\,r^2\Omega^2 \sin^2\theta\end{cases}$

motion $\quad\begin{cases} F_1 = -\,2\Omega r \cos\theta(u_\lambda \cos\lambda + u_\theta \cos\theta \sin\lambda + u_r \sin\theta \sin\lambda)\\ \qquad + r(-\,\dot{u}_\lambda \cos\theta \sin\lambda + \dot{u}_\theta \cos\lambda)\\ F_2 = 2\Omega r \cos\theta(-\,u_\lambda \sin\lambda + u_\theta \cos\theta \cos\lambda + u_r \sin\theta \cos\lambda)\\ \qquad + r(\dot{u}_\lambda \cos\theta \cos\lambda + \dot{u}_\theta \sin\lambda)\\ F_3 = -\,\Omega r \sin\theta\ u_\lambda.\end{cases}$

The torque can be written as a sum of two terms:

$$L_i = \int_V \rho\varepsilon_{ijk}x_j f_k\,dV + \int_S \varepsilon_{ijk}x_j p_{km}n_m\,dS.$$

The first part is due to a body force f_k; for example, gravitational

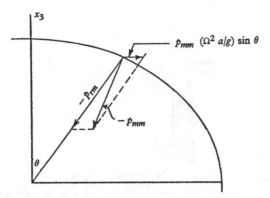

Fig. 6.5. The stress component, p_{mm}, normal to the geoid, has a component $p_{mm}(\Omega^2 a/g)\sin\theta$ in the plane normal to the x_3-axis.

attraction of the equatorial bulge. The second term is due to surface stresses p_{km} in the direction k on a surface element normal to n_m. As a special case we consider a smooth geoid whose surface is everywhere normal to local gravity. The radial components of surface stress do not produce a torque. The non-radial components (fig. 6.5) are

$$\left.\begin{aligned} p_{1m} &\approx p_{\theta m}\cos\theta\cos\lambda - p_{\lambda m}\sin\lambda + p_{mm}(\Omega^2 a/g)\sin\theta\cos\lambda \\ p_{2m} &\approx p_{\theta m}\cos\theta\sin\lambda + p_{\lambda m}\cos\lambda + p_{mm}(\Omega^2 a/g)\sin\theta\sin\lambda \\ p_{3m} &\approx -\,p_{\theta m}\sin\theta \end{aligned}\right\},$$

where p_{mm} is the component of stress along the normal to the geoid (the summation convention does not apply here). Hence

$$\left.\begin{aligned} L_1 &= a\int_S [-\,p_{\theta m}\sin\lambda - p_{\lambda m}\cos\theta\cos\lambda \\ &\qquad\qquad\qquad - p_{mm}(\Omega^2 a/g)\sin\theta\cos\theta\sin\lambda]\,\mathrm{d}S \\ L_2 &= a\int_S [p_{\theta m}\cos\lambda - p_{\lambda m}\cos\theta\sin\lambda \\ &\qquad\qquad\qquad + p_{mm}(\Omega^2 a/g)\sin\theta\cos\theta\cos\lambda]\,\mathrm{d}S \\ L_3 &= a\int_S [p_{\lambda m}\sin\theta]\,\mathrm{d}S \end{aligned}\right\}.$$

A graphical summary of the effect of various events on the excitation pole is given in fig. 6.6. It can be seen that a local counterclockwise circulation has a similar effect as a local defect of mass; both of these circumstances are characteristic of cyclones in the northern hemisphere.

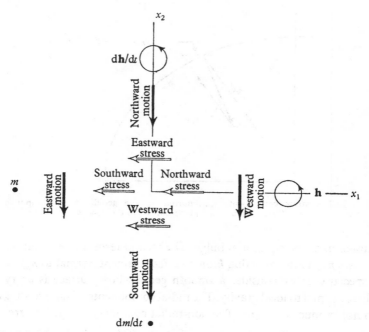

Fig. 6.6. Disturbances that will place the excitation pole on the meridian of Greenwich (ψ_1 positive, $\psi_2 = 0$). Perturbations in inertia due to a mass m at 180° has this effect, as will a positive rate of increase dm/dt at 270° east. The effect of relative angular momentum, $\mathbf{h}$ (from a horizontal vortex), and of its time rate of change, $d\mathbf{h}/dt$, are also shown. Motion toward $-x_2$ and stress toward $-x_1$ displace the excitation pole toward x_1, regardless in which quarter of the globe they occur.

9. Some idealized illustrations

Consider a merry-go-round rotating at a variable angular velocity on the Greenwich meridian at the equator. Then $h_2 = 0$, $h_3 = 0$, and from (6.1.5)

$$\psi_1 = \frac{h_1}{\Omega(C - A)}, \quad \psi_2 = -\frac{\dot{h}_1}{\Omega^2(C - A)}, \quad \psi_3 = 0.$$

This is the (non-modified) excitation computed from the point of view of the momentum approach.

The torque exerted on the support consists of the d'Alembert torque, $L_1 = -\dot{h}_1$, associated with the angular acceleration of the merry-go-round, and the gyroscopic torque $L_2 = -\Omega h_1$. When these components are substituted into (6.1.5) the foregoing expressions for ψ_i result, as of course they must.

Here the distinction between the momentum approach and torque approach is trivial. In geophysical cases it is not. Two investigations concerning the effect of motion in the core on the l.o.d., one using the momentum approach, the other the torque approach, obtained results that differ by a factor 10^5 (§ 11.12). The choice is essentially a matter of instrumentation: is it easier to measure the variable velocity of the merry-go-round, or the torque exerted on the foundation?

As a second example, consider the wobble induced by a hula dancer of mass m on the geographic north pole, moving her center of mass by an amount $l_1 = b \sin \sigma t$ (not necessarily small) along the meridian of Greenwich. Then $c_{23} = 0$, $h_1 = 0$, and

$$c_{13} = - \, mab \sin \sigma t, \quad h_2 = mab\sigma \cos \sigma t.$$

The excitation function is given by

$$\psi_1 = \frac{mab}{C-A}\left(1 - \frac{\sigma^2}{\Omega^2}\right)\sin \sigma t, \quad \psi_2 = \frac{2mab}{C-A}\frac{\sigma}{\Omega}\cos \sigma t. \quad (6.9.1)$$

The example illustrates two points. The three terms in (6.9.1) are in the ratio $1:(\sigma/\Omega):(\sigma/\Omega)^2$; for excitation frequencies small compared to Ω the term depending on the distribution of matter is dominant. Secondly, $\dot{c}_{13}$ and h_2 contribute equally to ψ_2, and this illustrates our remark that in (6.1.5) the effects of distribution in matter are not well separated from those due to relative motion. Here the wobble has been derived from considerations involving angular momentum of the Earth-dancer system. We can obtain identical results by considering the dancer as outside the planet Earth, and exerting a variable torque on the Earth by virtue of her motion.

As a third example, we compute the effect of firing a cannon ball of mass m with sufficient speed to escape the Earth's gravitational field. Suppose the cannon is located on the Greenwich meridian at some colatitude θ, firing south along a tangent to the surface. Then $\lambda = 0$, $u_\lambda = 0$, $u_r = 0$, and

$$c_{13} = + \, ma^2 \sin \theta \cos \theta H(t), \quad \dot{c}_{13} = + \, ma^2 \sin \theta \cos \theta \delta(t).$$

The quantity ρu_θ is zero just before the firing, when the resting cannon ball is part of the planet Earth, and zero after the firing when it is no longer a part. Hence

$$\dot{h}_2 = mau_\theta \delta(t).$$

Substituting term by term, the excitation function becomes

$$\psi_1 = -\frac{ma^2 \sin\theta\cos\theta}{C-A}H(t) + \frac{mau_\theta}{\Omega^2(C-A)}\delta(t);$$

$$\psi_2 = \frac{ma^2 \sin\theta\cos\theta}{\Omega(C-A)}\delta(t).$$

The ratio of ψ_2 to the second term in ψ_1 is $\Omega a \sin\theta\cos\theta/u_\theta$, and is small since the escape velocity $u_\theta = 11\cdot3$ km sec^{-1} far exceeds $\Omega a = 0\cdot46$ km sec^{-1}. Accordingly ψ_2 can be neglected. Each term in ψ_1 is associated with a circular motion of the pole of rotation. For the term containing $H(t)$ the pole revolves with a radius R about a position displaced by R from its original position; for the spike function the pole revolves with radius R' about the original position. According to (6.7.2, 3)

$$R = \frac{ma^2 \sin\theta\cos\theta}{(C-A)}, \quad R' = \frac{\sigma_0 mu_\theta a}{\Omega^2(C-A)}.$$

The ratio

$$\frac{R}{R'} = \frac{a\Omega}{u_\theta}\frac{\Omega}{\sigma_0}\sin\theta\cos\theta = (0\cdot041)(420)\sin\theta\cos\theta = 17\sin\theta\cos\theta$$

is large except very near the equator and the poles. Hence if we are to fire a projectile at escape velocity the most effective means of moving the pole is to place the cannon at $\theta = 45°$ and to fire the largest possible projectile that can be removed from the Earth's gravitational field.* But then it is easier and equally effective to fire

---

* Jules Verne has given a detailed discussion of turning the Earth by such means in an inaccessible book, 'Sans dessus dessous' (Verne, 1889, pp. 152, 187, 200) and we are indebted to Monsieur I. H. P. Eyries of the Service Central Hydrographique for tracing this book in the French National Library. Jules Verne reports great interest on the part of *le Gouvernement de Washington* in a report by the engineers of the 'North Polar Practical Association' to fire a projectile of 180,000 tons in order to displace the pole 23° and so to remove the obliquity of the ecliptic. Subsequently the French engineer Pierdeux discovered that the Earth's equatorial bulge had been neglected in the original calculations and, when this is allowed for, he obtained a polar displacement of 3 microns. Setting $m = 1\cdot8 \times 10^{11}$ g and $\theta = 45°$ gives

$$Ra = \frac{\frac{1}{2}ma^3}{C-A} = 0\cdot1 \text{ micron,}$$

the projectile to the equator (or poles), and there are cheaper ways of moving matter than by ballistics. The polar displacement (in radians) is of the order of the ratio of the displaced mass to the mass of the equatorial bulge, and presumably small.

or one thirtieth of Pierdeux's result. The story has a modern sequel. During the 1956 U.S. Presidential election the vice-presidential candidate, Senator Estes Kefauver, reported that the Earth's axis would be displaced by 10° as a result of hydrogen bomb tests. Suppose an energy of 10^{24} ergs could be available to fire a projectile at escape velocity. Its mass is $10^{24}/\frac{1}{2}u^2 = 1\cdot6 \times 10^{12}$ g, or roughly ten times that reported by Jules Verne, and the resulting displacement is then 1 micron. After seventy years, the government in Washington still refuses to recognize the existence of the equatorial bulge.

CHAPTER 7

OBSERVATIONS OF LATITUDE

1. A historical note*

Following Euler's original suggestion in 1765 that the Earth might undergo a free nutation with a period of $A/(C - A)$ sidereal days, Peters in 1841, Bessel in 1842, and Maxwell in 1851 all searched for changes in latitude with a ten-month period. The results were disappointing, indicating variations by less than $0.''1$ and of doubtful significance. Lord Kelvin suggested that the results might be significant after all. His optimism was based on geophysical considerations: he had estimated that shift in air mass alone should induce a wobble by something like $0.''05$ to $0.''5$. At Kelvin's request Newcomb analyzed the latitude of Washington, D.C., in the years 1862 to 1865 for a ten-month variation and obtained an amplitude $0.''05 \pm 0.03$. This result was announced by Kelvin in 1876 during his Presidential Address before the British Association as supporting evidence for the existence of a free nutation. Newcomb himself was skeptical.

As it turned out, Kelvin was right about the effect and wrong about the frequency. The correct solution was found, perhaps characteristically, during an investigation which was undertaken for a different purpose entirely and with no pre-conceived notion of a latitude variation of *any* frequency. In 1884 Küstner in Berlin began a short series of measurements to determine the constant of aberration, using small differences of stellar zenith distances in the manner devised by Talcott, of the Corps of Engineers, U.S. Army. He was surprised to find a quasi-annual variation of this constant. Having carefully examined all possible sources of error, he came to the conclusion that this was due to a latitude variation of $0.''2$. The result was announced at the Salzburg Congress in 1888, and the matter was promptly taken up by the International Geodetic Association.

* Much of the material is from Lambert *et al.* (1931); articles by Melchior (1957) and Larmor (1909) have been helpful. Sevarlic (1957) has compiled a bibliography comprising 1761 papers pertaining to latitude work.

A crucial test followed in 1891. Simultaneous measurements of latitude were conducted at Waikiki and Berlin. The two stations are almost exactly 180° apart in longitude so that the curves of variation should be opposite in phase. The expectation was confirmed in a striking manner. The recorded range was about $0''5$. Accordingly, Professor Förster, of Berlin, informed Lord Kelvin that Kelvin's prediction of 1876 had been confirmed, and 'that irregular movements of the Earth's axis to the extent of half a second may be produced by the temporary changes of sea level due to meteorological causes'.*

Meanwhile S. C. Chandler, a prosperous merchant in Cambridge, Massachusetts, had already begun his analysis of the variation. He was able to trace variations in latitude as far back as the time of Bradley, more than two hundred years earlier, and to show that many of the discouraging discrepancies encountered since that time were due to latitude variations. One of Chandler's first announcements (Chandler, 1891a) was to the effect that the observations showed, in addition to an annual term, a term with a period of 428 days, 40 per cent longer than Euler's classic value. This result was entirely unexpected and raised doubts concerning the validity of the observations. But only a year later Newcomb was able to demonstrate that the yielding of the Earth and Oceans could bring about just such an increase in period from ten to fourteen months (Newcomb, 1892). He attributed one-fourth of the increase to the mobility of the oceans, the remaining part to an elastic yielding of the Earth. With remarkable foresight Newcomb suggested† that the latitude observations may yet prove one of the best means of determining the rigidity of the Earth.

At the very start Chandler (1891b) discussed also the possibility of a secular term in latitude, but was unable to detect such a change. This is substantially the position today. Chandler also reported (1892) to have detected rather substantial fluctuations in the *period* of the non-annual term. Newcomb's (1892) reaction is still applicable to modern proposals of this concept of a variable period of nutation (§ 10.2): 'But the question now arises how far we are entitled to

---

* Lord Kelvin, Presidential Address, Royal Society, November 30, 1891.
† In a letter to Lord Kelvin, quoted in Kelvin's Presidential Address, Royal Society, November 30, 1892.

60 THE ROTATION OF THE EARTH

assume that the period must be variable. I reply that, perturbations aside, any variation of the period is in such direct conflict with the laws of dynamics that we are entitled to pronounce it impossible.'
The great interest in these discoveries led at once to the establishment of the International Latitude Service (ILS). To use the same stars, stations were chosen along a single latitude, 39° 08' North,

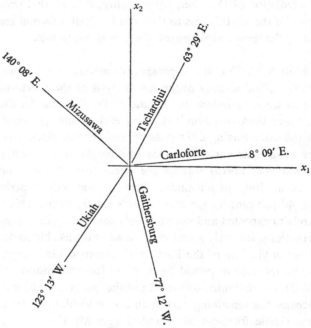

Fig. 7.1. The principal stations of the International Latitude Service. The stations lie along the indicated longitudes and in latitude 39° 08' North.

and distributed as well as possible in longitude (fig. 7.1). By the turn of the century all stations were reporting latitudes. Only four of the original stations are now operating: Mizusawa, Japan; Carloforte on a small island off Sardinia; and two U.S. stations, Ukiah, California, and Gaithersburg, Maryland. The latter was closed from 1914 to 1931 for reasons of economy. The station at Tschardjui, USSR (otherwise Charjui and various other transliterations of Russian) had to be shifted because of a change in the course of the river Amu Darya, and has been replaced by Kitab some 3° to the west. In the southern hemisphere observations are now made at La Plata, Argentina. The program has been noteworthy for its continuity in observations; during the first World War the stations were operated under the auspices of a 'Reduced Geodetic Association among Neutral Nations', organized to keep geodesy alive during the conflict.

Unfortunately, the program of the observations and the reduction of the data have lacked such a degree of continuity. There have been three major epochs: the German era from 1900·0 until 1922·7 under the direction of T. Albrecht, B. Wanach and H. Mahnkopf; the Japanese era from 1922·7 until 1935·0, under the direction of H. Kimura; and the Italian era from then until the present time under L. Carnera and G. Cecchini. Changes in the catalogue of appropriate zenith stars have introduced inhomogeneities into the observations which make the interpretation troublesome.

Until recently all ILS observations were conducted by means of visual zenith telescopes. In 1912 a Cookson floating zenith telescope was installed at Greenwich, and in 1915 a Ross photographic zenith tube* (PZT) at the U.S. Naval Observatory, Washington, and later in Richmond, Florida. Subsequently, a PZT was installed at Greenwich. The ILS station at Mizusawa now has a PZT, and it is hoped other ILS stations can be similarly equipped.

Danjon's 'impersonal astrolabe', which is of the same order of accuracy as the PZT, is now being used extensively. During the International Geophysical Year 22 visual zenith telescopes, 10 PZT's and 16 Danjon Astrolabes were in operation at over thirty stations, including Canberra (Australia); Uccle (Belgium); Ottawa (Canada); Tsientsin, Shanghai (China); Herstmonceux (England); Quito (Ecuador); Alger-La Bouzareah (France); Potsdam (GDR); Dehra Dun (India); Milan (Italy); Tokyo (Japan); Tananarive (Madagascar); Wellington (New Zealand); Borowiec, Poznan (Poland); Neuchâtel (Switzerland); Washington, Richmond, San Diego, State of Hawaii (USA); Blagoveschensk, Kazan, Gorky, Irkutsk, Poltava, Pulkovo, Moscow (USSR); Belgrade (Yugoslavia); in addition to the ILS stations.

---

* In 1896 photographic methods for the operation of the ILS were discussed; even then, photographic observations seemed preferable to visual.

2. Methods of observation

Determination of wobble involves precise measurements of latitude (§ 2.1). The meridian circle provides the fundamental method for determining latitude. This instrument consists of a refracting telescope that can be rotated about a horizontal axis oriented east-west. From the observed zenith distance, z_U, at the upper culmination

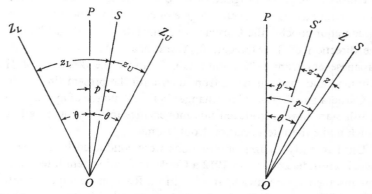

Fig. 7.2. The celestial pole is at P, the zenith at Z. In the fundamental method (*left*) the zenith distances, z_U and z_L of a star S are measured at upper and lower culmination. In the method of Talcott (*right*) the zenith distances, z and z', of two stars at north polar distances p and p' are measured.

(when the star is nearest the zenith), and the zenith distance, z_L at the lower culmination 12 sidereal hours later, one can obtain (fig. 7.2)

$$\theta = \tfrac{1}{2}(z_L + z_U), \quad p = \tfrac{1}{2}(z_L - z_U). \qquad (7.2.1)$$

The first expression gives colatitude θ (90°—latitude); the second equation the north polar distance p (90°—declination). This method is known as the *fundamental* method of determining latitude and declination, because the value of each of these is derived independently of a knowledge of the other.

Talcott's method consists of measuring the *difference* between the nearly equal zenith distances of two stars which pass the meridian within a few minutes of each other, one north and the other south of the zenith, and not very far from it. The angles associated with the two meridian passages are (fig. 7.2)

$$\theta = p' + z', \quad \theta = p - z,$$

whence, upon addition,

$$\theta = \tfrac{1}{2}(p' + p) + \tfrac{1}{2}(z' - z). \qquad (7.2.2)$$

The value of $(z' - z)$ is measured; the mean declination of the pair of stars, $\tfrac{1}{2}(p' + p)$, has to be known. Talcott's method is then not a fundamental method. Nevertheless it is the one usually employed because of two advantages: the use of zenith stars minimizes the error of optical refraction, and the measurement of the small angle $z - z'$ by means of a micrometer is far more accurate than measuring z_U and z_L on a graduated circle. In the Talcott method the classical instrument is the zenith telescope. By means of a precision level the telescope is set at the proper zenith distance for the star that first comes to the meridian. As the star passes the meridional (or N–S) cross hair, its distance north or south of the central E–W cross hair is measured by the micrometer. The telescope is then swung 180° about a vertical axis so that it points north of the zenith (if it was south before), and, if necessary, the telescope is readjusted so that the level is again horizontal. The essential point is that the angle between the telescope and the level remains undisturbed. The telescope is then positioned at the same zenith distance as previously, but on the opposite side of the zenith. The micrometer measurement is repeated for the second star, and the comparison of the two measurements give $z' - z$ directly without the need of depending upon a graduated circle.

There have been improvements over the zenith telescope. In the photographic zenith tube (PZT), the horizontal level has been replaced by a free surface of mercury from which the star images are reflected on photographic plates which are rotated about a vertical axis. Multiple exposures of a *single* star are taken before and after rotation, and the distance, $2z$, between images is obtained with a 'measuring engine'. The colatitude, θ, equals $p + z$ if the star is north of the meridian and $p - z$ if it is south. From a series of such exposures the time of meridian transit can also be determined.

The 'impersonal astrolabe' is based on a method altogether different from that of Talcott. A direct ray from a star, and a ray reflected by a mercury basin, enter the objective via a 60° prism, forming two images. When the images coincide, the zenith distance of the star is 30° (Danjon, 1958). The determination of time and latitude are interdependent (unlike the PZT), and a knowledge of the star's right ascension is required in addition to its declination.

3. Methods of reduction

The amplitude of the wobble is of the order of $0''1$. The probable error, based on internal consistency, of a single observation of one star is of the same order. Something like a thousand observations go into a monthly station average, so that such an average should be reliable to better than $0''01$ if the errors were random. A comparison of the observed latitude of Washington, D.C., with that inferred from the ILS indicates discrepancies as large as $0''1$ (see fig. 7.4). This and other evidence would indicate that there are important systematic errors.

The reduction of the ILS observations follows an elaborate scheme. The central difficulties involve the screw pitch of the micrometer and the declination of latitude stars. Neither is known *a priori* with sufficient accuracy for the reduction, and in the final analysis the latitude observations themselves have to provide the information for the various corrections. The corrections for the screw pitch include an annual term which may be as large as the annual motion of the pole itself! No wonder Melchior (1957) cautions that 'the geophysicist should be aware of the effects of inaccuracies in the screw pitch . . ., and should receive the provisional results with caution'.

To minimize errors due to screw pitch the group of 'latitude stars' (which are arranged in 'Talcott pairs') are so chosen that the *sum* of all micrometric measurements in one night is as small as possible. Unfortunately, the coordinates of the latitude stars are so affected by precession that after a decade or so some of the Talcott pairs are unsatisfactory and have to be replaced. Changes in the catalogue were made in 1912·0, 1922·7, 1935·0, and 1955·0.

In Cohn's star catalogue, which was used from 1899 to 1935, some declinations are known to be in error. The more accurate Boss's Catalogue has been used since 1935, but it also contains troublesome errors. Finally, the number of stations has varied from three (1922·7 to 1935·0) to six (1901·7 to 1906·0). All these changes seem to have introduced serious inhomogeneities into the observational material.

The *definitive* coordinates, which are published with a delay of some years after the *provisional coordinates*, include corrections for screw pitch and star declination; and according to Melchior these 'corrections deserve greater confidence than had been thought'. The reader is referred to Walker and Young (1957) and to Melchior's recent review for detailed information.

In the notation adapted by the International Latitude Service, the computed departure in colatitude, $\Delta\theta_u$, of a station u on *west* longitude ψ_u, is given by

$$- \Delta\theta_u(t) = x(t) \cos \psi_u + y(t) \sin \psi_u + z(t), \qquad (7.3.1)$$

where $x(= m_1)$ is the displacement of the pole of rotation towards Greenwich, $y(= - m_2)$ the displacement towards 90° *west* of Greenwich, and z a correction term introduced by Kimura. The coordinates x, y, z, are determined by a least square fit of $\Delta\theta_u$ to the observed latitudes at all stations.

The Kimura term represents a non-polar variation in latitude, as if all ILS stations increased or decreased their latitude simultaneously. It is of the order of 0″03. This term absorbs errors in the adopted values of the proper motions and declinations of the observed stars, the effect of neglected parallaxes, and errors in the fundamental constants (aberration and nutation). The term is not eliminated by replacing the zenith telescope with the PZT.

It has been noticed that $z(t)$ tends to have the same sign in the two hemispheres. (Kimura *latitude* departures are of opposite sign in the northern hemisphere as compared to the southern hemisphere.) This is the expected behavior from a shift in the Earth's center of mass, resulting, for example, from a seasonal flux of matter across the equator. The problem warrants further study.

The residual variation, which is uncorrelated between stations, is sometimes called the local Kimura term: it is simply the measured latitude minus that computed according to (7.3.1). This local variation is related to wind, pressure, and other meteorological variables.

Optical refraction is minimized by the use of zenith stars, but even 'zenith refraction' can introduce troublesome effects. It may be profitable to distinguish here between 'room refraction' associated with conditions immediately surrounding the telescope, and all the remaining atmospheric refraction. Przbyllok (1927) has compared the measured latitudes at the Naval Observatory, Washington, D.C., with those determined from the ILS. He finds the departure to be correlated with local wind direction; during northwest winds the Washington latitude is too large by $0''02$, during southeast winds too small by $0''02$. He attributes the effect to room refraction. Przbyllok suggests that monthly values at a station might deviate by as much as $0''25$, yearly values by $0''1$! During the last two years of operation of the original station at Tschardjui, the mean latitude was found to vary by $0''1$. Lambert (1922) attributes this anomalous observation to a change in refraction due to changed atmospheric conditions as the river Amu Darya shifted its course towards the station from its original channel some three kilometers away (surprisingly the effect on the local vertical is negligible). Under ordinary circumstances one might expect much of the seasonal variation in refraction to be absorbed in the Kimura z-term, and a correlation between room refraction and the z-term has in fact been found (Lambert *et al.* 1931, p. 269). But there is no assurance that a substantial error due to refraction does not remain in the computed annual variation in latitude. There also may be a drift in the annual mean of zenith refraction, associated with the change in climate during the twentieth century. This calls for further study.

Certain small astronomic effects are usually allowed for in the latitude determination: they include Battermann's effect in the aberration and small nutation terms due to perturbation of the Earth's orbit by Jupiter and Saturn. Declination is affected also by a sway of the rotation axis relative to the invariable axis (§ 6·6). Any geophysical event causing wobble must be associated not only with a variation in latitude but also with a variation in declination. In the calculation of latitude according to Talcott's formula (7.2.2) the declination is not corrected for sway. But the sway is of the order of the latitude variation times the precessional constant $H(= 0\cdot003)$ and certainly negligible. Chandler (1892) searched for empirical evidence of sway and found none: '... the comparison of the absolute and differential determination show that the phenomenon pertains entirely to a variation of the zenith, and in no part to a simultaneous variation of the zenith and the astronomical pole'.

4. Tidal disturbance in latitude

Tidal disturbances result in a deflexion of the vertical by as much as $0\rlap{.}''02$, and the tidal distortion of the solid Earth results in a movement of observatories relative to the pole by a few centimeters. Such disturbances have nothing to do with a wobble of the pole, yet they represent real changes in *astronomic* colatitude, this having been defined as the angle between the rotation axis and the local vertical. From the point of view of this subject a more appropriate definition of latitude would be one where the local vertical is replaced by a line from the Earth's center of mass through the observatory. This *geocentric* latitude is still subject to tidal disturbances because of the shift of the observatories relative to the pole, but the effect is much smaller, of the order of $0\rlap{.}''001$. In computing the position of the pole according to (7.3.1) most of the tidal disturbance is removed by the least square fit or absorbed in the Kimura z-term; still it might be safer first to subtract the latitude tides, and this can be done. The problem deserves further study.

The tidal effect on latitude is derived as follows: let U be the gravitational potential due to Sun or Moon. The tidal force has the component $(1/a)\,(\partial U/\partial\theta)$ in the direction θ (southward); the resulting tidal bulge has an additional component $(k/a)\,(\partial U/\partial\theta)$ (see § 5.1). The tide-producing body and tidal bulge pull a plum bob towards them (fig. 7.3) and, accordingly, the upward vertical is deflected away from body and bulge by an amount

$$\alpha = -\frac{1+k}{ga}\frac{\partial U}{\partial\theta}.$$

In the figure, $\partial U/\partial\theta$ is positive, and α is negative.

But the presence of a tidal bulge indicates that material has been drawn towards it. In the situation shown in fig. 7.3, the observatory has been displaced south by an amount $(l/g)(\partial U/\partial\theta)$, and the resulting increase in colatitude is $(l/ga)\,(\partial U/\partial\theta)$. The total increase in colatitude equals

$$\delta\theta = -\frac{1+k-l}{ga}\frac{\partial U}{\partial\theta}. \qquad (7.4.1)$$

Since $1 + k - l = 1 + 0.29 - 0.07 = 1.22$ is positive, the net deflexion of the upward vertical is away from the tide-producing body and bulge. A systematic derivation is given by Jeffreys (1952, § 7.02).

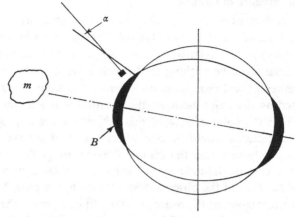

Fig. 7.3. The tide-producing body m and the resulting tidal bulge B both pull a plumb bob toward them, so that the upward vertical is deflected by an angle α in the opposite direction. For the situation shown, the colatitude is reduced by an amount α.

Convenient formulae for computing equilibrium tides can be found in the Admiralty Manual (Doodson and Warburg, 1941, table 7.1). The tide potential can be written

$$U = gK_{\mathbb{C}}bf(\theta) \cos [\beta(\lambda, t)], \qquad (7.4.2)$$

where
$$K_{\mathbb{C}} = \frac{3}{2} \frac{m_{\mathbb{C}}}{M} \frac{a^3}{r_{\mathbb{C}}^3} a = 53.7 \text{ cm} \qquad (7.4.3)$$

is the 'general lunar coefficient', $m_{\mathbb{C}}$ the Moon's mass, M the Earth's mass, $r_{\mathbb{C}}$ the Moon's distance, and a the Earth's radius. The convention is to refer solar as well as lunar tides to this coefficient $K_{\mathbb{C}}$. For some of the larger tidal components, table 7.1 lists the amplitude factor b, the latitude function $f(\theta)$ and the phase $\beta(\lambda, t)$, which depends on the east longitude λ of the station and the Greenwich mean time t.

For the M_2 component, the increment in colatitude has the amplitude

$$\delta\theta = (1 + k - l)K_{\mathbb{C}}(b/a) \sin \theta \cos \theta \cos 2\lambda$$
$$= 0\overset{''}{.}0192 \sin \theta \cos \theta \cos 2\lambda \qquad (7.4.4)$$

Table 7.1. Parameters for some equilibrium tides

Species	Symbol	Period	b	$f(\theta)$	$\beta(\lambda, t)$
Long period	Lunar	18·6 years	0·066	$\frac{3}{4}(\frac{1}{3} - \cos^2\theta)$	$(N_{☾} - N_{☾_0})$
	Sa	1 year	0·012		$(☉ - 1°.8)$
	Ssa	$\frac{1}{2}$ year	0·073		$2(☉ - 79°.8)$
	MSm	$31^d.85$	0·016		$(☾ - ☉ + p_{☾})$
	Mm	$27^d.55$	0·083		$(☾ - p_{☾})$
	MSf	$14^d.77$	0·014		$2(☾ - ☉ + 79°.8)$
	Mf	$13^d.66$	0·156		$2☾$
	$\cdots$	$13^d.63$	0·065		$2☾ + N_{☾} - N_{☾_0}$
Diurnal	O_1	$25^h.82$	0·377	$\sin\theta\cos\theta$	$(qt + ☉ - 2☾(- 169°.8 + \lambda)$
	P_1	$24^h.07$	0·176		$(qt - ☉ - 10°.2 + \lambda)$
	K_1	$23^h.93$	0·531		$(qt + ☉ + 10°.2 + \lambda)$
	N_2	$12^h.66$	0·174	$\frac{1}{2}\sin^2\theta$	$2(qt + ☉ - \frac{3}{2}☾ + \frac{1}{2}p_{☾}(- 79°.8 + \lambda)$
Semi-diurnal	M_2	$12^h.42$	0·908		$2(qt + ☉ - (- 79°.8 + \lambda)$
	S_2	$12^h.00$	0·423		$2qt$
	K_2	$11^h.97$	0·115		$2(qt + ☉ - 79°.8 + \lambda)$

☉ is the longitude of the 'mean Sun', increasing by $0°.0411$ per mean solar hour.

☾ is the mean longitude of the Moon, increasing by $0°.5490$ per mean solar hour.

$p_{☾}$ is the mean longitude of lunar perigee, increasing by $0°.0046$ per mean solar hour.

$N_{☾}$ is the mean longitude of the lunar ascending nodes, increasing by $- 0°.0022$ per mean solar hour.

q is the angular velocity of the Earth relative to the mean Sun, $15°$ per mean solar hour.

$\Omega = q + \dfrac{d☉}{dt}$ is the angular velocity of the Earth relative to the stars, $15°.0411$ per mean solar hour.

whereas, for the annual and semi-annual terms,

$$\delta\theta = 0\rlap{.}''0003 \sin\theta\cos\theta, \quad \delta\theta = 0\rlap{.}''0023 \sin\theta\cos\theta, \quad (7.4.5)$$

respectively. The annual and semi-annual terms are small, and no correction appears to be required. But there is a further complication. Suppose latitudes are measured each night at exactly the same time. Let the first observation coincide with the crest of the P_1 tide (period 24^h07). The observation on the following night falls 0·07 hours ahead of the crest. In roughly 180 days the observations coincide with low tide, and in a year they coincide again with high tide. The measurements look exactly as if there had been a one-year tide with amplitude of the P_1 component.

This is the phenomenon of 'aliasing' (see Appendix A.2). In the analysis for an annual tidal disturbance (frequency f) from observations taken at daily intervals, Δt, we have

$$f = \dot{\odot}/2\pi, \quad (\Delta t)^{-1} = q/2\pi,$$

where $\dot{\odot} = \mathrm{d}\odot/\mathrm{d}t$ and q are the angular velocities of the mean Sun and of the Earth relative to the Sun (table 7.1). According to (A.2.14) the effect of the following tides is then entirely included in an analysis for the annual tide:

$$K_1, \quad \text{angular frequency} \quad q + \dot{\odot},$$
$$P_1, \quad \text{angular frequency} \quad q - \dot{\odot}.$$

Thus K_1 and P_1 are 'synodic' with the annual tide. For the semi-annual tide, frequency $2\dot{\odot}$, the semi-diurnal component

$$K_2, \quad \text{angular frequency} \quad 2q + 2\dot{\odot}$$

is synodic, and for the MSf tide, frequency $2(\dot{\mathbb{C}} - \dot{\odot})$,

$$M_2, \quad \text{angular frequency} \quad 2q + 2(\dot{\mathbb{C}} - \dot{\odot})$$

is synodic. Allowance has to be made for the synodic tides in addition to the long-period tides; and since the synodic tides are larger, their effect is more important. The fact that the latitude program does not consist of single nightly observations but a series of observations centered at a fixed time reduces the aliasing effect but does not eliminate it. The amplitude of the annual component is of the order $0\rlap{.}''1$. The aliased variation might conceivably amount to $0\rlap{.}''01$. A larger water tide in the vicinity of a station would further contribute to this effect. The aliased amplitude for the semi-annual tide is much smaller, but so is the observed variation in latitude. The problem deserves further study.

There have been a number of attempts to detect short-period variations in latitude as a means of evaluating $1 + k - l$. An analysis by Nishimura (1950) for the M_2 tide at ILS stations gave the following values:

Carloforte, $1 \cdot 08 \pm 0 \cdot 06$ Tschardjui, $1 \cdot 31 \pm 0 \cdot 19$

Ukiah, $1 \cdot 06 \pm 0 \cdot 06$ Cincinnati*, $1 \cdot 66 \pm 0 \cdot 18$

The average is $1 \cdot 20 \pm 0 \cdot 10$. There is an intriguing suggestion of a difference between the oceanic and continental stations. Table 7.2 gives the results of a similar analysis at Greenwich and Washington. In the Greenwich observations allowance has been made for the gravitational attraction by the tide in the river Thames. The Washington observations have been corrected for the ocean tide as well as some minor effect due to tides in the Potomac river. The table gives the amplitude of latitude variation due to the water tide only (neglecting load deformation), and the amplitude of the total observed variation, including water tide.

Table 7.2

Location	Source	Interval	Instrument	Observed	Water tide	$1 + k - l$
Greenwich	Spencer Jones, 1939b,	1911–1936	Cookson Telescope	$0{\cdot}^{\prime\prime}0050$	$0{\cdot}^{\prime\prime}0021$	$0 \cdot 92 \pm 0 \cdot 17$
Washington	Markowitz and Bestul, 1941	1916–1940	PZT	$0{\cdot}^{\prime\prime}010$	$0{\cdot}^{\prime\prime}0014$	$1 \cdot 4 \pm 0 \cdot 2$

The foregoing values are similar to the value $1 + k - l = 1 \cdot 22$ computed by Takeuchi (1950) from altogether different considerations (§ 5·1). The latitude tides are believed to furnish the most accurate estimates of the number l (assuming k is known). Measurements of the variable distance between two points have led to rather inaccurate determination of l but as far as they go they are not inconsistent with the latitude results.

* Cincinnati Observatory is at latitude 39° 08′ and has sometimes collaborated with stations of the International Latitude Service.

Table 7.2 emphasizes the relative importance of water tides on latitude observations. A discussion by Robinson in 1804 concerning the deflexion of the vertical by the great tides in the Bay of Fundy is referred to by Thomson and Tait (1883, § 818). The effect depends critically on the distance from the water line. Consider the gravitational attraction due to an infinitely long strip of water of density ρ, thickness h, extending from the coast line $(x = a)$ to a distance b offshore:

$$\rho G h \int_{-\infty}^{\infty} \int_{a}^{a+b} \frac{x}{(x^2 + y^2)^{\frac{3}{2}}}\, dx\, dy = 2\rho G h \ln \left(\frac{a + b}{a} \right).$$

The angular deflexion of the vertical is this quantity divided by g. For $a = 10$ km, $b = 1000$ km, and a tidal amplitude of $h = 100$ cm, the deflexion is 6.3×10^{-8} radians, or $0''.013$.

Finally, it may be noticed that variations in rotation also produce changes in the local vertical and hence in latitude. The potential arising from rotational deformation is of the same type as the tidal potential. Any wobble will therefore introduce additional changes in latitude which can be computed according to (7.4.1) by setting U equal to the rotational potential (5.2.2):

$$U = -\,\omega_1\omega_3 x_1 x_3 - \omega_2\omega_3 x_2 x_3 = -\tfrac{1}{2}a^2\Omega^2|\mathbf{m}|\sin 2\theta,$$

where $|\mathbf{m}| = m_1 \cos \lambda + m_2 \sin \lambda$ is the instantaneous displacement of the rotation pole towards the station at θ, λ. The correction

$$\delta\theta = (1 + k - l)\frac{a\Omega^2}{g}|\mathbf{m}|\cos 2\theta = 0.00422\,|\mathbf{m}|\cos 2\theta \quad (7.4.6)$$

is negligible compared with the variation in latitude $|\mathbf{m}|$ of the station.* For a fractional variation m_3 in the l.o.d. we have

$$U = \Omega^2 a^2 m_3 \sin^2 \theta, \quad (7.4.7)$$

$$\delta\theta = -(1 + k - l)\frac{a\Omega^2}{g} m_3 \sin 2\theta, \quad (7.4.8)$$

or $\delta\theta = 0''.000010 \sin 2\theta$ for $m_3^{-1} = -0.865 \times 10^8$, corresponding to an increase in the l.o.d. by 1 ms.

---

* But this minute disturbance associated with the annual wobble of about 10 ft is actually larger than that associated with annual variation by 2 per cent in the Sun's distance! The latter effect gives rise to the annual tide Sa and is included in all textbooks on the subject (see Munk and Haubrich, 1958).

5. Summary of observations

The most detailed analysis of the ILS observations is the one by Walker and Young (1957). The best results are obtained from the unsmoothed monthly values (their table 1a) in accordance with (7.3.1). The first three curves in fig. 7.4 are based on these values. Earlier analyses by Ledersteger (1949), Jeffreys (1940) and Pollak (1927) are based on smoothed values summarized in table 1b of Walker and Young (1957).

The only other comparable series of data are those from Greenwich (systematically reduced by Spencer Jones*) and Washington (Markowitz, 1942, and subsequent papers in the *Astronomical Journal*). The superior instruments at these observatories together with the uniformity in the program makes up for the fact that these are not located at the same latitude. The last two curves of fig. 7.4 show the component $- m_2$ from the unsmoothed ILS observations (corresponding to latitude variation at 90° W longitude) and the latitude of Washington (located at 77° W), both after removal of the seasonal terms.† The main features are duplicated, but there are considerable differences in detail.

The first two curves of fig. 7.4 show that the latitude variations consist of a seasonal term plus a fourteen-month variation of variable amplitude. Rudnick's (1956) power spectrum (fig. 7.5) may serve to summarize the essential features of any of the latitude observations.‡ Rudnick computed harmonics 40–62 of the 54·4 year record. The principal results are:

(1) 98·5% of the power is associated with positive (west-to-east) motion.

(2) 93% of the power is contained in the frequency range 0·74 to 1·14 cycles per year.§

(3) 22% of the power in this range is in an annual line without

---

* Royal Observatory, Greenwich, 1939. Observations made with the Cookson floating telescope.
† This is accomplished by subtracting from each January the mean of all Januaries etc.
‡ Rudnick uses Kulikov's (1950) summary of the latitude observations from 1891 to 1945 at 0·1 year intervals. These have been reproduced by Struve (1952). The material is considered inferior to that used by Walker and Young.
§ Henceforth designated by the abbreviation c/year.

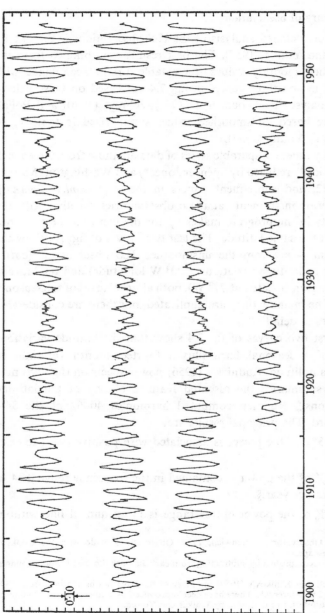

Fig. 7.4. The component, m_1, of the unsmoothed ILS observations, before (*top*) and after (*second curve*) removal of the seasonal variation; the component, $-m_2$, of the unsmoothed ILS observation after removal of the seasonal variation (*third curve*) and the corresponding non-seasonal variation in the latitude of Washington, as obtained with the PZT (*bottom*).

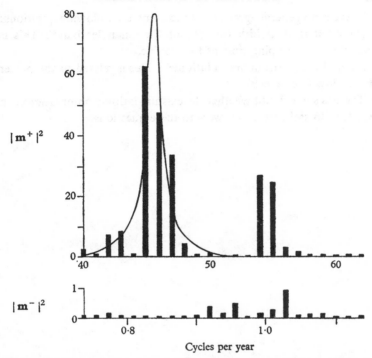

Fig. 7.5. The spectrum of variation in latitude, according to Rudnick (1956). The upper figure refers to the positive (west-to-east) motion, the lower figure to the negative motion of the pole of rotation (see § 6.7). For both motions the harmonics 40 to 62 are shown, with the corresponding frequency scale, in cycles per year, indicated below. The length of the spectral lines gives the contribution per harmonic toward the mean square radius arm (in units of $(0\overset{..}{.}01)^2$). The scale for negative motion is ten times that for positive motion. The curve has been fitted by the method of maximum likelihood (see Appendix A.2).

recognizable structure (in the analysis it happens to fall between harmonics 54 and 55).

(4) 78% of the power in this range is contained in the Chandler peak which is centered near 0·85 c/year and has a noticeable band-structure.*

All analyses since the day of Chandler are in agreement concerning the virtual absence of negative motion and the predominance of an annual line and a Chandler peak, but the terminology differs widely.

* Some analyses of 19-year series refer to the harmonic with frequency (15/19) c/year as the 'Witting D term' as distinct from the Chandler term at the neighboring harmonic, with a frequency (16/19) c/year. Such a distinction can serve no useful purpose.

There is no general agreement concerning the following questions:
(1) What is the width (or 'Q') of the Chandler band? This is related to the damping time of the wobble.
(2) Is there a drift in 'mean latitude'? This is related to the power at very low frequencies.

There is some doubt whether the existing latitude observations can ever lead to satisfactory answers to these questions.

CHAPTER 8

OBSERVATIONS OF THE LENGTH OF DAY

The times when stars of accurately known right ascension transit across the meridian are measured in terms of the oscillation of quartz crystals. If the crystals maintained their frequency precisely and the length of day were invariable, then consecutive transits of a star should occur after some fixed number of quartz oscillations, and always at the same time of the quartz clock when expressed in proper sidereal units. In fact, transits are noted to occur later by $0\overset{s}{.}03$ (30 ms) in spring and earlier by about 20 ms in late summer, as measured by a crystal clock calibrated to keep the best possible sidereal time throughout the year. This seasonal departure indicates a variable l.o.d., or a variable frequency of the quartz crystal, or both. The fact that different quartz clocks give nearly the same result is evidence for a seasonal variation in the l.o.d. On the other hand, different quartz clocks give quite different results concerning departures from one year to the next and are therefore unsuitable as a time standard for periods longer than a year. For measuring such long-period variations a suitable time standard has been provided by Ephemeris Time (based largely on the motion of the Moon) and, more recently, by atomic frequency-standards.

1. A historical note

By 1930 the transit of a star could be determined to an accuracy of better than $0\overset{s}{.}02$, and the detection of a seasonal variation in the l.o.d. was awaiting the perfection of a clock. The first significant report of a seasonal variation came from Stoyko (1936, 1937) at the Bureau de l'Heure, Paris, based largely on the performance of pendulum clocks. Stoyko found that the l.o.d. in January exceeded that in July by 2 ms. By 1950 determinations based on quartz crystal clocks had been published by Scheibe and Adelsberger (1950), Finch (1950), and Stoyko (1950). The reported double-amplitudes of the annual variation in the l.o.d. were 2·6, 1·8, and 2·8 ms, respectively, confirming Stoyko's remarkable achievement with the use of pendulum clocks.* We now have reason to believe that the fore-going values are too large by a factor of two or three. Perhaps because these results were independently reported by French, German, and British investigators, they received more credence than they deserve. Thus, up to 1950, the amplitude of the annual term was reported around 1 ms. After 1950 the Greenwich observations gave only 0·53 ms, and this was reduced further to 0·38 ms as a result of corrections for errors in the FK3 catalogue of the right ascension of stars (Smith and Tucker, 1953). In 1951 the seasonal variation was determined also in Washington: for 1951 to 1954 Markowitz (1955) reported 0·52 ms. In Markowitz's program the observed stars form a complete band around the sky. This permits the determination of a 'closing error' so that the star positions are internally consistent. No reference to a catalogue of right ascensions is required. A value of roughly ½ ms is now adopted by all parties. The reduction in the annual term beginning in 1950 remains somewhat of a mystery. It has been suggested that the change is real, but there is little evidence to support this (§ 9.4). The earlier observations include more random errors and perhaps some systematic seasonal clock error; such errors would enhance the amplitude of any particular harmonic.

The semi-annual terms were initially determined by Smith and Tucker (1953) as 0·20 ms, and this value was increased to 0·38 ms after correction for the FK3 errors. Markowitz (1955) obtained 0·34 ms. An amplitude of about ⅓ ms is now accepted by all parties.

Some calculations by Andersson (1937), Stoyko (1951), and by Mintz and Munk (1951; 1954) had indicated that the semi-annual term was due, at least in part, to bodily tides. This suggested the existence of even shorter period terms in the l.o.d., corresponding to the lunar monthly and fortnightly tides. The amplitudes in these variations of the l.o.d. were computed to be 0·23 and and 0·43 ms,

---

* The disadvantage of pendulum clocks as compared to crystal clocks is not just a matter of precision: pendulum clocks have to be corrected for variable gravity (Jeffreys, 1928; Stoyko, 1951). A typical value of the annual potential is $U = 500 \text{ cm}^2\text{s}^{-2}$, and this is due about equally to the annual wobble and the annual tide Sa (Munk and Haubrich, 1958). The variation in g has the amplitude $\Delta g = 2a^{-1}(1 - \tfrac{3}{2}k + h)U$ (see Jeffreys, 1952; § 7.02 to 7·11); for $U = 500 \text{ cm}^2\text{s}^{-2}$ this gives $\Delta g = 2$ microgals, and $(\Delta g/g) = 0.2 \times 10^{-8}$. The resulting variation in the l.o.d., as determined by the swings of a pendulum (frequency $\sim \sqrt{g}$), is

$$\tfrac{1}{2}(0.2 \times 10^{-8})\, 0.865 \times 10^8 \approx 0.1 \text{ ms}$$

as compared to the observed variation of 0·5 ms.

respectively, and thus of the same order as the annual and semi-annual variation. But what is observed are the cumulative amounts by which the Earth is slow, and these are very small when the periods are short. It turns out that the maximum cumulative time effect is about 1 ms for both tidal terms. As late as 1953 any detection seemed out of the question because of the limits imposed by astronomic time determination. But within two years observations made with the PZT at Washington, D.C., and Richmond, Florida, on a total of 1500 nights indicated monthly and fortnightly terms of the expected amplitude and phase (Markowitz, 1955).

There have been two recent instrumental developments which increase by an order of magnitude the accuracy to which variations of periods longer than one year can be determined. The development of the moon camera (Markowitz, 1954) should make it possible to determine Ephemeris Time during one lunation to an accuracy of $0^\text{s}1$. Atomic frequency standards were established by Essen at the National Physical Laboratory, G.B., in 1955 (Essen and Parry 1955; Essen, Parry, Markowitz and Hall, 1958) and by Harris Hastings at the U.S. Naval Research Laboratory late in 1956. The drift of the quartz clocks can be checked against these standards with an accuracy of a few parts in 10^{10}. The same order of accuracy has been achieved by a group at Neuchâtel using an Ammonium Maser (Blaser and Bonanomi, 1958; Blaser and DePrius, 1958).

2. Methods of observation

In our subject the Earth is a geophysical laboratory, not a time-keeper. There are no errors in its variable rate of rotation, $m_3(t)$; there are only errors in 'reading' $m_3(t)$ from the astronomic observations. The l.o.d. is determined from a comparison with some independent time standard. The precision in this determination depends both on the precision of the astronomic observations and of the time standard. In turn, the precision of the time standard depends on two things: the precision in the clock rate and the precision to which it can be read. One or the other is usually limiting. The time of day as read on a wrist watch would not be more accurate if the point on the second hand were sharpened.

The precision in the determination of the l.o.d. depends then on

(1) the reading error in $m_3(t)$, (2) the reading error, and (3) the clock error in the time standard. It is impossible to express this precision in terms of a single number. The precision is strongly frequency dependent. Thus one speaks of the short-term (month-to-month) and long-term (year-to-year) performance of crystal clocks. A convenient (but unconventional) display is in terms of the power spectrum of 'noise' inherent in the various methods of determining

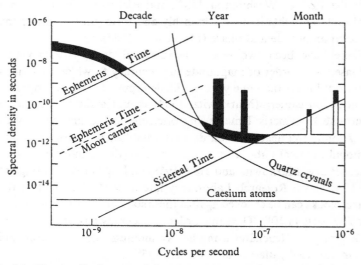

Fig. 8.1. The contribution toward the mean square error in m_3 per unit frequency band (i.e., the error spectrum in the l.o.d.) plotted against frequency (in c/s) for the four basic types of observation. The observed spectrum is indicated by the broad band and vertical 'lines', filled where these rise above instrumental noise level. The presentation is schematic.

the l.o.d. (fig. 8.1). Only if the observed spectrum is comfortably above this noise level can there be any hope of accurate determination.

We start with the reading error in $m_3(t)$. The probable error, ε, of a time sight (from one night's observations) is about 5 ms (Markowitz, 1955). The mean square error from N nights is then ε^2/N, provided the errors are independent (there is some indication of errors persisting over several days). Values of sidereal time are given at intervals of M days, and each reported value is an average of observations from N nights (N is not necessarily equal to M). At this sampling-rate we can study fluctuations in time error whose periods exceed $2M$ days, i.e., the spectrum in the frequency range 0 to the Nyquist frequency $(2M\Gamma)^{-1}$, where Γ is the length of day (see *aliasing*, Appendix A.2). The mean square error per unit frequency band, i.e. the spectral *density*, is then

$$\frac{\varepsilon^2/N}{(2M\Gamma)^{-1}} = \frac{2M\Gamma\varepsilon^2}{N}.$$

For illustration we assume $N = M$ (e.g., weekly averages at weekly intervals). The spectral density then equals

$$2\Gamma\varepsilon^2 = 2 \times 86{,}400^s \times (0^s\!005)^2 \approx 5 \text{ sec}^3. \qquad (8.2.1)$$

This is the spectral density of the time error, $\tau(t)$. The spectral density of the error in the l.o.d. is Γ times the spectrum of $d\tau/dt$; the error spectrum of m_3 just equals that of $d\tau/dt$, and this is given by (A.2.13)

$$(2\pi f)^2 2\Gamma\varepsilon^2 = (2\pi f)^2 \times 5 \text{ sec} \qquad (8.2.2)$$

where f is the frequency under consideration (see fig. 8.1: Sidereal Time). At the annual frequency, $f = 0.32 \times 10^{-7}$ c/s, the noise spectrum is 2×10^{-13} sec. These values are representative of all modern methods.

The earliest time standard against which sidereal time could be compared was provided by Ephemeris Time (ET). This is based on the orbital motion of the planets, including the Earth, about the Sun, or of satellites about their primaries. The position of these bodies is observed against the background of stars; gravitational theory gives the celestial coordinates of these bodies as a function of ET. The reader is referred to § 11.2 for a further discussion of ET. The reading error severely limits the usefulness of ET as a time standard. The Moon has the most rapid motion in celestial longitude (0"55 per sec) and is therefore the most favorable object for determining ET. The reading error has been estimated at 2^s per lunation. This gives, for $N = M$,

$$(2\pi f)^2 2\Gamma\varepsilon^2 = (2\pi f)^2 \times 1.9 \times 10^7 \text{ sec} \qquad (8.2.3)$$

for the error spectrum in m_3; this lies 4×10^6 above the error spectrum (8.2.2) due to sidereal time determination.

The development of the moon camera (Markowitz, 1954) may reduce the probable error of a time sight to $0^s\!5$, and thus reduces the noise level considerably, as shown on fig. 8.1. In this dual-rate camera the Moon and surrounding stars are exposed simultaneously for about 20 seconds. The Moon is held fixed relative to the stars by means of a tilting filter, so that both Moon and stars are sharp on the photograph. Even so, the error spectrum of ET lies orders of magnitude above that for sidereal time and constitutes the limiting factor in any determination of the l.o.d. from a comparison between ephemeris and sidereal times.

In the discussion so far of ET as an independent time standard, it has been tacitly assumed that the accuracy is limited by the reading error. If ET (for example, the Moon's position) could be read with enormous precision, then the accuracy would be limited by clock errors (e.g., defects in gravitational theory due to tidal friction, § 11.2). It is not impossible that the reading errors of ephemeris and sidereal times could be made commensurate by a rapidly revolving satellite, but then clock errors of various sorts are sure to become bothersome.

Next we turn to quartz crystal clocks. The outstanding feature is the individuality in their performance. Fig. 8.2 illustrates the performance of the two most successful clocks in the American time service (Markowitz, 1959). The curves give the clock correction, that is the amount $\tau(t)$ by which the Earth is slow. $\tau(t)$ has been split into two terms: a drift of the type $A + Bt + C\,e^{-\kappa t}$, plus a residual. The drift-curves for the two clocks are seen to separate by tens of seconds in a few years. From a comparison of sidereal and ephemeris time we know that the variable rotation of the Earth produces changes in τ by only a fraction of a second in a few years. This removes all hope for detecting year-to-year changes with quartz crystals. The two residual curves, on the other hand, differ by an amount small compared to the residual itself. This is evidence that the residual is due to variations in the l.o.d. rather than to clock errors.

The important feature is that month-to-month variation in crystal frequency is sufficiently regular to permit the detection of the annual variation in the l.o.d., whereas year-to-year variation of the crystal completely mask the long-term variation in the l.o.d. In other words, the spectrum of the clock rate, in appropriate units, falls below that of the l.o.d. for frequencies higher than 1 c/year, but is far above it for frequencies of less than 0·1 c/year, say. In fig. 8.2 we have guessed what this clock spectrum may look like, based on some performance records presented by Smith (1953). To isolate the seasonal variation from the low-frequency noise calls for a good high-pass (anti-drift) filter. The subtraction of a clock drift is one method of high-pass filtering. Some authors use an exponential

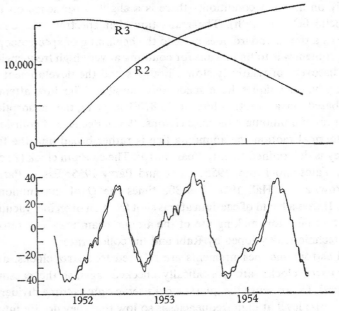

Fig. 8.2. The performance of clocks R2 and R3 (according to Markowitz, 1959). The curves give the clock correction in ms consisting of a drift (upper curves) plus a residual term (lower curves).

drift; others use power series. There is no evidence from the known physical structure of quartz crystals to favor either type of equation (Mason, 1958). In all events the constants have to be adjusted every few years. There may be better methods of high-pass filtering than subtracting a particular form of clock drift.*

Three basic shortcomings of quartz clocks are: (1) in the course of time, alterations in the internal structure (poorly understood) lead

---

* The spectrum of pendulum clocks is far richer in high frequencies than the spectrum of quartz clocks, and for this reason pendulum clocks are not well suited for measuring the annual term. A detailed comparison of the errors of pendulum and quartz clocks has been made by Greaves and Symms (1943).

to a drift in the resonance frequency; (2) stringent temperature controls are required for adequate performance; and (3) no two quartz clocks can be made precisely alike, so that if a primary standard were destroyed it would be impossible to reproduce it precisely. An entirely different time standard is provided by atomic resonances. The frequency is invariant, reproducible, and depends hardly on ambient conditions (there is a slight dependence on the magnetic field strength). The realization that spectral lines could serve as a time standard goes back to the beginning of spectroscopy. The problem was to find means for counting at very high frequencies. The discovery of relatively slow vibrations and the development of microwave techniques have made this possible. The first attempt was based on a spectral line at 23,870·1 mc/s in the absorption spectrum of ammonia molecules (Lyons, 1952). Because of collisions and thermal motion the ammonia line is rather broad and the frequency is determined to only 1 part in 10^7. The caesium clock (Zacharias, Yates and Haun, 1955; Essen and Parry 1955; Essen, Parry, Markowitz and Hall 1958) has 300 times the Q of the ammonia clock. It uses a beam of caesium atoms shot from an oven in a vacuum tank to a detector, making use of the atomic beam magnetic resonance technique developed by Rabi and his collaborators.

All astronomic measurements are referred to quartz clocks, and the quartz clocks are periodically checked against the caesium standard. In this sense the quartz clocks serve only as 'time dividers'. Their noise level at high frequencies is so low that they do not introduce an appreciable error over and above that inherent in the caesium standard. There is no need to operate the caesium oscillator on a continuous basis. Failure to appreciate this fact may have been responsible for a delay by several years in applying atomic frequency standards to astronomic measurements.

The caesium frequency standard can be used to calibrate the quartz clocks with a precision approaching $\pm$ 1 part in 10^{10}. This performance is based on a comparison of two caesium beams for an interval of the order of one hour. If the interval is much shorter, then the calibration is less accurate; however, little is gained by making it much longer. The latter statement requires some elaboration. If the uncertainty were due only to random processes, such as those associated with the finite width of the absorption line, then the difference in counts between the two caesium oscillators during an interval t would increase as $t^{\frac{1}{2}}$ on the average, and the error of the measured frequency would then decrease as $t^{-\frac{1}{2}}$ (Townes, 1951). However, the counting circuitry inevitably reacts back on the system, and ultimately limits the overall precision of the frequency standard. If the caesium atom is linked through a servo to the quartz clock, the back reaction is more severe and the precision drops to something of the order of 5 parts in 10^{10}.

If the caesium frequency standard were the only source of uncertainty, then successive daily measurements (assumed unrelated) of $m_3(t)$ would be scattered about the 'true' value by 10^{-10} parts, and the mean square error would be of the order 10^{-20}. The power spectrum of this time series is then 'white' (§ A.2) and equals

$$\frac{\varepsilon^2/N}{(2M\Gamma)^{-1}} = 2\Gamma\varepsilon^2 = 2 \times 86{,}400^s \times 10^{-20} \approx 1\cdot7 \times 10^{-15} \text{ sec,}$$

$$(8.2.4)$$

as shown.

Fig. 8.1 has been plotted for the special case $M = N$, i.e., the interval between recorded values equals the averaging time. This is not necessarily the case. One could use weekly averages at monthly intervals or monthly (overlapping) averages at weekly intervals. In the former case precision is lost; in the latter case, periods much shorter than a month are smoothed from the record, and little is gained by taking such frequent readings.

The level of the observed spectrum is indicated by the broad band in Fig. 8.1. For frequencies lower than one cycle per decade, the estimate is based on the discussion of 'turning points' (§ 11.5); for frequencies around one cycle per year and higher we refer to § 9.9. The four spectral lines which rise out of the continuum represent the annual, semi-annual, monthly and fortnightly terms. Ideally these lines are infinitely high and infinitely narrow. For any finite length of record, T, we must allow for the smearing due to the resolution, T^{-1}, imposed by the length of record. A harmonic $m_3 = M_3 \cos 2\pi f_a t$ of amplitude M_3 and frequency f_a has a mean square value of $\frac{1}{2}M_3^2$ which is smeared over a frequency interval T^{-1}. Thus $ST^{-1} = \frac{1}{2}M_3^2$, or

$$S(f) = \tfrac{1}{2}M_3^2 T \text{ from } f = f_a - \tfrac{1}{2}T^{-1} \text{ to } f + \tfrac{1}{2}T^{-1}, \quad (8.2.5)$$

where $S(f)$ is the equivalent power density of the smeared line. The annual variation of m_3 equals 0.5 ms per day, or roughly 6×10^{-9}. The value of $S(f)$ for four years of record is 2.1×10^{-9} sec, as shown on fig. 8.1. For the semi-annual term the corresponding spectral density is 0.9×10^{-9} sec. These values are comfortably above noise level. Markowitz's determination of the monthly and fortnightly lines from two years of observation corresponds to approximately the same S-level, and both lie just above the error spectrum (imposed principally by sidereal time determination).

Before 1950 the seasonal S-levels were one-fourth the present values, being based on one year (say) instead of four years of record; at the same time the combined spectral density of the errors of the quartz crystals and the sidereal time determination could well have been 1000 times above the present level. Then a frequency analysis would include about as much noise as the annual effect itself, and the apparent amplitude would be too large. Over the years, the resolution was increased and the noise level reduced, and this might account for the continuous decrease in the annual term after its early discovery.

Fig. 8.1 serves to summarize the following known features:

(1) For periods somewhat shorter than one year the accuracy in sidereal time determination is the limiting factor. This prevents the determination of the irregular variations (the continuum) for periods shorter than six months.

(2) For periods longer than one year the error spectrum of the quartz clock rises sharply, and has prevented observations of the irregular variations in that frequency range.

(3) This low-frequency noise is enormously reduced by use of the caesium frequency standard.

(4) For periods much shorter than a decade, Ephemeris Time provides a time base less accurate than crystal clocks; for periods longer than a decade Ephemeris Time is more accurate and (as it turns out) adequate for detecting changes in the l.o.d.

(5) Before the advent of atomic standards the peak in the error spectrum at about one decade (falling between the frequencies for which Ephemeris Time and quartz clocks could provide adequate precision) has prevented the study of variations in the l.o.d. with

periods between a decade and a year. The caesium standard now provides the opportunity of studying this unknown part of the spectrum.

3. Tidal disturbance in the l.o.d.

Motion of the pole of rotation along the meridian of an observatory causes a variation in latitude. Motion at right angles to the meridian displaces the meridian and has an effect on timekeeping. Accordingly, sidereal time has to be corrected for wobble.

In consequence of any polar displacement $\mathbf{m} = |\mathbf{m}|e^{il}$, a station at colatitude θ and east longitude λ is displaced northward and westward (relative to the stars) through the angles

$$-\,\delta\theta = |\mathbf{m}|\cos(l-\lambda), \quad -\,\delta\lambda = |\mathbf{m}|\sin(l-\lambda)\cos\theta. \quad (8.3.1)$$

Since

$$1 \text{ sec of time} \quad \text{and} \quad 15\sin\theta \text{ sec of arc} \quad (8.3.2)$$

are equivalent, then with $\mathbf{m}$ given in second of arc the number of seconds of time by which the stars are late (the Earth is slow) equals

$$\tau = \tfrac{1}{15}\,|\mathbf{m}|\,\sin\,(l-\lambda)\,\cot\,\theta. \quad (8.3.3)$$

This corresponds to an apparent change in the l.o.d. by

$$\Delta(\text{l.o.d.}) = \frac{\mathrm{d}\tau}{\mathrm{d}t} \times (\text{l.o.d.}) \quad (8.3.4)$$

which must be subtracted from the observed variation.

The annual wobble (frequency σ) consists almost entirely of positive (west-to-east) motion, and we may write

$$\left.\begin{array}{l} \mathbf{m} = \mathbf{m}^+\,e^{i\sigma t} = |\mathbf{m}^+|\,e^{i(\sigma t + l+)} \\[4pt] \tau = \tfrac{1}{15}\cot\theta\,|\mathbf{m}^+|\,\sin\,(\sigma t + l^+ + \lambda) \\[4pt] \dfrac{\mathrm{d}\tau}{\mathrm{d}t} = -\tfrac{1}{15}\cot\theta\sigma|\mathbf{m}^+|\,\cos\,(\sigma t + l^+ + \lambda) \end{array}\right\}. \quad (8.3.5)$$

Setting $|\mathbf{m}^+| = 0''084$ (table 9.1) and $2\pi/\sigma = 365\cdot24$ mean solar days gives $(5\cdot6\cot\theta)$ ms for the amplitude of τ, and $(0\cdot096\cot\theta)$ ms for the amplitude of $\Delta(\text{l.o.d.})$. At $\theta = 45°$ wobble accounts for one-fourth the observed annual variation in the l.o.d.! The Chandler wobble introduces a 14-month term of about twice the amplitude of the annual wobble. In fact, Markowitz (personal communication) has found a 14-month term of 10 ms from a comparison of Richmond and Washington PZT observations. As a result, corrections are now applied individually at the two stations. For the various reasons given in ch. 7, it will be difficult to determine latitude to better than $0''01$. This implies an uncertainty in timekeeping by roughly 1 ms.

Bodily tides deflect the vertical and shift stations relative to one another, hence affect the sidereal time of a place, this having been defined in terms of the time when stars of known right ascension transit across the local meridian. The situation is very similar to the one discussed in § 7.4 with regard to tidal disturbances of latitude. A tidal potential, U, is associated with a westward deflexion of the upward vertical and a westward displacement of the station, by the amounts

$$\frac{1 + k}{ga \sin \theta} \frac{\partial U}{\partial \lambda}, \quad -\frac{l}{g \sin \theta} \frac{\partial U}{\partial \lambda},$$

respectively. The amount by which the Earth is slow is then

$$\tau = \frac{\xi}{2\pi} \text{(l.o.d.)}, \quad \xi = \frac{1 + k - l}{ga \sin \theta} \frac{\partial U}{\partial \lambda}. \tag{8.3.6}$$

If ξ is in seconds of arc, τ equals $\xi/(15 \sin \theta)$ seconds of time (8.3.2).

Long-period tides have no effect because they are independent of longitude. But here again (as in the case of latitude) the effect of aliasing must be taken into account. An annual effect (frequency σ) due to the synodic tides K_1 and P_1 is computed as follows, assuming nightly measurements (see § 7.4):

$$b = \sqrt{[(0·531)^2 + (0·176)^2]} = 0·56,$$

$$\xi = \frac{bK}{a} (1 + k - l) \cos \theta = 0''0119 \cos \theta \cos \sigma t,$$

$$\tau = (0·8 \cot \theta \cos \sigma t) \text{ ms},$$

$$\Delta \text{ (l.o.d.)} = (d\tau/dt) \times \text{(l.o.d.)} = (0·0114 \cot \theta \sin \sigma t) \text{ ms}.$$

This compares to the observed variation in l.o.d. by 0·5 ms. A semi-annual effect due to the synodic tide K_2 gives

$$\Delta \text{ (l.o.d.)} = (0·0056 \cos 2\sigma t) \text{ ms},$$

as compared to the observed amplitude, 0·3 ms. The effect of ocean tides might be larger, but in all events the aliased tides can account for only a few per cent of the observed variation in the l.o.d.

Errors in τ due to uncertainty in wobble and to the east-west deflexion of the vertical by tides and zenith refraction are all of the order of one millisecond and comparable with the errors in measurement achieved by modern methods. In the final analysis it is the lack of knowledge concerning these geophysical events that limits the accuracy of timekeeping (and of measuring wobble).

CHAPTER 9

SEASONAL AND OTHER SHORT-PERIOD VARIATIONS

This chapter is principally concerned with annual, semi-annual, monthly and fortnightly variations in rotation. There is also a brief discussion of the continuous spectrum which is superimposed on these four spectral lines. Circumstances favor the detection of lower frequencies for both wobble and l.o.d., other factors remaining the same. The annual wobble is amplified because of its proximity to the resonant period of 14 months. The detection of variations in l.o.d. depends on the integrated discrepancies in the l.o.d., and this again favors the lower frequencies. The annual variation in rotation is due to meteorologic events. These are treated by considering the balance of momentum and mass for the planet Earth. The semi-annual change in the l.o.d. appears to be caused by the atmosphere and by a tidal distortion of the solid Earth. In the monthly and fortnightly variations in l.o.d. the tidal effects seem predominant.

1. The astronomical evidence

Wobble.—Table 9.1 summarizes the evidence with respect to the annual and semi-annual terms in the latitude observations. $\odot$ is the longitude of the 'mean sun' measured from the beginning of the year. The conversion from components to circular vectors is in accordance with (6.7.6). The Jeffreys–Markowitz estimate of the annual term is taken from observations at Greenwich and Washington; unfortunately these do not cover the same years. The Greenwich observations give m_1; the latitude variation at Washington (on 77° west longitude) equals

$$m_1 \cos 77° - m_2 \sin 77°,$$

and this determines m_2. The results by Walker and Young based on unsmoothed values, fig. 7.4, give larger annual coefficients.

Values for the semi-annual term are very much in doubt. Rudnick finds comparative amplitudes at a period of 0·8 years, for example.

Table 9.1. Seasonal components of the rotation pole, in units of 0″.01. Standard deviations in the last column refer to all six amplitudes appearing on the same line.

		Annual				
Source	Interval	m_1	m_2	m^+	m^-	
Jeffreys, 1952	1892–1938	$-3{\cdot}6\cos\odot - 8{\cdot}5\sin\odot$	$7{\cdot}0\cos\odot - 2{\cdot}9\sin\odot$	$8{\cdot}4\ \exp i(\odot + 113°)$*	$0{\cdot}83\ \exp - i(\odot + 114°)$*	± 0·8
Pollak, 1927	1890–1924	−3·7 −8·9	7·0 −3·9	8·8 116°**	0·96 84°	
Rudnick, 1956	1891–1945	−3·2 −8·2	6·7 −2·8	8·0 112°	0·78 108°	
Walker and Young, 1957	1899–1954†	−6·4 −7·1	7·0 −4·6	8·9 128°	0·90 177°	± 1·1
Walker and Young, 1957	1900–1934	−5·5 −7·0	7·5 −4·6	8·8 125°	0·52 209°	± 1·1
	1900–1920	−4·8 −6·0	6·6 −3·7	7·6 124°	0·57 195°	± 1·6
Jeffreys, 1940 Markowitz, 1942	1912–1935 1916–1940	−3·2 −7·8	5·6 −1·6	7·1 110°	1·36 126°	
		Semi-annual				
Walker and Young, 1957	1899–1954†	$-0{\cdot}1\cos 2\odot + 0{\cdot}6\sin 2\odot$	$-0{\cdot}5\cos 2\odot + 0{\cdot}0\sin 2\odot$	$0{\cdot}56\ \exp 2i(\odot - 48°)$	$+0{\cdot}07\ \exp - 2i(\odot - 67°)$	± 0·2
Walker and Young, 1957	1900–1934	−0·2 0·7	−0·6 −0·3	0·70 − 55°	0·07 − 22°	± 0·3
	1900–1920	−0·3 0·8	−0·8 −0·6	0·92 − 59°	0·15 0°	± 0·3
Rudnick, 1956	1891–1945			0·46	0·26	

* Correcting Rudnick's results. This does not alter Rudnick's correction of Jeffreys's values.
† The annual and semi-annual coefficients for the 1899–1954 series have been checked by us and found to be correct.

All that can be said is that the semi-annual terms are much smaller than the annual term.

Fig. 9.1 shows the annual ellipse described by the rotation pole (solid) based on the analysis by Jeffreys. Characteristic parameters are readily computed from the formulae following (6.7.6): the semi-major axis has a magnitude $0''092$ which happens to lie on the

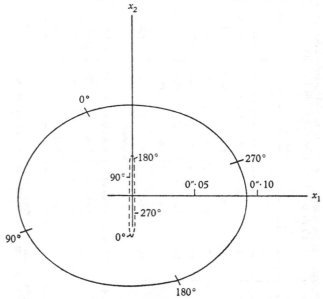

Fig. 9.1. The annual motion of the poles of rotation (solid) and excitation (dashed) with respect to the reference system centered at the mean pole, with x_1 toward Greenwich, x_2 toward 90° east of Greenwich. The positions of the pole at $\odot = 0°$ (Jan. 1), $\odot = 90°$ (April 2), $\odot = 180°$ and $\odot = 270°$ are indicated. The amplitude scale in seconds of arc is given along the x_1-axis; $0''01$ is equivalent to 1 ft of polar displacement.

longitude of Greenwich and is occupied when $\odot = -113°, +67°$, that is, early in September and March. The preference of the polar motion for the Greenwich coordinate system is remarkable.

Table 9.2 and fig. 9.1 give the corresponding values of the modified excitation function (6.3.3). This is obtained from $\mathbf{m}$ according to the equation (6.3.6)

$$i\frac{1}{\sigma_0}\frac{d\mathbf{m}}{dt} + \mathbf{m} = \mathbf{\Psi}.$$

But σ_0^{-1} is the period of the Chandler motion, $1 \cdot 20$ years per cycle, and

Table 9.2. Seasonal components of the modified excitation pole, in units of $0''.01$.

Annual

Source	Interval	Ψ_1	Ψ_2	Ψ^+	Ψ^-
Jeffreys, 1952	1892–1938	$-0{\cdot}1\cos\odot - 0{\cdot}1\sin\odot$	$-3{\cdot}2\cos\odot + 1{\cdot}4\sin\odot$	$1{\cdot}68\exp i(\odot + 293°)$	$1{\cdot}82\exp -i(\odot + 114°)$
Rudnick, 1956	1891–1945	0·2 −0·2	−3·1 1·0	1·60 292°	1·72 108°
Jeffreys, 1940 ⎱ Markowitz, 1942 ⎰	1912–1935 ⎱ 1916–1940 ⎰	−1·3 −1·1	−3·2 2·2	1·42 290°	2·99 126°
Walker and Young, 1957	1899–1954	−0·9 +1·3	−1·5 3·1	1·78 308°	1·98 177°

Semi-annual

Source	Interval	Ψ_1	Ψ_2	Ψ^+	Ψ^-
Walker and Young, 1957	1899–1954	$-0{\cdot}1\cos 2\odot - 0{\cdot}6\sin 2\odot$	$0{\cdot}9\cos 2\odot + 0{\cdot}2\sin 2\odot$	$0{\cdot}78\exp 2i(\odot + 42°)$	$0{\cdot}24\exp -2i(\odot - 67°)$
	1900–1934	0·5 −0·7	1·1 0·2	0·98 45°	0·24 −22°
	1900–1920	1·1 −1·1	1·1 0·1	1·29 31°	0·51 0°

$d/dt = (d/d\odot) \times (d\odot/dt)$, where $d\odot/dt$ is the rate of change in the longitude of the mean sun, 1 cycle per year. The foregoing equation then becomes

$$1 \cdot 20 \, i \frac{d\mathbf{m}}{d\odot} + \mathbf{m} = \mathbf{\Psi} \qquad (9.1.1)$$

and the components of $\mathbf{\Psi}$ can readily be derived.

The positive motion of the rotation pole has an amplitude, $|\mathbf{m}^+|$, about ten times that of $|\mathbf{m}^-|$. But this does not imply a similar ratio in the excitation function. In fact, $|\mathbf{\Psi}^+|$ and $|\mathbf{\Psi}^-|$ are about the same, and the amplification of $|\mathbf{m}^+|$ over $|\mathbf{m}^-|$ is in the expected ratio (6.7.8)

$$\frac{|\mathbf{m}^+|}{|\mathbf{m}^-|} = \frac{\sigma + \sigma_0}{\sigma - \sigma_0} = \frac{1 + 0 \cdot 85}{1 - 0 \cdot 85} = 12 \cdot 3.$$

But, although $\mathbf{\Psi}^+$ and $\mathbf{\Psi}^-$ are of the same magnitude, the west-to-east term $\mathbf{\Psi}^+$ is derived from a much more prominent part of the $\mathbf{m}$ spectrum and therefore much more accurately established than its companion $\mathbf{\Psi}^-$.* In other words, as pointed out by Rudnick, the *length* of semi-major axis of the annual excitation ellipse (fig. 9.1) is much better known than its orientation and the eccentricity. This limitation must be kept in mind when interpreting the excitation ellipse in terms of geophysical events.

Length of day.—Table 9.3 lists recent determinations of the l.o.d.† Observed monthly and fortnightly terms are listed in the left column of table 9.4.

From the observed seasonal variation in the time τ, by which the Earth is slow, the variation in the l.o.d. is computed from the expressions

$$\Delta \, (\text{l.o.d.}) = \frac{d\tau}{d\odot} \frac{d\odot}{dt} \times (\text{l.o.d.}),$$

$$\frac{d\odot}{dt} = 0 \cdot 0172 \, \frac{\text{radians}}{\text{mean solar day}} = 1 \cdot 99 \times 10^{-7} \, \text{sec}^{-1}.$$

* Jeffreys gives a standard error of $\pm 0''\!0076$ for amplitudes of m_1 and m_2. The same value applies to $\mathbf{m}^+$ and $\mathbf{m}^-$. Then $|\mathbf{\Psi}^+| = 1 \cdot 68 \pm 0 \cdot 15$, $|\mathbf{\Psi}^-| = 1 \cdot 82 \pm 1 \cdot 67$.

† By international agreement all time services now correct for the seasonal effect according to the annual and semi-annual BIH values given in the table. These rounded-off values are in close agreement with those given by Markowitz (1959). In accordance with the notation adopted by commission 31 of the International Astronomical Union, UT 0 is Universal Time; UT 1 is UT 0 corrected for observed polar motion (in accordance with the *Rapid Latitude Service*); and UT 2 is UT 1 corrected for the seasonal variation (in accordance with the BIH formula, table 9.3).

Table 9.3. Variation in the amounts, τ, the Earth is slow, and in the l.o.d., both in ms.*

Annual

Source†	Location	Interval	τ	Δ(l.o.d.)
S and T	Greenwich	1951-2	$-8.1 \cos ⊙ + 16.6 \sin ⊙ = 18.5 \sin(⊙ - 26°)$	$0.285 \cos ⊙ + 0.139 \sin ⊙ = 0.318 \cos(⊙ - 26°)$
	Canberra	1951-2	$0.4 \quad 23.9 \quad 23.9 \quad (⊙ - 359°)$	$0.411 \quad -0.007 \quad 0.411 \quad 359°$
	Washington	1951-2	$-12.8 \quad 18.3 \quad 22.3 \quad 35°$	$0.315 \quad 0.218 \quad 0.384 \quad 35°$
	Richmond	1951-2	$-13.5 \quad 18.6 \quad 23.0 \quad 36°$	$0.320 \quad 0.232 \quad 0.396 \quad 36°$
M	Wash. and Richmond	1951-8	$-17.5 \quad 21.5 \quad 27.7 \quad 39°$	$0.370 \quad 0.301 \quad 0.477 \quad 39°$
BIH	Richmond	1956-8	$-17 \quad 22 \quad 28 \quad 38°$	$0.39 \quad 0.31 \quad 0.50 \quad 38°$

Semi-annual

Source†	Location	Interval	τ	Δ(l.o.d.)
S and T	Greenwich	1951-2	$-1.5 \cos 2⊙ - 10.4 \sin 2⊙ = 10.5 \sin 2(⊙ - 86°)$	$-0.358 \cos 2⊙ + 0.052 \sin 2⊙ = 0.361 \cos 2(⊙ - 86°)$
	Canberra	1951-2	$-2.1 \quad 8.2 \quad 8.5 \quad 83°$	$-0.282 \quad -0.072 \quad 0.292 \quad 83°$
	Washington	1951-2	$-8.9 \quad 6.5 \quad 11.0 \quad 117°$	$-0.224 \quad -0.306 \quad 0.378 \quad 117°$
	Richmond	1951-2	$-6.5 \quad 8.3 \quad 10.5 \quad 109°$	$-0.286 \quad -0.224 \quad 0.361 \quad 109°$
M	Wash. and Richmond	1951-8	$-6.8 \quad 6.8 \quad 9.4 \quad 112°$	$-0.232 \quad -0.223 \quad 0.332 \quad 112°$
BIH	Richmond	1956-8	$6 \quad 7 \quad 9.2 \quad 110°$	$-0.24 \quad -0.21 \quad 0.32 \quad 110°$

* All values are corrected for polar motion according to the *definitive* coordinates provided by the International Latitude Service.

† S and T: Smith and Tucker, 1953; M: Markowitz, 1959; BIH.: Bureau Int. de l'Heure.

Table 9.4. Observed and computed variations in the time, τ, by which the Earth is slow, in milliseconds (according to Markowitz, 1959).

	Observed		Computed	
	Monthly	Fortnightly	Monthly	Fortnightly
	$-1{\cdot}0\cos(\mathbb{C}-p_{\mathbb{C}}) + 0{\cdot}4\sin(\mathbb{C}-p_{\mathbb{C}})$	$-0{\cdot}2\cos 2\mathbb{C} + 1{\cdot}0\sin 2\mathbb{C}$	$0\cos(\mathbb{C}-p_{\mathbb{C}}) + 0{\cdot}7\sin(\mathbb{C}-p_{\mathbb{C}})$	$+0{\cdot}2\cos 2\mathbb{C} + 1{\cdot}0\sin 2\mathbb{C}$
1951·7 to 52·7				
1952·7 to 53·7	0·6 0·9	0·5 1·5	0 0·7	0·2 0·9
1953·7 to 54·7	−0·2 0·5	1·3 0·7	0 0·8	0·3 0·8
1955	0·4 1·1	0·0 0·5	0 0·8	0·3 0·7
1956	0·5 2·7	0·3 0·3	0 0·8	0·3 0·6
1957	0·1 1·6	1·0 − 0·4	0 0·9	0·2 0·5
1958	− 1·1 0·2	1·5 + 0·3	0 0·9	0·2 0·5

If τ and Δ (l.o.d.) are in the same units (milliseconds in table 9.3) then

$$\Delta \text{ (l.o.d.)} = \frac{d\tau}{d\odot} \times 0.0172.$$

For the monthly and fortnightly terms, the appropriate 'speeds' are 0·228 and 0·230 radians per day, respectively, so that

$$\Delta \text{ (l.o.d.)} = \frac{d\tau}{d(\mathbb{C} - p_{\mathbb{C}})} \times 0.228; \quad \Delta \text{ (l.o.d.)} = \frac{d\tau}{d\mathbb{C}} \times 0.230.$$

2. Tides

There is a slight additional protuberance, on the average, in the equatorial region at the expense of the polar regions, and a resulting

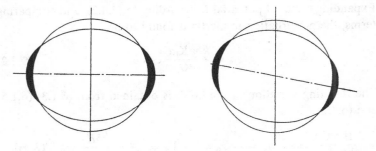

Fig. 9.2. The tidal bulge during equinox (*left*), solstice (*right*). (From Mintz and Munk, 1954.)

increment in the Earth's moment of inertia over what it would be if the Sun were absent. The essential effects here are *variations* in the increment due to (1) variations in the Sun's distance from the Earth and (2) changes in the Sun's declination because of the obliquity of the ecliptic. The former gives rise to a small annual term; the latter to a semi-annual term with maxima in the moment of inertia (and thus the l.o.d.) during spring and fall equinoxes and minima during summer and winter solstices (fig. 9.2). There are analogous effects due to the motion of the Moon. Jeffreys (1928) first drew attention to the effect of obliquity on the l.o.d. long before it was discovered. There have been further calculations by Andersson (1937), Stoyko (1951) and by Mintz and Munk* (1951, 1954). The most detailed

---

* They use a derivation based on the Love number h, rather than k. This leads to the correct values, but the procedure is poor, as pointed out by Melchior (1957, p. 439).

calculations are due to Woolard (1959) based on his own systematic development of the tidal potential (Woolard, 1953).

The effect is most easily derived from MacCullagh's formula (§ 5.2). The distortion of the Earth due to a perturbing tidal potential, U, gives rise to variations, by c_{ij}, in the inertia C_{ij}. On the average $c_{11} = c_{22}$; furthermore $c_{11} + c_{22} + c_{33} = 0$. Using these conditions and equations (5.2.1) we obtain, for $r = a$,

$$\left.\begin{array}{l} kU = V = \dfrac{G}{2a^5}\,(c_{33} - c_{11})\,(x_1^2 + x_2^2 - 2x_3^2) \\[2mm] = \dfrac{9G}{4a^3}\,c_{33}(\tfrac{1}{3} - \cos^2\theta) \end{array}\right\}. \qquad (9.2.1)$$

Expanding the tidal potential U according to (7.4.2), and comparing terms, the perturbation in inertia is found to be

$$c_{33} = \frac{kbgK_{\mathbb{C}}a^3}{3G}\cos\beta. \qquad (9.2.2)$$

The resulting variation in the l.o.d. is obtained from (6.1.3) (6.1.5) and (6.4.2):

$$\frac{\Delta\,(\text{l.o.d.})}{\text{l.o.d.}} = -m_3 = -\phi_3 = -\Psi_3 = \frac{c_{33}}{C} = \frac{kbK_{\mathbb{C}}}{a}\cos\,[\beta(\lambda, t)]. \qquad (9.2.3)$$

We have made use of $3g = 4\pi Ga\bar{\rho}$ and $C = \tfrac{1}{3}Ma^2$, $M = \tfrac{4}{3}\pi a^3\bar{\rho}$.

Computed values for the long-period of tides are summarized in table 9.5. The phase β is given in the last column of table 7.1. Amplitudes are based on the tidal effective Love number $k = 0.29$

Table 9.5. Computed variations in τ and Δ (l.o.d.) due to tides.

Symbol	Period	τ in ms	Δ l.o.d. in ms
Lunar	18·6 years	149 $\sin\beta$	0·14 $\cos\beta$
Sa	1 year	1·41 $\sin\beta$	0·025 $\cos\beta$
Ssa	½ year	4.33 $\sin\beta$	0·15 $\cos\beta$
Mm	27^d55	0·762 $\sin\beta$	0·17 $\cos\beta$
MSf	14^d55	0·062 $\sin\beta$	0·027 $\cos\beta$
Mf	13^d66	0·775 $\sin\beta$	0·33 $\cos\beta$
. . .	13^d63	0·296 $\sin\beta$	0·14 $\cos\beta$

(§ 5.5) as determined from the frequency of the Chandler wobble. At first glance this appears to be the appropriate choice. The tidal distortion which varies the l.o.d. and the rotational distortion which lengthens the Chandler wobble are both spherical harmonics of degree two, and both involve in a similar way the Earth as a whole, including oceans. But there are important differences. The frequencies are not the same. Furthermore the wobble distortion is of the type p_2^1 and involves a tilting of the axis; all long-period tidal distortions are of the type p_2^0 and do not tilt the axis. For the fourteen-month wobble a dynamic theory is required; for the fourteen day tide an equilibrium argument may be adequate. Jeffreys and Vicente (1957a, b) have shown that a fluid core reduces the value of k appropriate to the wobble by 20 per cent compared to the value appropriate to the fortnightly tide.

Comparison of tables 9.3 and 9.5 shows that the observed semi-annual term is somewhat larger, and the annual term very much larger than the corresponding computed terms. One suspects that meteorologic factors enter here. For the monthly and fortnightly terms the observations are not of sufficient duration to resolve neighboring frequency lines, such as MSf, Mf, and the $13\overset{d}{.}63$ tide. Accordingly Markowitz (1959) has lumped these into a single term of variable amplitude, and the comparison between theory and observation is made year by year (table 9.4). The observed fortnightly term agrees roughly with the computed value; the observed monthly term is larger.* There is tolerable agreement with respect to phase. The computed and observed variations in the l.o.d. would show the best agreement if (Markowitz, 1959)

$$k = 0\cdot30 \pm 0\cdot07, \qquad k = 0\cdot38 \pm 0\cdot07$$

for the fortnightly and monthly terms, respectively. A detailed discussion of the Love numbers derived by various observations is postponed until after the treatment of the Chandler wobble.

So far we have been concerned with the effect of tides on the l.o.d. This would be the only effect if the oceans were uniform over the entire Earth or varied only with latitude. The asymmetry in the distribution of land and sea must be associated with a wobble. We

---

* Perhaps because it is just above 'noise level', see fig. 8.1.

consider only equilibrium tides. The ocean is then lifted relative to the ground by an amount

$$\xi(\theta, t) = (1 + k - h)(U/g) = \xi_0(t)(\tfrac{1}{3} - \cos^2 \theta) \qquad (9.2.4)$$

to which we must add a correction tide $\xi'(t)$ (introduced by Sir George Darwin) in order that the total tidal volume

$$\int\limits_{\text{ocean}} \xi(\theta, t)\, dS + \xi'(t) \int\limits_{\text{ocean}} dS$$

integrated over all of the ocean surface vanishes in spite of the irregular distribution of land and sea. The integral can be expressed explicitly in terms of the harmonics of the ocean function (see Appendix A.1). When this is done, $(\tfrac{1}{3} - \cos^2 \theta)$ in equation (9.2.4) is replaced by

$$\frac{\tfrac{1}{3}a_0^0 + \tfrac{2}{15} a_2^0}{a_0^0} - \cos^2 \theta = 0.31 - \cos^2 \theta. \qquad (9.2.5)$$

We do not allow for the load deformation and self attraction of the tide. This could be done by the method of the generalized Love numbers (§ 5.12).

The excitation function is found directly from § 6.8:

$$\Phi = -\frac{a^4}{C - A}\, \rho_w \int \mathscr{C}(\text{oceans})\, (\xi + \xi')\cos\theta \sin\theta\, e^{i\lambda}\, ds$$

$$= -\frac{9(1 + k - h)}{H}\, \frac{\rho_w}{\bar{\rho}}\, \frac{K_{\mathbb{C}}}{a}\, b \cos\beta \left[\left(\frac{1}{75}\, \frac{a_2^0}{a_0^0} - \frac{1}{105} \right) c_2^1 - \frac{1}{63}\, c_4^1 \right],$$

$$(9.2.6)$$

where $\rho_w = 1.025$ g cm^{-3} is the density of sea water and $\bar{\rho} = 5.53$ g cm^{-3} the mean density of the Earth. Setting $k = 0.29$, $h = 0.59$, $H = 0.00327$, $K_{\mathbb{C}} = 53.7$ cm, we obtain

$$\phi_1 = 0''.010\, b \cos\beta, \quad \phi_2 = 0''.002\, b \cos\beta.$$

The tidal amplitude factor, b, is of the order 0.1 at most, and the excitation pole less than $0''.001$ or less than one inch of polar displacement.

3. The inverted barometer problem

Here it will be shown that the ocean responds nearly like an inverted barometer to seasonal variations in atmospheric pressure.

For every millibar (mb) of atmospheric pressure over and above the mean pressure over the entire ocean surface, the sea surface is locally depressed by approximately 1 cm.

Let $\rho_w g z$ be the impressed atmospheric pressure disturbance; z is the equivalent water barometer. In response the ocean yields by an amount ξ. The total pressure disturbance just beneath the surface is $\rho_w g (z + \xi) = \rho_w g \zeta$. For a frozen ocean $\zeta = z$ (fig. 9.3a); if the ocean responds as an inverted barometer, $\xi = -z$ and $\zeta = 0$ (fig. 9.3b). But there are other possible steady-state configurations: for example, anti-cyclonic geostrophic circulation (clockwise in the northern hemisphere, as shown in fig. 9.3c). There can be combinations of (b) and (c). A layered ocean permits additional classes of equilibrium response, such as the one shown in (d). Situations like (b) and (c) are known as *barotropic*, (d) as *baroclinic* (surfaces of equal pressure and density intersect).

We can proceed along the lines of the investigations by Charney (1955) and by Stommel and Veronis (1956) into the effect of a variable wind stress. Let u_x, u_y designate the velocity components in the directions x (eastward) and y (northward), respectively. The equations of motion for a homogeneous ocean are

$$\left. \begin{aligned} \frac{\partial u_x}{\partial t} - f_C u_y &= -g \frac{\partial \zeta}{\partial x} \\ \frac{\partial u_y}{\partial t} + f_C u_x &= -g \frac{\partial \zeta}{\partial y} \end{aligned} \right\}, \tag{9.3.1}$$

where $f_C = 2\Omega \cos \theta$ is the Coriolis frequency. Conservation of mass requires that in water of depth h

$$\frac{\partial \xi}{\partial t} = -h \left(\frac{\partial u_x}{\partial x} + \frac{\partial u_y}{\partial y} \right). \tag{9.3.2}$$

Cross-differentiation of the equations of motion gives

$$\frac{\partial^2 u_x}{\partial y \partial t} - \frac{\partial^2 u_y}{\partial x \partial t} - \beta u_y + \frac{f_C}{h} \frac{\partial \xi}{\partial t} = 0, \tag{9.3.3}$$

with $\beta = df_C/dy$. If we confine our attention to frequencies σ, small compared to the Coriolis frequency f_C, then the acceleration terms

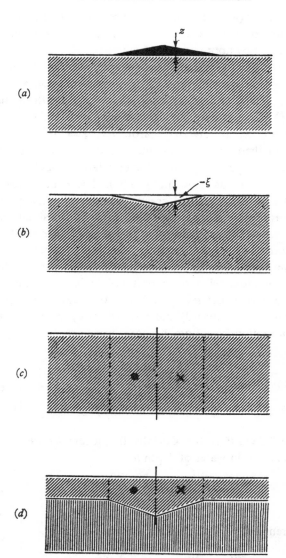

Fig. 9.3. Response of the ocean to a conical pressure spot applied to the sea surface. ● designates flow into the paper, ✗ out of the paper.

in (9.3.1) are small compared to the Coriolis terms, fu. Accordingly (9.3.3) can be written

$$\lambda^2 \frac{\partial}{\partial t} \nabla^2 \zeta + \beta \lambda^2 \frac{\partial \zeta}{\partial x} - \frac{\partial \zeta}{\partial t} = -\frac{\partial z}{\partial t},$$

where $\lambda^2 = gh/f_C^2$. For cellular oscillations of the type

$$z, \zeta = (z_0, \zeta_0) \cos l_y y \cos (l_x x + \sigma t),$$

we obtain

$$\zeta_0 = \frac{z_0}{1 + \lambda^2(l_x^2 + l_y^2) - \lambda^2 \beta l_x/\sigma}. \tag{9.3.4}$$

For extended disturbances (l small) of high frequency (σ large) this gives $\zeta_0 = z_0$: the ocean does not respond. For sufficiently low frequencies, $\zeta_0 = 0$: the response is that of an inverted barometer. This is the expected asymptotic behavior.

We set $f_C = 10^{-4}$ sec^{-1}, $h = 5 \times 10^5$ cm, so that $\beta = 2 \times 10^{-13}$ cm^{-1} sec^{-1}, $\lambda^2 = 5 \times 10^{16}$ cm^2. Consider cellular pressure spots with dimensions of 6000 km. Then $l_x = l_y = 10^{-8}$ cm^{-1}, and $\lambda^2(l_x^2 + l_y^2) = 10$. At the critical frequency

$$f_1 = \frac{\sigma_1}{2\pi} = \frac{1}{2\pi} \frac{\lambda^2 \beta l_x}{1 + \lambda^2(l_x^2 + l_y^2)} \approx \frac{1}{8} \text{ days}^{-1}, \tag{9.3.5}$$

the denominator of (9.3.4) vanishes. For frequencies between f_1 and f_C (periods less than a week and more than two days, say) the sea surface responds nearly as an inverted barometer, giving about 90 per cent of the inverted barometer response. At the critical frequency there is a change in phase, and for frequencies much less than f_1 the response is again nearly that of an inverted barometer, but slightly larger. Groves (1957) has presented empirical evidence that the sea-level response is roughly two-thirds of inverted barometer response for pressure disturbances from individual storms (2 cycles per week is a typical frequency). This finding is in agreement with the foregoing model. Thus it would appear that for the annual frequency the inverted barometer model is justified.

But there are difficulties with this model. The critical period can be associated with the time required for a barotropic Rossby (or planetary) wave to travel across the pressure spot. For a two-layer ocean there is another critical period associated with a baroclinic Rossby wave. Compared with the barotropic wave, the phase velocity is reduced in the ratio of the density difference between the two water layers to the density difference between water and air. This ratio is of the order 10^{-3}. The critical baroclinic period is 8000 days! Thus for decade variations in the pressure pattern the expected response is the one shown in fig. 9.3 (*d*).

We have ignored the troublesome mathematical features that appear when the isobars intersect the coastline. This problem has not been solved. The work of Charney, Stommel and Veronis suggests that the presence of boundaries would favor a response of the inverted barometer type.

4. Geostrophic and ageostrophic motion

Here it will be shown that geostrophic motion has no effect on wobble (with some reservations), and that the effect of ageostrophic motion on wobble and l.o.d. is small compared to the effect of the associated mass transport.

According to the geostrophic approximation the motion is along isobars with the Coriolis force balanced by the horizontal pressure gradient. All other forces, centrifugal, inertial, frictional, are neglected. Seasonal variations in angular momentum are associated with winds and currents that obey the geostrophic rule to a first approximation.

For horizontal motion near the Earth's surface, the excitation function (§ 8·2) can be written

$$\left.\begin{aligned}
\Phi &= -\frac{2a}{\Omega(C-A)} \int \rho \cos\theta \, (u_\lambda - iu_\theta \cos\theta) \, e^{i\lambda} \, \mathrm{d}V, \\
\phi_3 &= -\frac{a}{\Omega C} \int \rho \sin\theta \, u_\lambda \, \mathrm{d}V
\end{aligned}\right\}. \qquad (9.4.1)$$

In geostrophic motion, the pressure departure p and the velocity are related according to

$$\rho u_\lambda = \frac{1}{2a\Omega\cos\theta}\frac{\partial p}{\partial\theta}, \quad \rho u_\theta = -\frac{1}{2a\Omega\cos\theta\sin\theta}\frac{\partial p}{\partial\lambda} \qquad (9.4.2)$$

and, when this is substituted above, we obtain

$$\begin{aligned}
\Phi &= -\frac{a^2}{\Omega^2(C-A)} \int \left[\sin\theta\frac{\partial p}{\partial\theta} + i\cos\theta\frac{\partial p}{\partial\lambda} \right] e^{i\lambda} \, \mathrm{d}\theta \, \mathrm{d}\lambda \, \mathrm{d}r \\
&= -\frac{a^2}{\Omega^2(C-A)} \int \left[\frac{\partial}{\partial\theta}(p\sin\theta\,e^{i\lambda}) + i\frac{\partial}{\partial\lambda}(p\cos\theta\,e^{i\lambda}) \right] \mathrm{d}\theta \, \mathrm{d}\lambda \, \mathrm{d}r \\
&= 0.
\end{aligned}$$

The first expression vanishes upon integration with respect to θ, $\sin \theta$ being zero at both limits, $\theta = 0, \pi$; the second expression vanishes upon integration with respect to λ. Two circumstances have been ignored: (1) there is an equatorial singularity in the geostrophic winds (but not in the integrand for $\boldsymbol{\phi}$); (2) the intersection of level surfaces with mountains excludes certain ranges in θ, λ from integration. The latter feature is pertinent to motion in the lower atmosphere; the application to ocean currents is questionable.

With respect to ϕ_3 we wish to compare the role played by a variable distribution of matter with that played by the associated variation in geostrophic motion. Returning to § 6.8 we find that the integrands of the volume integrals are

$$\rho F_3 \text{ (motion)} = -\rho \Omega r \sin \theta \, u_\lambda$$

$$\Delta\rho F_3 \text{ (matter)} = -\Delta\rho \Omega^2 r^2 \sin^2 \theta,$$

respectively. Setting $dP = -\rho g \, dr$ and integrating vertically from the ground ($P = P_0$) to the top of the atmosphere ($P = 0$) we obtain

$$\int_a^\infty \rho F_3 \text{ (motion)} \, dr \approx -\frac{\Omega a \sin \theta}{g} \int_0^{P_0} u_\lambda(P) \, dP = -\frac{\Omega a \sin \theta \, P_0}{g} \overline{u_\lambda}$$

$$\int_a^\infty \Delta\rho F_3 \text{ (matter)} \, dr \approx -\frac{\Omega^2 a^2 \sin^2 \theta}{\rho g} \int_0^{P_0} \Delta\rho(P) \, dP = -\frac{\Omega^2 a^2 \sin^2 \theta}{g} p,$$

where $\overline{u_\lambda}$ is the vertically integrated motion (weighted by density) and p the pressure departure on the ground.

To compare the two terms we introduce the ratio

$$\gamma(\theta) = \overline{u_\lambda}(\theta)/u_\lambda(P_0, \theta) \tag{9.4.3}$$

between the average motion and that on the ground. For winds limited to the lower atmosphere $\gamma \ll 1$; for a high-level disturbance $\gamma \gg 1$. As a very rough approximation we ignore 'tilt', that is, we presume that $\overline{u_\lambda}(\theta)$ and $u_\lambda(P_0, \theta)$ have an identical dependence on θ so that γ is independent of θ. Introducing the geostrophic rule (9.4.2) the ratio of the effect of motion to that of matter becomes

$$\left[\frac{P_0/\rho_0}{\frac{1}{2}a^2\Omega^2}\right] \left[\frac{1}{2\sin 2\theta} \frac{1}{p} \frac{\partial p}{\partial \theta}\right] \gamma. \tag{9.4.4}$$

The first term is the ratio, at the ground, of the thermal energy per unit mass to that of kinetic energy of rotation. This ratio equals 0·8.

The remaining terms, except for γ, are likewise of the order of unity for the large-scale motion here under consideration. The relative role of motion and matter is then indicated by the value of γ.

Next we consider the motion normal to the isobars. The divergence of this ageostrophic component gives rise to the changes in the distribution of matter. In the ocean and atmosphere, this ageostrophic component is small compared to the geostrophic component for the scale here under consideration.

Again referring to § 6.8 we note that any of the three integrands are proportional to $\Omega u x$ and $\Omega^2 x^2$ for the case of motion and matter, respectively. Now if the displacement, x, of any particle varies in time according to $e^{i\sigma t}$, its velocity, u, varies as σx, and the ratio of the effects of motion to matter is σ/Ω. In the case of acceleration the ratio is $(\sigma/\Omega)^2$. The problem and result is analogous to that of the hula dancer (see 6.9.1).

5. Distribution of air and water

The seasonal variation in the distribution of air mass can be estimated from observations of pressure on the ground. The effect of winds is taken up in the next section. It turns out that the least motion required to accomplish the observed variation in atmospheric pressure is altogether negligible compared with the observed winds (which are roughly along isobars). This justifies a separate discussion of the effects of wind and air mass. Similar remarks apply to ocean currents and water mass.

For changes in the distribution of matter near the Earth's surface, $|r - a| \ll a$, the excitation function (§ 6.8) can be written*

$$\psi = -\frac{a^4}{C - A} \int q(\theta, \lambda; t) \sin \theta \cos \theta \, e^{i\lambda} \, ds, \qquad (9.5.1a)$$

$$\psi_3 = -\frac{a^4}{C} \int q(\theta, \lambda; t) \sin^2 \theta \, ds,$$

where
$$q(\theta, \lambda; t) = \int [\rho(r, \theta, \lambda; t) - \bar{\rho}(r, \theta, \lambda)] \, dr \qquad (9.5.2)$$

is the seasonal *departure* in surface load (g cm^{-2}), and $ds = \sin \theta \, d\theta \, d\lambda$.

---

* In accordance with our working rule (§6.4) we first compute the excitation function, ψ_i, as if the Earth were rigid. The equations for ψ_i are the same as those for the total excitation ϕ_i (§6.8) provided we neglect rotational and load deformations.

Information concerning $q(\theta, \lambda; t)$ comes from a variety of sources. The quality of the observational material is uneven, as is to be expected, but not so poor as to make an effort to evaluate ψ_i worthless. It is advisable to put certain restraints on the observations in order to assure that for the seasonal variation the following quantities are conserved: (1) the total mass; (2) the amount of dry air; (3) the amount of water.

We can always separate the load into the terms $q_E(t) + q(\theta, \lambda; t)$, where $q_E(t)$ is the average over the Earth. The first condition is automatically fulfilled in the evaluation of ψ, for the integral vanishes for that part, q_E, of the load which does not depend upon θ, λ. This is not the case for ψ_3. We can make it so by writing

$$\psi_3 = -\frac{a^4}{C} \int q(\theta, \lambda; t) \, (\sin^2 \theta - \tfrac{2}{3}) \, \mathrm{d}s. \qquad (9.5.1b)$$

We shall estimate the contribution towards ψ_i from the atmosphere, the oceans and the solid Earth. With regard to any seasonal *net* transfer of matter from one to the other of these three portions of the planet Earth, we could proceed in two ways: (1) by ignoring this net transfer for all three portions, thus presuming that the effect on rotation of any loss of water from the oceans (say) is cancelled by an equal and opposite effect arising from a gain in total ground water; or (2) by including these positive and negative contributions in the tabulated values of ψ_i for each of the three portions. We have selected the former alternative.

Air mass.—Ten years after the annual wobble had been determined by Chandler, Spitaler (1901) attempted to explain it in terms of the observed seasonal shift of air mass. Some difficulties are connected with this pioneering effort, as pointed out by Jeffreys (1916), but the principal conclusion is still the only conclusion we dare make today: that the shift in air mass associated with the high pressure in winter over Siberia has the phase and magnitude to account for the observed wobble. A second calculation by Jeffreys (1916) was based on more complete meteorological charts from *Bartholomew's Meteorological Atlas*. Jeffreys developed the complete theory and made allowance for snow, vegetation and the mobility of the oceans. A calculation by Schweydar (1919) leads substantially to the same result as Jeffreys's. Rosenhead (1929) made use of further observations as compiled in Sir Napier Shaw's *Manual of Meteorology*. Our calculations are based on Rosenhead's values. Huaux (1951) made a calculation based on Laurent's mean monthly pressures between 65° N. and 65° S.

Consider the observed *departure*, $p(\theta, \lambda; t)$, in station-level (not sea-level) pressure from the annual mean at that station. Then

$$p \approx g \int_a^\infty \Delta\rho \, dr = gq \qquad (9.5.3)$$

in accordance with the hydrostatic approximation. In terms of the ocean function (§ A.1) we can write

$$q_E(t) = (1 - a_0^0)q_L(t) + a_0^0 q_O(t) \qquad (9.5.4)$$

for the seasonal departure in atmospheric load averaged over the entire Earth, land and oceans, respectively. Jeffreys and Rosenhead set $q_E = 0$. But the conservation principle applies only to dry air (seasonal variations in CO_2 and O_2 associated with photosynthesis are negligible). If the atmosphere were saturated, it would contain about 15×10^{18} g of water; the actual content is not much less, $12 \cdot 2 \times 10^{18}$ g on the average (Reichel, 1949), and the excess in July over that in January is $1 \cdot 4 \times 10^{18}$ g. From a seasonal compilation of water vapor by Bannon and Steele (1957) we find

$$q_E(t) = \frac{1}{4\pi} \int q(\theta, \lambda; t) \, ds = -0 \cdot 17 \cos \odot - 0 \cdot 08 \sin \odot \text{ g cm}^{-2},$$
$$(9.5.5)$$

which gives a January departure of $0 \cdot 19$ g cm^{-2} as compared to Reichel's $0 \cdot 27$ g cm^{-2}.

Following Jeffreys's original paper, we find it useful to allow immediately for the adjustment of the oceans to the superficial pressure.* The inverted barometer rule applies to the seasonal variation (§ 9·3). The ocean thus yields so as to annul horizontal pressure gradients; seasonal departures of the total load on the sea floor are everywhere the same but vary in time because of the

---

* The need for this correction was pointed out by Lamp (1891).

seasonal variation in the fraction of the atmosphere that lies above oceans. The integrals

$$
\left.
\begin{aligned}
\psi(t) &= -\frac{a^4}{C-A}\left\{\int_{\text{land}} q(\theta, \lambda; t)\sin\theta\cos\theta\, e^{i\lambda}\, ds \right. \\
&\qquad\qquad\qquad + q_O(t)\int_{\text{oceans}} \sin\theta\cos\theta\, e^{i\lambda}\, ds\Big\} \\
\psi_3(t) &= -\frac{a^4}{C}\left\{\int_{\text{land}} q(\theta, \lambda; t)(\sin^2\theta - \tfrac{2}{3})\, ds\right. \\
&\qquad\qquad\qquad + q_O(t)\int_{\text{oceans}} (\sin^2\theta - \tfrac{2}{3})\, ds\Big\}
\end{aligned}
\right\}
\tag{9.5.6}
$$

assume that dry air and water mass are conserved. The integrals taken over the entire Earth vanish; hence we can write them in the form

$$
\left.
\begin{aligned}
\psi(t) &= -\frac{a^4}{C-A}\int_{\text{land}} [q(\theta, \lambda; t) - q_O(t)]\sin\theta\cos\theta\, e^{i\lambda}\, ds \\
\psi_3(t) &= -\frac{a^4}{C}\int_{\text{land}} [q(\theta, \lambda; t) - q_O(t)](\sin^2\theta - \tfrac{2}{3})\, ds
\end{aligned}
\right\}
\tag{9.5.7}
$$

which, together with (9.5.4) and (9.5.5), permits us to evaluate the excitation function solely from observed pressures over land.

Nearly all meteorological compilations involve sea-level pressure P_0. This is computed from the observed station-level pressure P according to Laplace's formula (or some equivalent rule)

$$
z = 1{\cdot}8 \times 10^6 \left[1 + \frac{2(T + T_0)}{1000}\right]\left[\log_{10}\frac{P_0}{P}\right],
\tag{9.5.8}
$$

where z is the height above sea-level in cm, T the (variable) temperature at height z in °C, and T_0 the corresponding temperature at sea-level. At any given time, contours of station-level pressure, $P = $ constant, differ so little from contours of height, $z = $ constant, that no interest is attached to them, and all charts show instead sea-level isobars, $P_0 = $ constant. We require the departures, $P - \bar{P}$, of station-level pressures* from their annual mean, and not $P_0 - \bar{P_0}$.

---

* The error resulting from the use of sea-level pressures, as they are found on climatic charts, has already been pointed out by Angot. See *Annuaire de la Société Météorologique*, T. 35, 1887.

Jeffreys and Rosenhead applied the correction (rather, the removal of a correction already made) on a seasonal basis: first ψ was computed from seasonal charts of P_0, and then a correction excitation function, ψ', was computed in accordance with (9.5.8) using seasonal charts of temperature and allowing for the known elevation of continents.

On the basis of Rosenhead's tabulation* (fig. 9.4) we obtain (in parts per 10^8)

$$\psi_1 = -1\cdot58 \quad \cos \odot -1\cdot56 \quad \sin \odot, \quad \psi_1 + \psi_1' = -1\cdot68 \quad \cos \odot -0\cdot94 \quad \sin \odot,$$
$$\psi_2 = -25\cdot40 \qquad -6\cdot77 \qquad , \quad \psi_2 + \psi_2' = -16\cdot28 \qquad -1\cdot61 \qquad ,$$
$$\psi_3 = 0\cdot0433 \qquad +0\cdot0248 \qquad , \quad \psi_3 + \psi_3' = 0\cdot0246 \qquad +0\cdot0186 \qquad ,$$

$$(9.5.9)$$

for the excitation function associated with sea-level and station-level pressures, respectively. All components are now expressed in the same dimensionless ratios. Most of the contribution to ψ comes from Asia, where the difference between P and P_0 is very large. At $\theta = 50°$, $\lambda = 80°$, for example, $P_0 = 25$ mb and $P = 2$ mb! This accounts for the large difference between the corrected and uncorrected excitation function. The largest wobble term is reduced from 25·4 to 16·3; for the l.o.d. from 0·043 to 0·025. A new calculation is needed based directly on the raw pressure observations as they are entered into the logs of observatories.† Under the circumstances

---

* Our values for ψ_1, ψ_2 are not directly comparable with Rosenhead's values (his $\lambda_0 \ \mu_0$) for the following reasons: (1) the time argument was changed from 0° on 21 March, 90° on 21 June etc., to 0° on 1 January, 90° on 1 April etc.; (2) values were multiplied by $4\cdot85 \times 10^{-6}$ to change units from seconds of arc to radians (numerical values in (9.5.9) are in units of 10^{-8} radians); (3) values were divided by 1·41 to remove amplification due to rotation deformation. Rosenhead works out the theory for the load deformation and accordingly multiplies his (modified) values by 0·60 to obtain λ_1, μ_1, for comparison with astronomically derived λ_2, μ_2. But we have shown (§ 6.6) that the effects of rotational and load deformation cancel to a first order and that one might as well work with the excitation function, ψ, derived for a rigid Earth; (4) a number of errors have been corrected; in particular, Rosenhead treated the Arctic Ocean as land.

† This has now been done. Hassan (1960) has computed mean monthly values of ψ_i for the period 1873 to 1950, using station-level pressures tabulated in Clayton's World Weather Records, published by the Smithsonian Institution, Washington, D.C., and subsequently by the U.S. Weather Bureau. On the basis of these time series the seasonal variation and irregular variation (footnote, § 10.6) have been rediscussed by Munk and Hassan (1961). Our best estimate for the seasonal variation now is

$$\psi_1 = -1\cdot8 \cos \odot + 0\cdot2 \sin \odot + 0\cdot4 \cos 2 \odot - 0\cdot8 \sin 2 \odot$$
$$\psi_2 = -12\cdot9 \qquad -1\cdot0 \qquad +1\cdot8 \qquad +1\cdot4$$
$$\psi_3 = -0\cdot0028 \qquad +0\cdot0085 \qquad -0\cdot0017 \qquad -0\cdot0006$$

and this takes the place of $\psi_i + \psi_i'$ in (9.5.9).

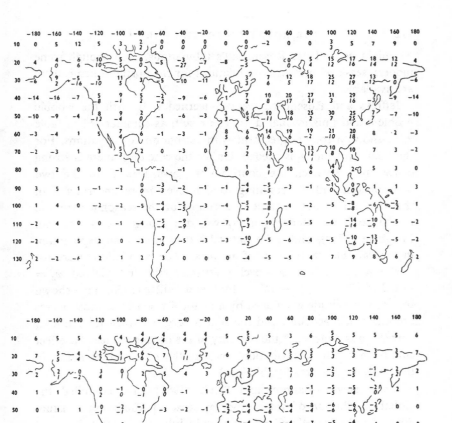

Fig. 9.4. Surface pressure differences, in millibars, for January minus July (*top*) and April minus October (*bottom*) according to Rosenhead (1929). Values refer to average conditions in a 10° × 20° grid centered at $\theta = 10°, 20°, \ldots, 130°$ colatitude, and $\lambda = 180°, 200°, \ldots, 340°, 0°, 20°, \ldots, 180°$ east longitude. Station-level difference can be interpreted as $2q$, twice the January (*top*) and April (*bottom*) load departures, respectively, in units of g cm^{-2}. For areas that are predominantly over land, the station-level pressure, P, (lower number, in italics) is given as well as the sea-level pressure, P_0.

we might as well use Rosenhead's tabulations, even though some modern sea-level pressure compilations for the northern hemisphere are available.

Jeffreys and Rosenhead were concerned only with the wobble problem and thus evaluated ψ. The result, as summarized in table 9.10, shows a rough agreement with astronomic observations. For the l.o.d. problem ψ_3 is needed. Unlike the case of ψ there is strong cancellation between the contribution toward ψ_3 from the two hemispheres, and the resulting total is even more uncertain than that of ψ. The first attempt to account for the seasonal variation in the l.o.d. was undertaken by van den Dungen, Cox and van Mieghem (1949), and quite naturally they sought an explanation in the variable distribution of air mass. The calculations were for northern hemisphere sea-level pressures only and yielded $\psi_3 = - 0.2 \times 10^{-8} \cos (\odot - 15°)$. Munk and Miller (1950) have shown that this amplitude is reduced by a factor 4 when both hemispheres are taken into account, and the sign is reversed when the sea-level correction is removed;* the mobility of the oceans reduces the effect still further. The contribution towards the observed variation in the l.o.d. is quite small (table 9.10).

Water vapor.—Bannon and Steele (1957) have compiled the global distribution of water vapor by season.† Maximum moisture appears to be over the Ganges Valley in July (6.5 g cm^{-2}). Seasonal variations up to $3–4 \text{ g cm}^{-2}$ are observed over the northwestern Atlantic and Pacific Oceans. From this compilation we have computed $q_E(t)$ as given in (9.5.5) and the following values of excitation function (in parts per 10^8):

$$\left.\begin{array}{l} \psi_1 = - 0.75 \cos \odot + 0.39 \sin \odot, \\ \psi_2 = 1.50 \cos \odot + 0.17 \sin \odot, \\ \psi_3 = - 0.0058 \cos \odot - 0.0026 \sin \odot \end{array}\right\}. \qquad (9.5.10)$$

Measurement of atmospheric pressure on the ground includes the weight of water vapor. Consequently the foregoing values are already included in (9.5.9). Water vapor accounts for about 10 per cent of the effect of air mass.

---

* This was discussed in "Letters to the Editor," *Tellus*, 2. 319–321, 1950.

† Independent compilations for the northern hemisphere have been made by Starr, Peixoto and Livadas (1957) in connexion with their study of the flux in water vapor.

Snow.—Jeffreys's (1916) remarks concerning the inadequacy of information for estimating the global distribution of snow still apply. His estimates are based in large measure on a remark by Sir Napier Shaw that snow in Siberia commences to accumulate in October and reaches a maximum of about 1 m in March. The mean density of snow is taken at $\frac{1}{3}$ g cm^{-3}. The snow load is assumed

Table 9.6. q_0 in g cm^{-2} due to snow.

					λ				
	0°	20°	40°	60°	80°	100°	240°	260°	280°
30°		30	35	50	40	20	10	20	45
θ									
40°	15	25	35	45	35	15		25	30

to increase at a uniform rate from 0 on the September equinox to $q_0(\theta, \lambda)$ g cm^{-2} on the March equinox and then to diminish uniformly to 0 on 21 May. Fourier analysis of this assymetric saw-tooth pattern yields

$$q(\theta, \lambda; t) = q_0(\theta, \lambda)$$
$$\left[\frac{7}{2\pi^2} \cos (\odot - 79°8) - \frac{3\sqrt{3}}{2\pi^2} \sin (\odot - 79°8) \right]. \quad (9.5.11)$$

The distribution of q_0 given in table 9.6 was chosen to give Jeffreys's values

$$\int q_0 \sin \theta \cos \theta \, e^{i\lambda} \, ds = (3 \cdot 06 + i \, 1 \cdot 63) \text{ g cm}^{-2}.$$

Furthermore $\int q_0 (\sin^2 \theta - \frac{2}{3}) \, ds = - 5 \cdot 6 \text{ g cm}^{-2}.$

We have allowed for removal of water from other sources, hence the negative sign (see remarks following 9.5.1*b*). The resulting excitation function equals (in parts per 10^8):

$$\left. \begin{array}{l} \psi_1 = - 6 \cdot 14 \cos \odot - 5 \cdot 77 \sin \odot, \\ \psi_2 = - 3 \cdot 27 \cos \odot - 3 \cdot 07 \sin \odot, \\ \psi_3 = 0 \cdot 036 \cos \odot + 0 \cdot 034 \sin \odot \end{array} \right\} \quad (9.5.12)$$

If the snow load were distributed uniformly over the entire Earth, this would give a mean load

$$q_E = 0.44 \cos \odot + 0.41 \sin \odot \text{ g cm}^{-2}. \qquad (9.5.13)$$

Our results differ somewhat from those given by Jeffreys because we have not allowed for rotational deformation. The summary in table 9.10 shows that the effect of snow is about one-third that of atmospheric mass. Accordingly it plays an important role in determining wobble and a negligible role with regard to the l.o.d. A new calculation based on modern observations is desirable. In the summary table 9.10 the effect of snow is included in the entry for ground water. The ground water values are obtained from other considerations (see below) and do not depend on the preceding estimates.

Polar ice has been neglected. Floating ice has no effect because the mass per unit area remains unchanged. Ice attached to the margins of continents may play a slight role. The variable ice sheet over Antarctica and Greenland ought to be taken into account, but there is no information concerning the seasonal variation. In all events these polar variations play a minor role because the latitude is so high.

Vegetation.—There is an additional source of variable storage in the form of plant material. According to Jeffreys (1916),

During the summer the vegetative parts of plants increase in mass in two ways. In trees, large quantities of sap rise from the ground, and thus the woody parts become heavier. Leaves are also produced in deciduous trees. In herbs, the whole of the subaerial portion is regenerated annually in the earlier part of the summer. Later in the year, usually in late summer or early autumn, the subaerial parts of all terrestrial plants partially dry up, and ultimately herbs wilt and fall to the ground, while deciduous trees cast all their leaves. The dead portions continue to lose weight until decomposition is complete. We may say then that there is a periodic part of the mass of trees, shrubs, and herbs which has a maximum in summer and a minimum in winter.

Jeffreys estimates this variable load to amount to 2 g cm^{-2} over England, and sets

$$q = 3.3 \cos \theta \cos (\odot - 194°) \text{ g cm}^{-2} \qquad (9.5.14)$$

over land surfaces. The factor $\cos \theta$ assures that the variation vanishes at the equator, where the vegetation seems to have no annual period.

The poles produce little effect on account of the terms $\sin^2 \theta$ in the expressions for ψ and $\sin^3 \theta$ in the expression for ψ_3. For the induced wobble Jeffreys obtains somewhat less than $0''01$. The effect is small but just large enough to deserve consideration. This intriguing calculation is probably the only geophysical application Sir Harold has been able to make of his early career as a Botanist.

An independent estimate can be made from considerations involving photosynthesis:

$$CO_2 + H_2O \rightarrow \tfrac{1}{6} C_6H_{12}O_6 + O_2. \qquad (9.5.15)$$

The first line of table 9.7 gives Riley's (1944) estimate of the total annual production of organic carbon on land (20 billion tons) and

Table 9.7. Annual production (+) and consumption (−) by photosynthesis in units of 10^{18} g

	Atomic ratio	Ocean	Continents	Total
1. Organic carbon	12/12	0·12	0·02	0·14
2. Dry plant tissue	30/12	0·30	0·05	0·35
3. Wet plant tissue		4·80	0·50	
4. Difference		remains in sea	− 0·45	
5. Water	− 18/12	remains in sea	− 0·03	
6. Total			− 0·48	− 0·48
7. CO_2	− 44/12	withdrawn from sea	− 0·07	− 0·07
8. O_2	33/12	0·32	0·05	0·37

sea (120 ± 80 billion tons). But for every atom of carbon (atomic weight 12) one molecule of $C_6H_{12}O_6$ is produced (gram molecular weight 30), and thus the total production of carbohydrate is found by multiplying by 30/12 (line 2). This represents grams of dry plant tissue. The most abundant plants in the ocean (phytoplankton) weigh roughly 16 times their dry weight. For potatoes the ratio is 10:1. The third line is found by multiplying by these ratios (net weight/dry weight). The difference between dry and wet weights is then the amount of water taken up by the plants from their surroundings (line 4). An additional molecule of water is consumed for each carbon atom produced as part of the photosynthetic process (line 5).

The sum of lines 4 and 5 give therefore the total water withdrawn by the plants each year (line 6). The last two lines give the corresponding figures for CO_2 and O_2 in the *atmospheric* budget. CO_2 is plentiful in the sea, and only the amount withdrawn by land plants need be considered. In the case of O_2 most of the amount produced in the sea is passed on to the atmosphere.

The total annual consumption of 0.48×10^{18} g of water by plants is distributed largely in the temperate zones:

$$20° \text{ N to } 60° \text{ N, land area } 0.66 \times 10^{18} \text{ cm}^2$$
$$20° \text{ S to } 60° \text{ S, land area } 0.19 \times 10^{18} \text{ cm}^2$$
$$\overline{0.85 \times 10^{18} \text{ cm}^2}$$

This gives $0.48 \times 10^{18} \, g/0.85 \times 10^{18} \text{ cm}^2 \approx 0.6 \text{ g cm}^{-2}$ per year on the average. The same amount must be released by the plants back to the surroundings. The seasonal *variation* must be less than this amount, with the above figure being approached as an upper limit if photosynthetic production were to take place all at once, and the release of water were spread evenly over the rest of the year. To allow for this we set

$$q(t) = 0.5 \cos \theta \cos (\odot - 194°) \text{ g cm}^{-2}$$

over land, in the place of Jeffreys's (9.5.14). The resulting excitation function is very roughly (in parts per 10^8)

$$\left. \begin{aligned} \psi_1 &= 0.2 \cos \odot + 0.1 \sin \odot, \\ \psi_2 &= 0.3 \cos \odot + 0.1 \sin \odot, \\ \psi_3 &= 0.006 \cos \odot + 0.002 \sin \odot \end{aligned} \right\}, \qquad (9.5.16)$$

and accounts for somewhat less than 2 per cent of the wobble and less than 1 per cent of the variation in the l.o.d. The vegetative load averaged over all of the Earth's surface is

$$q_E(t) = \frac{1}{4\pi} \int_{\text{land}} q(\theta, \lambda; t) \, ds = -0.033 \cos \odot - 0.008 \sin \odot \text{ g cm}^{-2}. \qquad (9.5.17)$$

As in the case of snow, the preceding estimate does not affect the totals in line 5, table 9.10.

Ground water.—The term is here used to include moisture stored on the surface (snow, vegetation) as well as in the ground. Seasonal variation in the water stored beneath the surface is of the order 10 g cm^{-2}. Clearly this is an important consideration. Detailed studies for many areas are available, and these show a marked degree of local variability. Global estimates are hard to come by.

Many authors have considered separately the storage (1) between the surface and the shallowest position of the water table, (2) between the shallowest and the instantaneous positions, and (3) beneath the instantaneous water table. For the seasonal variation (1) is the most important. But in any attempt to estimate global variations one appears to be limited to certain budgetary considerations in which the soil moisture at *all* depths is lumped together with the water on the ground (snow, vegetation, lakes) into a single storage term,* q. The 'hydrological equation' is then

$$P - E = R + \dot{q} \qquad (9.5.18)$$

which has a counterpart in the atmosphere,

$$- (P - E) = R' + \dot{q}'. \qquad (9.5.19)$$

P is precipitation and E evaporation, including 'evapotranspiration' from plants. R is the loss of water (other than to the atmosphere), due principally to river discharge. The corresponding term R' is the divergence of the advected water vapor. q and q' designate the water content of the ground and atmosphere, respectively, and $\dot{q} = dq/dt$, $\dot{q}' = dq'/dt$. All terms in the equations are in units of g cm^{-2} sec^{-1}.

We desire to solve (9.5.18) for $q(t)$. The term most difficult to determine from observations is E. Two methods have been used to get around this difficulty. By adding (9.5.18) and (9.5.19) one obtains

$$- \dot{q} = R + R' + \dot{q}' \qquad (9.5.20)$$

and, from the known wind vector $\mathbf{u}$ and vapor content v (in g cm^{-3}), the divergence term is evaluated according to

$$R' = \int_a^\infty \text{divergence } (\mathbf{u}v) \, dr. \qquad (9.5.21)$$

In practice monthly (or seasonal) averages are used for each of the terms in (9.5.20). It is then essential to interpret R' as the appropriate mean of the divergence of $\mathbf{u}v$, and not as the divergence of the mean $\mathbf{u}v$. Daily weather maps appear to give an adequate sampling rate

* Here q is in units of g cm^{-2}. Hydrologists use cm, and this is to be interpreted as cm^3 of water/cm^2 of area. Storage in cm must not be confused with the level of the water table, also in cm. If the ground is saturated beneath the table, and dry above, then a rise in the table by z cm will increase q by z times the porosity.

to include all of the important contributions from eddy processes. This method has been used by Benton and Estoque (1954). They find that, between May and August, 4.5 g cm^{-2} are lost from the North American continent to the atmosphere, whereas 16 g cm^{-2} are gained between September and April. The difference, 11.5 g cm^{-2}, is lost by run-off during the year. If the run-off were uniform, then the excess storage in April over that in September is 8.3 g cm for North America. A detailed study for the northern hemisphere has recently been completed by Starr and his colaborators (Starr and Peixoto, 1958; Starr, Peixoto and Livadas, 1958). There is a surprisingly large divergence of vapors from the atmosphere above some deserts. The source of this water remains unexplained. There is a marked convergence in the vicinity of headwaters and drainage basins of many large rivers, with some surprising exceptions (e.g., the Indus river). This method provides perhaps the most promising means for assessing the role of ground water. An extension to the southern hemisphere is most desirable.

The second method is based on Thornthwaite's work (1948). The underlying assumption is that unless the soil is dry the plant cover adjusts itself such that the evaporation is a function only of temperature and latitude. This 'potential evapotranspiration' $E_P(T, \theta)$ is given by an empirical formula. The integration of (9.5.18) proceeds according to the following rules (with some simplifications):

(1) if the soil is saturated,
$$q = q_S \text{ and } P \geqslant E_P, \text{ then } E = E_P \text{ and } \dot{q} = 0;$$
(2) if the soil is dry,
$$q = 0 \text{ and } P \leqslant E_P, \text{ then } E \leqslant E_P \text{ and } \dot{q} = 0;$$
(3) if the soil is wet,
$$0 \leqslant q \leqslant q_S, \quad \text{then } E = E_P \text{ and } \dot{q} = P - E.$$

Thus $E = E_P$ unless the soil is dry. The applicability of such a hypothesis depends heavily on comparison with observations and other methods. For North America, Thornthwaite's method is in remarkably close accord with the calculations by Benton and Estoque (1954).

Hylckama (1956) has computed q for the whole Earth per $10°$ square per month, using Thornthwaite's method (fig. 9.5). Harmonic

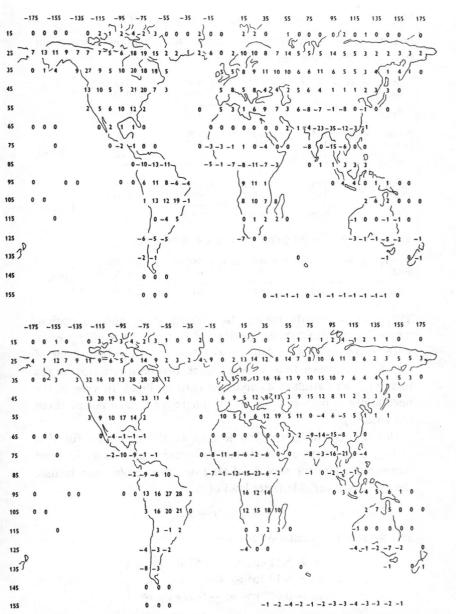

Fig. 9.5. Ground water storage, in g cm⁻², for January minus July (*top*) and April minus October (*bottom*). Values refer to average conditions in a 10° × 10° grid centered at $\theta = 15°, 25°, \ldots, 155°$ colatitude, and $\lambda = 185°, 195°, \ldots, 355°, 5°, 15°, \ldots, 175°$ east longitude. Hylckama's monthly values of grams of water in each 10° × 10° square were divided by the area of the square and rounded to the nearest g cm⁻². For comparison with air mass, see fig. 9.4.

analysis gives the following values for the excitation function (in parts per 10^8) in the two hemispheres:

		cos ⊙	sin ⊙	cos 2⊙	sin 2⊙	
North	$\psi_1 =$	− 1·44	− 2·85	+ 0·58	− 0·40	
South		0·73	2·73	0·09	0·16	
Total		− 0·71	− 0·12	0·67	− 0·24	
North	$\psi_2 =$	6·15	6·58	− 0·28	− 1·20	
South		0·26	− 0·67	− 0·17	0·28	(9.5.22)
Total		6·41	5·91	− 0·45	− 0·92	
North	$\psi_3 =$	0·032	0·076	− 0·005	− 0·014	
South		− 0·005	− 0·031	0·000	− 0·003	
Total		0·027	0·045	− 0·005	− 0·017	

The mean storage (in g cm^{-2}) has the components

		cos ⊙	sin ⊙	cos 2⊙	sin 2⊙	
North	$q_E =$	0·15	+ 0·46	− 0·08	− 0·05	
South		0·04	0·31	0·00	0·02	(9.5.23)
Total		0·19	0·77	− 0·08	− 0·03	

These terms include contributions from snow and vegetation. According to (9.5.23) the continents contain $7·5 \times 10^{18}$ g more water in spring than in fall. Previously a loss from the oceans of 6×10^{18} g between October and March had been reported on the basis of Miss Pattullo's compilation (Munk, 1956). In view of the uncertainty of both compilations this agreement must not be taken too seriously.

Oceans.—In the calculation leading to the excitation function (9.5.9), it has been assumed that the ocean yields as an inverted barometer under the superficial air load. The results then include the effect of a variable water load of

$$- q(\theta, \lambda; t) - q_0(t)$$

and the resulting excitation function*

$$\left.\begin{array}{l} \psi_1 = 5.11 \cos ⊙ + 1·57 \sin ⊙, \\ \psi_2 = 5·10 \cos ⊙ + 0·55 \sin ⊙, \\ \psi_3 = 0·012 \cos ⊙ - 0·019 \sin ⊙ \end{array}\right\}. \qquad (9.5.24)$$

* For this computation it is slightly better to compute $q_0(t)$ directly from oceanic pressures rather than by subtracting $q_L(t)$ from $q_E(t)$ as we had done previously (9.5.4).

Mobility of the oceans is responsible for an appreciable reduction in the values of $\psi_i + \psi_i'$ (9.5.9) and merits careful consideration. Rosenhead's (1929) procedure is to add three times the foregoing values of ψ to the excitation function already corrected for the inverted barometer adjustment. Thus, if p is the pressure departure on the sea surface, Rosenhead assumes that the pressure departure on the sea floor equals $- 3p$. This procedure is based on a comparison of tide-gauge and barograph records in Japan which indicated that the sea-level is depressed when the atmospheric pressure is high by an amount four times the response of an inverted water barometer. There are two difficulties with this interpretation of the Japanese records (Nomitsu, 1932). An examination of the seasonal variation of sea-level based on all available tide-gauge records reveals that this 4:1 porportionality between sea-level and pressure is restricted, with one exception, to seas surrounding Japan (Pattullo, Munk, Revelle and Strong, 1955, fig. 5). But, even if the Japanese situation were typical, a recorded variation in water-level by ξ cm does not imply a variable load $q = \rho_w\xi$ g cm^{-2}. It is important to distinguish between the *recorded* departures of sea-level, as obtained from tide-gauge records, and the *steric* departures which are inferred from the measured changes in specific volume of the water column. The latter depend almost entirely on temperature, and considerable data have been amassed since 1940 by means of the bathythermograph (salinity is important in the Gulf of Bengal and may play a role in high latitudes). The excess of recorded over steric departures is a measure of the load departure. At low and middle latitudes the recorded and steric variations are roughly the same, and the mass per unit area remains fairly constant throughout the year. If it were exactly constant, the effect of sea-level on the Earth's wobble enters only to a second order (§ 5.9) and should be dropped altogether. In a similar manner Lawford and Veley (1951) have gone astray in computing the effect of sea-level on the l.o.d. Rosenhead's correction amounts to two-thirds of the observed wobble! The agreement within 10 per cent between computed and observed values must be regarded as accidental.

The question remains concerning seasonal departures in oceanic load over and above that required by the inverted barometer response to atmospheric pressure. Variations in wind stress must be an important cause. Only a very rough guess can be made. Table 9.8

Table 9.8. Seasonal departures in ocean volume (10^{18} cm^3) as derived from the recorded and steric levels.

	θ	January departures			April departures		
		recorded	steric	diff.	recorded	steric	diff.
N. Subpolar	30°	1	0	1	$-1{\cdot}5$	$-0{\cdot}5$	-1
N. Subtropical	65°	-8	-6	-2	$-7{\cdot}5$	-5	$-2{\cdot}5$
S. Subtropical	115°	0	0	0	4	4	0
S. Subpolar	150°	-1	$-0{\cdot}5$	$-0{\cdot}5$	1	$0{\cdot}5$	$0{\cdot}5$
Total				$-1{\cdot}5$			$-3{\cdot}0$

is based on the global compilation by Pattullo *et al.* (1955). In April the oceans in the northern hemisphere have a deficit of water mass and those in the southern hemisphere a surplus, but the deficit is larger by 3×10^{18} g. This is the time of year when the terrestrial water content is at a maximum. Hylckama's (1956) estimate is 4×10^{18} g for the terrestrial spring surplus. The uncertainty of the oceanic estimate is such that the spring deficit could be twice as large as given by the table, or that there could be none at all. In January the subtropical and subpolar zones are out of phase in both hemispheres and the deficit smaller. The corresponding water load (averaged over the entire Earth) equals

$$q_E(t) = - {\cdot}14 \cos \odot - {\cdot}66 \sin \odot \text{ g cm}^{-2}.$$

We have computed ψ_i in accordance with table 9.8 assuming that the load departures are concentrated at the four latitudes indicated. But the recorded departures in the table include the effect of yielding to atmospheric pressure, and accordingly the values (9.5.24) have been subtracted. The result is

$$\left. \begin{array}{l} \psi_1 = - 9{\cdot}8 \cos \odot - 5{\cdot}4 \sin \odot, \\ \psi_2 = - 4{\cdot}6 \cos \odot - 8{\cdot}2 \sin \odot, \\ \psi_3 = 0{\cdot}009 \cos \odot + 0{\cdot}030 \sin \odot \end{array} \right\} . \qquad (9.5.25)$$

To balance the global water budget from all sources we require an additional term (table 9.10)

$$q_E(t) = 0\cdot12 \cos \odot - 0\cdot03 \sin \odot \text{ g cm}^{-2}. \qquad (9.5.26)$$

The water budget closure is better than we have any right to expect. The balance term (9.5.26) could be associated with a uniform ocean tide of 2 mm amplitude whose excitation function is (in parts per 10^8):

$$\left.\begin{array}{l} \psi = -\dfrac{2\pi}{5}\dfrac{a^4}{C-A}\dfrac{c_2^1}{a_0^0}q_E(t) \\[2mm] \quad = (0\cdot13 + 0\cdot21i)\cos\odot + (-0\cdot03 - 0\cdot05i)\sin\odot \\[3mm] \psi_3 = -\dfrac{8\pi}{15}\dfrac{a^4}{C}\dfrac{a_2^0}{a_0^0}q_E(t) \\[2mm] \quad = -0\cdot0008 \cos\odot + 0\cdot0002 \sin\odot \end{array}\right\} . \quad (9.5.27)$$

6. Winds and currents

Prompted by a remark by Starr (1948) that seasonal variations in the angular momentum of the atmosphere should produce 'practically undetectable inequalities in the rate of the Earth's rotation', Munk and Miller (1950) computed the component h_3 and found that the l.o.d. in January should exceed that in July by 1·8 ms. It turned out that an inequality of this magnitude and phase had in fact been reported (but not interpreted) by Stoyko, Finch, and by Scheibe and Adelsberger (§ 8.1). Most of the atmospheric momentum was found to be associated with the jet stream and surrounding westerlies which had been properly recorded only in the last two decades.

A recomputation by Mintz and Munk (1951), incorporating new data for the southern hemisphere, and extending the calculation in the northern hemisphere to *all* longitudes, led to a reduction in the computed January to July change from 1·8 ms to about 0·6 ms. Munk and Miller had used only the published sections along 100° E and 80° W where, as it turned out, the variations are substantially larger than at all other longitudes. On the basis of the new calculations it was concluded that only one-third to one-half of the observed variation could be attributed to winds. Similar results were obtained by Pariyski and Berlyand (1953). Extending the calculation from two to four seasons (Mintz and Munk, 1954) did not materially

124   THE ROTATION OF THE EARTH

change the numerical results. But by now estimates based on astronomical measurements had been substantially reduced, partly on account of errors in the FK 3 catalogue. With the reduction by a factor of three of both the atmospheric and astronomical terms (each revision for its own peculiar reason) the conclusion now seems

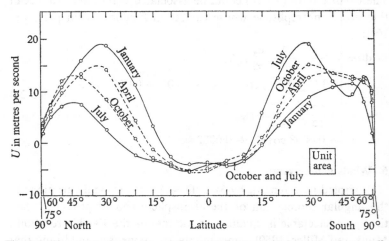

Fig. 9.6. The curves denote the relative westerly zonal winds, averaged over all longitudes and over the lower 90 per cent of the mass of the atmosphere, as a function of $\int_0^\theta \sin^2 \theta \, d\theta$. For this function the area under each each curve is proportional to relative westerly angular momentum; unit area equals 10^{32} g cm^2 sec^{-1}. An increase by this amount corresponds to an increase by 0·16 ms in the l.o.d. (from Mintz and Munk, 1954).

to be what it was in the first place, namely that the annual change in the l.o.d. is due largely to winds.

Let
$$U(\theta) = \frac{1}{2\pi} \frac{1}{P_0} \int_0^{2\pi} \int_0^{P_0} u_\lambda(\theta, \lambda, P) \, d\lambda \, dP \qquad (9.6.1)$$

designate the mean zonal wind averaged over all longitudes and between the surface ($P = P_0$) and the top of the atmosphere ($P = 0$). Fig. 9.6 shows pole to pole plots of $U(\theta)$. Between 20° N and 20° S the winds are based on rather scanty pilot balloon and radar wind observations; elsewhere the winds were computed from the observed pressure pattern according to the geostrophic rule (Mintz, 1954; Mintz and Munk, 1954). Observations up to the 100 mb level were available, and the calculation includes therefore the lower 90 per cent of the mass of the atmosphere.

The excitation function

$$\psi_3 = -\frac{2\pi a^3 P_0}{\Omega g C}\int_0^\pi U \sin^2\theta\,\mathrm{d}\theta, \qquad (9.6.2)$$

and the associated relative westerly angular momentum $h_3 = -\psi_3 C$, have been obtained by numerical integration (table 9.9). The agree-

Table 9.9. Relative westerly angular momentum of the atmosphere, h_3, and its seasonal departure, δh_3, in units of $10^{32}\,g\,\mathrm{cm}^2\,\mathrm{sec}^{-1}$, together with the seasonal departure of ψ_3, in parts per 10^8.

		Jan.	April	July	Oct.
90° N to 20° N	h_3	10·5	8·1	3·0	6·6
20° N to 20° S	h_3	− 1·0	− 2·2	− 1·7	− 2·8
20° S to 90° S	h_3	6·5	9·0	10·5	9·7
90° N to 90° S	h_3	16·0	14·9	11·8	13·5
90° N to 90° S	δh_3	1·95	0·85	− 2·25	− 0·55
90° N to 90° S	ψ_3	0·33	0·14	− 0·38	− 0·08

ment is good with the Greenwich observations and poor with the BIH values (table 9.10).

It is clear that the atmospheric contribution to the annual change in the l.o.d. owes its existence to the lack of cancellation between the two hemispheres. The principal difference is the larger momentum in the southern hemisphere during the southern summer, as compared to the northern hemisphere during the northern summer (6·5 vs 3·0 × $10^{32}\,g\,\mathrm{cm}^2\,\mathrm{sec}^{-1}$).

To obtain a more detailed picture of the seasonal variation, Mintz and Munk computed the angular momentum north of 20° N month-by-month for the 700 and 500 mb levels, the two levels for which the data are most reliable. The results show a maximum in early February and a minimum in late July. Generalizing from these two levels, one obtains (in parts per 10^8)

$$\psi_3 = -0.31\cos\odot - 0.20\sin\odot + 0.03\cos 2\odot + 0.01\sin 2\odot. \qquad (9.6.3)$$

There is a discrepancy between the computed annual amplitude,

0.37×10^{-8}, and the observed amplitude, 0.58×10^{-8}, even after allowing for Earth tides and other effects (table 9.10). There are several possibilities. Jenkinson (1955) and Palmer (personal communication) find substantial seasonal variations in the momentum of the equatorial atmosphere, and these have the order of magnitude and phase to account for the discrepancy. The equatorial wind observations leading to the numerical values (9.6.3) are inadequate. Furthermore, wind speeds of 100 m sec^{-1} have been observed at the 25 mb level indicating that a significant portion of the atmospheric momentum may be associated with stratospheric westerlies. These were not taken into account by Mintz and Munk. The problem of the annual variation in the l.o.d. is not yet closed.

With regard to the semi-annual term it is found that bodily tides could account for about half the observed BIH value (table 9.10). The semi-annual terms in (9.6.3) are much too small to fill the gap. Professor Palmer has drawn our attention to recent compilations of equatorial winds at the 300 mb level. These are of the same order as the annual winds at that level and of the required phase to account for the semi-annual discrepancy. Stratospheric winds may also play a role. The seasonal reversals have been noted to be so abrupt that one expects a relatively important contribution towards the semi-annual term and other higher harmonics.*

In § 8.1 we have referred to a possible reduction in the observed annual variation since 1950. To examine this possibility, Mintz and Munk computed the westerly angular momentum of the atmosphere north of 20° N at the 700 mb level, taking five-day means, from January 1949 until April 1950. When these curves are compared with the corresponding *normal* monthly values at the levels, as derived from U.S. Weather Bureau normal charts, they are found not to differ significantly.

The importance of winds with regard to the seasonal change in the l.o.d. led Munk and Groves (1952) to examine the effect of winds on the seasonal wobble.† They investigated numerically the role of

* Schwerdtfeger (1956, 1960) has analyzed antarctic and other southern hemisphere observations and finds a pronounced semi-annual variation in the mass of the atmosphere south of 60° S. The phase is the one required to account for the observed discrepancy, and the amplitude may be of the required magnitude.

† Such an effect was already considered by Volterra (1895).

geostrophic winds, not being aware of the fact that the integrated effect over the whole Earth must vanish (§ 9.4). The results proved inconclusive, as in fact they must.

Oceans.—The essential elements of oceanic circulation are: (1) the antarctic circumpolar current, which is largely a zonal current; (2) the equatorial circulation; (3) the subtropical and subpolar gyres in the 'confined' basins of the North Atlantic, North Pacific, South Atlantic, South Pacific and Indian Oceans. A discussion of these features can be found in Munk (1950) and Stommel (1957).

The antarctic circumpolar current is the only important around-the-globe motion (the arctic ocean has negligible zonal circulation). The current resembles somewhat the atmospheric jet stream. Its angular momentum can be estimated from

$$h_3 = 2\pi a^3 \, \overline{\sin^2 \theta} \times \text{transport.} \tag{9.6.4}$$

Sverdrup *et al.* (1946, pp. 164, 710) estimate the total transport at 10^{14} g sec^{-1}. A representative value of $\overline{\sin^2 \theta}$ is sin^2 (35°). This gives $h_3 \approx 10^{32}$ g cm^2 sec^{-1}, or about 10 per cent of the angular momentum of either atmospheric jet stream. The seasonal range is probably of the order of 20 per cent, and this gives 0.02×10^{-8} for the magnitude of ψ_3 as compared to 0.38×10^{-8} for the winds.

It may seem surprising that the atmosphere should contribute so much more than the oceans, especially since the mass of the atmosphere corresponds to only 10 m of water compared to a mean ocean depth of 4000 m. However, the mean zonal velocity of the ocean water is probably less than 2 cm sec^{-1}, whereas the mean zonal wind-speed in the westerlies (weighted for density stratification) is 20 m sec^{-1}. Thus the ratio in masses of 400–1 is more than offset by a velocity ratio of 1–1000. An additional factor is the larger *percentage fluctuations* of the atmospheric circulation about its mean value.

The equatorial circulation has an effect smaller than that of the antarctic circumpolar current. It has only one quarter of the transport, and much of this is offset by an east-flowing undercurrent. Moreover, seasonal variations in equatorial regions are small.

The effect of the continental barriers is to break up the remaining currents into a series of gyres. By far the most important are the subtropical gyres of the North Atlantic, including the Gulf Stream system, and of the North Pacific, including the Kuroshio system. The effect must be relatively small as compared to the unobstructed zonal circulation in the southern oceans. The effect on wobble is small (Munk and Groves, 1952).

7. Torque approach

The excitation function from a torque due to surface stresses can be written (§ 6.8)

$$\psi = - \frac{a^3}{\Omega^2(C - A)}$$

$$\int [p_{\theta m} + i p_{\lambda m} \cos \theta + p_{mm}(\Omega^2 a/g) \sin \theta \cos \theta] \, e^{i\lambda} \, ds \quad (9.7.1a)$$

$$\psi_3 = - \frac{1}{\Omega C} \int L_3 \, dt, \quad L_3 = a^3 \int p_{\lambda m} \sin \theta \, ds. \quad (9.7.1b)$$

Stress on the surface of land and sea provides an alternate method for computing the effect of the atmosphere, including winds and shift in air mass. Stress on the land surface and sea bottom gives the combined effect of atmosphere and ocean.

We consider the effect of the normal stress, p_{mm} (or simply p) acting on the smooth geoid in a direction opposite to gravity. Then $\psi_3 = 0$, and

$$\psi = - \frac{a^4}{C - A} \int \frac{p}{g} \sin \theta \cos \theta \, e^{i\lambda} \, ds, \quad (9.7.2)$$

which is precisely the excitation due to the distribution of matter, subject to the hydrostatic approximation (9.5.1, 9.5.3). Again this illustrates the equivalence of the momentum approach and torque approach. But what is measured is pressure on the ground, and the present interpretation is more direct.

The total stresses give the wobble due to changes in air mass and in winds. The normal stress on the geoid corresponds to the changes in air mass. The remaining stresses therefore correspond to the effect of winds. These remaining stress components are not known with sufficient accuracy to permit a direct evaluation of ψ_i. But ψ must be negligible because the effect of winds on wobble is so small.* On

---

* This invalidates the suggestion by Munk and Groves (1952) that the pressure of the monsoon winds against the Himalayas may produce a measurable wobble.

the other hand ψ_3 is by no means negligible. This is discussed in § 9.8.

8. Discussion of the seasonal variation

Summary of empirical results.—Table 9.10 summarizes the astronomical and geophysical evidence regarding the annual and semi-annual variations, respectively. Astronomical observations are given in the same dimensionless units (10^{-8} radians) as the geophysical observations; we no longer care whether the measurements were in centiseconds of arc or milliseconds of time. The modified excitation function Ψ'_i has been derived from ψ_i according to the transfer function (6.3.3). In the case of bodily tides the deformations due to rotation and load are already included so that $\Psi'_i = \psi_i$. The summary may be regarded as a seasonal budget for the planet Earth, including its atmosphere, oceans and interior, of four time-variable quantities: the components of momentum and the mass of water (expressed as the mean surface density, q_E).

Consider first the annual wobble. The astronomic observations leave no doubt that Ψ'_2 is larger than Ψ'_1, but the phase is not well-determined. According to Walker and Young (1957) the maximum in Ψ'_2 is reached 40° (40–1/2 days) earlier than according to Jeffreys (1952). Other investigations, including those based on the independent Greenwich–Washington observations, agree much more nearly with Jeffreys.

To a first order the annual wobble is accounted for by the shifts in air mass. The values in line 2 already contain, among other things, the items separately listed in lines 2·1, 2·2, 2·3. Without the pressure correction, and without allowance for the isostatic yielding of the ocean, the computed wobble would be too large. The effects of ground water (including snow and vegetation, as given) and of non-isostatic movements in the oceans throw the momentum budget further out of balance. We consider (4.2) as the most uncertain of the items; it turns out that the agreement would be better without this term.* The water budget, reflecting the seasonal flux between

---

* *Note added in press.* The analysis of observations during the IGY by Lisitzin (in preparation) and by Pattullo (1960) does, in fact, indicate that the term (4.2) is too large. In the subpolar oceans in particular, the steric departures are small and the recorded departures very nearly in accord with the inverted barometer rule.

Table 9.10. The modified excitation function, Ψ_i, in parts per 10^8, and the mean water load, q_E, in g cm^{-2}.

Annual

	Ψ_1	Ψ_2	Ψ_3	q_E
Astronomical Observations				
Jeffreys	$-0.5\cos\odot - 0.5\sin\odot$	$-15.5\cos\odot + 6.8\sin\odot$	$-0.45\cos\odot - 0.36\sin\odot$	
W & Y	−4.8 / 6.3	−7.3 / 15.0	−0.33 / −0.16	
BIH				
S & T				
Geophysical Observations				
1. Tides	order (0.05)	order (0.01)	order (0.01)	
2. Air Mass	−1.7 / −0.9	−16.3 / −1.6	−0.03 / −0.00	$-0.17\cos\odot - 0.08\sin\odot$
2.1 Pressure correction	−0.1 / 0.6	9.1 / 5.2	−0.02 / 0.01	0.77
2.2 Water vapor	−0.8 / 0.4	1.5 / 0.2	−0.01 / −0.00	0.41
2.3 Isostatic oceans	5.1 / 1.6	5.1 / 0.6	−0.00 / 0.01	−0.01
3. Ground water	−0.7 / −0.1	6.4 / 5.9	0.02 / 0.03	0.19
3.1 Snow	6.1 / −5.8	−3.3 / −3.1	0.02 / 0.02	0.44
3.2 Vegetation	0.2 / 0.1	0.3 / 0.1	0.00 / 0.02	−0.03
4. Oceans	−9.6 / −5.5	−4.4 / −8.3	0.01 / 0.00	
4.1 Isostatic (2:3)	(5.1) / (1.6)	(5.1) / (0.6)	(0.01) / (−0.01)	−0.14
4.2 Non-isostatic	−9.8 / −5.4	−4.6 / −8.2	0.01 / 0.02	−0.66
4.3 Balance	0.2 / −0.1	0.2 / −0.1	−0.00 / 0.00	0.12
5. Winds			−0.31 / −0.20	−0.03
6. Currents			order (0.02)	0.00
1 + 2 + 3 + 4 + 5 + 6	−12.0 / −6.5	−14.3 / −4.0	−0.29 / −0.14	0.00

Semi-annual

	Ψ_1	Ψ_2	Ψ_3
Astronomical Observations			
W & Y	$-0.5\cos 2\odot - 2.9\sin 2\odot$	$4.4\cos 2\odot + 1.0\sin 2\odot$	$0.28\cos 2\odot + 0.24\sin 2\odot$
BIH			0.41 / −0.06
S & T			
Geophysical Observations			
1. Tides	order (0.3)	order (0.1)	0.16 / −0.06
2. Ground water	0.7 / −0.2	−0.4 / −0.9	−0.00 / −0.01
3. Winds			0.03 / 0.01

atmosphere, oceans, and water stored on and in the ground, comes out better than one has a right to expect. The 'balance tide' 4·3 has no appreciable effect on the remaining calculations. The effect of ocean currents is negligible. The most obvious cause is then still the one pointed out by Spitaler in 1901: the loss in winter of air mass over the North Atlantic and Northeast Pacific Oceans and the corresponding gain over the Asiatic continent. The new observations have only confused the issue. The essential problems now are the seasonal variation in ground water and the shift in ocean mass induced by wind stress. With regard to the semi-annual wobble the astronomic and geophysical evidence is too uncertain to permit any conclusions.

One can be more definite about the variation in the l.o.d. (component Ψ'_3) even though this variation was discovered only in relatively recent times. The BIH values are based on all observations and are considered as representative. Certainly the seasonal variation in the distribution of matter has only a small effect. It would be surprising if this should amount to more than 20 per cent of the observed variation. Ocean currents are negligible. But the computed effect of winds comes close to the observed variation, and this gives us some confidence in the southern hemisphere winds as determined from few scattered stations. The wind calculation does not include recently discovered stratospheric winds of such magnitude that their effect may be considerable, even though the air density is small. It seems likely that equatorial winds have a larger effect than the values in table 9.10 indicate. Perhaps the combined effects of the stratospheric and equatorial winds can account for the fact that the observed variation is 1·5 times the computed variation.

With regard to the semi-annual term the discrepancy is even larger. According to the table the only important term is due to bodily tides, and this gives only half the observed value. Here again it seems likely that equatorial and stratospheric winds will alter the picture when properly taken into account. It is conceivable that a fictitious semi-annual term is introduced by way of correcting sidereal time for the polar motion (§ 8.2).

In the interpretation of these results it is instructive to consider the reverse problem: to inquire as to the simplest meteorological pattern that could account for the observed variation in rotation.

Air mass.—Suppose the pressure departure, p, is expanded in spherical surface harmonics. Then

$$\tfrac{3}{2} \sin \theta \cos \theta \, (a_2^1 \cos \lambda + b_2^1 \sin \lambda), \quad a_2^0 \, (\tfrac{3}{2} \cos^2 \theta - \tfrac{1}{2}), \quad (9.8.1a, b)$$

are the only terms in this expansion for which (9.5.1a) and (9.5.1b), respectively, do not vanish, and these give

$$\Psi = - \frac{2\pi}{5} \frac{a^4}{g(C - A)} c_2^1, \quad \psi_3 = \frac{8\pi}{15} \frac{a^4}{gC} a_2^0. \quad (9.8.2a, b)$$

From the observed excitation function (Ψ according to Jeffreys; Ψ_3 according to BIH, table 9.10) we have computed c_2^1, a_2^0, and upon substitution into (9.8.1a, b) these give

$$p = \sin \theta \cos \theta \, [0\cdot14 \cos \lambda \cos (\odot - 45°)$$
$$+ \, 3\cdot20 \sin \lambda \cos (\odot - 336°)] \Big\} \quad (9.8.3)$$
$$p = - 16\cdot6 \, (\tfrac{3}{2} \cos^2 \theta - \tfrac{1}{2}) \cos (\odot - 38°)$$

for the simplest pressure patterns (in millibars) to achieve the observed wobble and variation in the l.o.d., respectively (fig. 9.7). For the wobble the magnitude of the requisite pressure range, 3·2 mb from summer to winter at latitude 45°, is not excessive; local pressure ranges over Asia go up to 20 mb. The similarity with the observed high-pressure cell over Asia during the winter is at once apparent (fig. 9.4). The northwestern quadrant, centered over America, is dominated in summer by the east Pacific and Bermuda high-pressure areas and in winter by the Aleutian and Iceland low pressures. These features, which here overshadow a relatively weak opposite effect of the American continent, are again in agreement with the required pattern. There is no resemblance with observed features in the southern hemisphere. The observed variation in the l.o.d. calls for an equatorial high pressure and polar low pressure during northern winter, with an annual range of 33 mb at the equator! This is clearly ruled out by the observations.

Geostrophic winds.—A similar procedure can be followed with respect to winds. For geostrophic winds we need be concerned only with ψ_3. Consider an atmospheric shell contained between the pressure surfaces $P - \tfrac{1}{2}\delta P$ and $P + \tfrac{1}{2}\delta P$. In an expansion into spherical harmonics, the only components of zonal wind for which

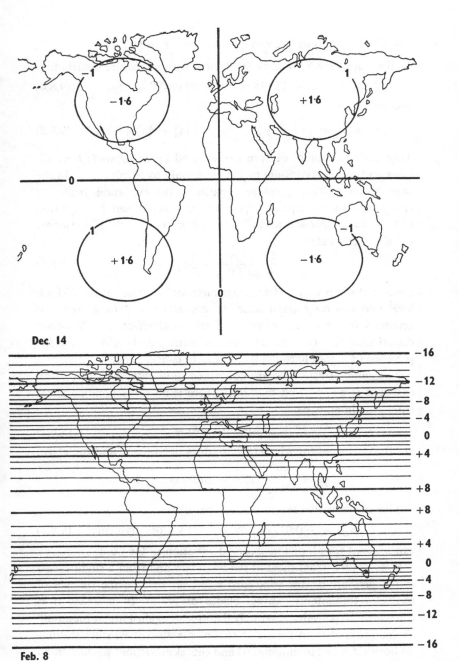

Dec. 14

Feb. 8

Fig. 9.7. Departure of surface pressure (millibars) from the annual mean which could account for all of the observed wobble (*top*) and variation in the l.o.d. (*bottom*). The dates are those for which the pressure departures are a maximum or minimum. Zero isobar on Dec. 14 is at 1° W (not zero) longitude.

Ψ_3' does not vanish are those associated with a pressure distribution

$$\tfrac{1}{2}a_2^0(3\cos^2\theta - 1) + \tfrac{1}{8}a_4^0(35\cos^4\theta - 30\cos^2\theta + 3) + \ldots \quad (9.8.4)$$

and these give

$$\psi_3\delta P = \left(\frac{8\pi}{15}\frac{a^4}{gC}\right)\left(\frac{15}{2}\frac{\delta P}{a^2\Omega^2\rho_0}\right)(a_2^0 - \tfrac{4}{3}a_4^0 + \ldots). \quad (9.8.5)$$

Thus only harmonics of even degree need be considered; both air mass and winds contribute to p_2^0, winds only to p_4^0, p_6^0, ... In the case of 'non-tilting' pressure departures the excitation from the entire atmosphere equals $\gamma P_0\psi_3$, where γ is defined by equation (9.4.3). On comparison with (9.8.2) we note that for the p_2^0 harmonic the effect of winds is

$$\frac{15}{4}\frac{P_0\gamma}{\tfrac{1}{2}a^2\Omega^2\rho_0} = 2\cdot 8\gamma \quad (9.8.6)$$

times that of air mass. The annual variation is largest at the 200 mb level and relatively small near the ground; $\gamma = 3$ to 4, and this accounts for the predominance of the wind effect. For low-level disturbance the two effects may be more nearly alike, and the presence of higher harmonics can lead to cancellation.

Tangential stress.—We are concerned only with the effect of an eastward stress, $p_{\lambda m}$, on the l.o.d. There is no corresponding effect from $p_{\theta m}$. In an expansion of $p_{\lambda m}$, the terms

$$a_0^0 + a_2^0(\tfrac{3}{2}\cos^2\theta - \tfrac{1}{2}) + \ldots \quad (9.8.7)$$

are the only ones for which (9.7.1b) does not vanish, and these give

$$\psi_3 = \frac{a^3\pi^2}{\Omega C}\int[a_0^0(t) - \tfrac{1}{8}a_2^0(t) + \ldots]\,\mathrm{d}t.$$

Taking only the first term, we find that a stress

$$p_{\lambda m} = -\,0\cdot 017\cos\odot + 0\cdot 021\sin\odot \quad \mathrm{dyn\ cm^{-2}} \quad (9.8.8)$$

can account for all of the observed variation in the l.o.d.

Stresses arise from shear at the boundary and from a slight excess in the pressure on the windward side over that on the lee side of obstacles (mountains, waves, blades of grass). In practice, it is convenient to consider separately the stresses against major mountain chains (derived from measurements of pressure at weather stations on both sides of the mountains) and the 'skin friction' against small obstacles (derived from surface winds by statistical considerations).

These normal stresses are here replaced by equivalent tangential stresses on the geoid.

We consider the American continent for illustration. The mountain effect is due largely to pressure against the Rocky Mountains and the Sierra Nevadas. Mintz (1951) has shown that we shall not be far wrong if we replace the Rocky Mountains by a wall, 3 km high, extending along 110° W between 30° N and 60° N, with pressure against its western face minus pressure against its eastern face in January exceeding that in July by 3.5 mb. This yields, approximately, in parts per 10^8,

$$\psi_3 = 1 \cdot 2 \sin (\odot - 15°), \qquad (9.8.9)$$

or twice the value derived from the BIH observations! From general consideration of atmospheric circulation, we know that stresses of the order of 1 dyne cm^{-2} are representative over large portions of the Earth's surface. Mintz (1951) finds that the zonal stress between 35° N and 90° N averages 1 dyne cm^{-2}, of which one-third to one-half is due to skin friction, the remaining part to the mountain effect.

We conclude that the *magnitude* of the wind stress on the surface of the Earth is larger by a factor 10^2 than the minimum values (9.8.8) required to account for the observed variation in the l.o.d. This means that there must be cancellation between nearly equal eastward and westward stresses.

Fig. 9.8 gives a schematic presentation of the *mean* flux of relative westerly angular momentum. In accordance with the suggestion made originally by Jeffreys (1926, 1933) the northward flux across OB is associated with the tilted crest and trough lines of atmospheric waves: $\overline{\rho u_\lambda u_\theta}$ is negative. In the region of the westerlies, north of B, there is a flux into the ground; in the easterlies the flux is from the ground into the atmosphere. The solid Earth is twisted so that a point north of B is slightly eastward, and a point south of B westward of what it would be in the absence of wind stress. The displacements amount to a small fraction of a cm.

The westerly angular momentum of the atmosphere is of the order 10^{33} g cm^2 sec^{-1}, the torque due to wind stress on AB is 10^{27} g cm^2 sec^{-2}; upon division we obtain 10^6 sec for the time constant of atmospheric circulation. If the flux across BC suddenly vanished (or across *AB* doubled) the momentum of the atmosphere would be greatly reduced in ten days. A 10 per cent variation in stress would produce accordingly large variations in momentum in 100 days. This latter case is essentially the nature of the annual variation. To come within 25 per cent, the eastward and westward stresses would each have to be known to an accuracy of one part in a hundred. Under the circumstances the momentum approach is inherently more accurate than the torque approach for determining the annual variation in

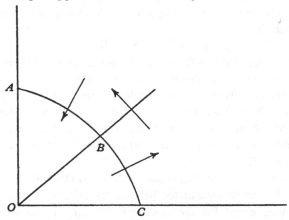

Fig. 9.8. Schematic presentation of the mean flux of relative westerly angular momentum.

the l.o.d. A calculation by the torque method has been given by van den Dungen, Cox and van Mieghem (1950).

We now return to (9.8.9). Suppose the winds over the Rocky Mountains (or any other chain) were to increase suddenly. There would be an immediate and pronounced effect on the l.o.d., but its persistence is limited by the total momentum content of the atmosphere to a few weeks, at most. We may therefore expect high-frequency variations in the l.o.d. by an amount of the same order as the annual term. But what is measured is the integrated time error, $\int \Delta$ (l.o.d.) dt, and the effect of a high-frequency variation is relatively small.

Distribution of continents.—In the final analysis the seasonal variation in rotation must depend on the distribution of land and sea. If the Earth were covered entirely by continents or oceans, then there would be no seasonal variation in rotation. What arrangements in the distribution of continents could be expected to lead to a maximum seasonal wobble and variation in l.o.d.?

Consider first the effect of air mass. It is a fact that in winter (northern and southern) pressure is high over continents. The top and left diagrams in fig. 9.9 appear to represent the distribution that would lead to maximum wobble and variation in l.o.d. respectively. Both diagrams refer to January, and the phase is then roughly in agreement with the observed. It should be noted that the winds

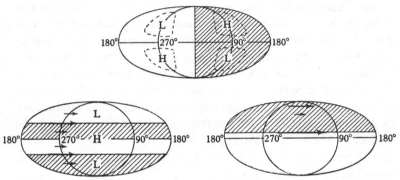

Fig. 9.9. The shaded areas represent arrangements of continents that would be associated with seasonal atmospheric variations believed to cause maximum variations in rotation. The distribution in matter is indicated by areas of high (H) and low (L) pressures. The arrows designate around-the-world winds.

associated with the pressure distribution of the left diagram further lengthen the day in January.

But it has been shown that pressure plays a minor role in the variation of the l.o.d. The distribution of continents in the right diagram might be more effective. Here it is assumed that the ground pressure is uniform and that the winds aloft increase according to the 'thermal wind equation'

$$\frac{\partial u_\lambda}{\partial r} = \frac{1}{2\Omega \sin \theta} \frac{1}{T} \frac{\partial T}{\partial \theta}. \tag{9.8.10}$$

The continent extends not quite to the equator and an eastward jet is associated with the sharp increase in temperature at the continental boundary. There are additional easterlies associated with the drop in temperature over the polar cap.

The actual distribution of land and sea is in a sense a mixture of the eastern continent (top) and the northern continent (right). In the former case the resemblance is only in the northern hemisphere. The southern hemisphere plays a passive role with respect to wobble.

Response of core.—The work of Poincaré (1910), Bondi and Lyttleton (1948) and of Jeffreys and Vicente (1957*a*, *b*) has demonstrated that a dynamic treatment for the core is required. For the annual wobble we have the advantage of a frequency so close to the 14-month nutation that we may expect the same tidal-effective Love numbers to be applicable. To allow for the response of the core, whether static or dynamic, we need only to write the equations in terms of the *observed* frequency of nutation, σ_0, and this has been done. For the semi-annual wobble the response of the core requires a different set of Love numbers (Jeffreys and Vicente, 1957*a*, p. 161).

The situation is altogether different with respect to the variation in the l.o.d. Electromagnetic coupling between mantle and core (viscous coupling is negligible) depends on the electric conductivity in the mantle (§ 11.12). The core may not fully partake in the annual variation. The core's moment of inertia, C_C, accounts for 10 to 12 per cent of the Earth's inertia, C, and the variation in the l.o.d. would be increased in the ratio

$$I/(I - I_C) = 1 \cdot 1 \qquad (9.8.11)$$

if the core were not to participate at all. The annual slippage (frequency σ) between core and mantle at the equator is then $(2\pi/\sigma)a_C\Psi_3$ ≈ 23 m, where a_C is the radius of the core. The problem is open, and we have made no corrections.

Non-equilibrium response.—Bondi has suggested that if a phase lag between computed and observed wobble could be determined, it would provide information concerning anelastic properties of the mantle.* The problem can be treated by replacing σ_0 by the complex frequency

$$\sigma_0 = \sigma_0 + i\alpha$$

in equations (6.7.7) and (6.7.8). The amplitude of the free motion (Chandler wobble) then diminishes according to $e^{-\alpha t}$, and the forced motion shows a phase lag

$$\phi = \frac{\alpha}{\sigma_0} \frac{\sigma}{\sigma - \sigma_0} = \frac{1}{2Q} \frac{\sigma}{\sigma - \sigma_0} \text{ rad}, \qquad (9.8.12)$$

* *The Observatory.* No. 902, February 1958, p. 12.

where Q is assumed to be a large number. From the study of the Chandler wobble we find $Q \approx 30$ (table 10.1), but there is some evidence from the pole tide that Q may be 100 (§ 10.4). The corresponding phase lags are $\phi = 6°$ and $2°$, respectively. The lags are uncertain and small compared to the discrepancy between observed and computed phases.

Variation of the gravitational constant.—Dicke (1957, 1958) has suggested the possibility of an annual variation in the Earth's radius due to a variation in the gravitational constant, G. Mach's principle suggests that inertial forces are a gravitational interaction with distant matter. It follows that G varies with the velocity relative to distant matter. Dicke estimates an annual variation in G as measured on the Earth by 1 part in 10^7 to 1 part in 10^8, and a resulting variation by the same order of g at a fixed point. A direct experimental check is not yet possible. G can be determined to only 1 part in 10^4, and g to not quite 1 part in 10^6.

Hess (1958) has investigated the possible effect on the l.o.d. by such a variation in G. Let G_0 designate the gravitational constant in a special coordinate system from which the Universe as a whole appears isotropic. Then according to Dicke

$$G = G_0[1 + 2(v/c)^2],$$

where v is the velocity relative to this system and c the velocity of light. It is now assumed that the galaxy is at rest relative to this special coordinate system. Hess writes

$$v = v_0 + v_1 \cos (\odot + 29°),$$

where v_0 is the appropriate galactic component of the Sun's velocity and v_1 the average orbital speed of the Earth. Setting $v_0 = 100 \text{ km sec}^{-1}$ and $v_1 = 30 \text{ km sec}^{-1}$ gives $G = G_0[1 + 1\cdot47 \times 10^{-7} \cos (\odot + 29°)]$. On the basis of various elastic models for the Earth, Hess estimates that

$$\frac{\Delta C}{C} = -0\cdot12 \frac{\Delta G}{G}$$

and obtains $\quad \Delta \text{ (l.o.d.)} = -1\cdot5 \cos (\odot + 29°)$ ms

compared with the BIH value $+0\cdot5 \cos (\odot - 38°)$. There is no support for Dicke's proposal in the observed variation of the l.o.d.

However, Hess's choice of phase and of v_0 depends on the arbitrary assumption that our galaxy is at rest with respect to the special coordinates. There would be no conflict with observations of the l.o.d. if v were smaller by an order of magnitude and this may well be the case.

Other considerations.—Seasonal variations in the distribution of solid matter has been considered only with regard to bodily tides. There may be other factors. A brief discussion of seasonal oscillations of the Earth's surface as derived from frequent levelling can be found in the IUGG News Letter (January, 1953, p. 177). Measurements in Germany by Kneissl (1955) did not indicate any short-period movements.

Compared to the observed values any effect of human activity is negligible. If all American automobiles were driven from Alaska to Mexico, the change in moment of inertia would be by 1 part in 10^{14}.

Landsberg (1933, 1948) has suggested that the strain associated with the annual wobble may trigger earthquakes. He presents some evidence that the frequency of deep-focus earthquakes is at a maximum during equinox and a minimum during solstices. The evidence is marginal. The annual wobble is smaller than the Chandler wobble so that a further study might be based on the combined effect of the 12- and 14-month terms.

9. The continuous spectrum

The discussion so far has been limited to the discrete spectrum; in particular, to the fortnightly, monthly, semi-annual and annual variation. These spectral lines are superimposed on a continuous spectrum arising from the irregular motions of the atmosphere and ocean. The *continuum* is the more interesting in many ways. It may provide pertinent information concerning persistent anomalies in the weather.

There have been no trustworthy records of the continuous wobble spectrum for frequencies higher than 1 c/year. The development of the caesium frequency standard (§ 8.1) may permit the detection of the continuum for the l.o.d. Fig. 9.10 shows the variation of the l.o.d. after removal of the annual, semi-annual, monthly and fortnightly terms. The most important feature is the mean drift in 1956 and 1957.

This represents the first detailed look at the historical variation (ch. 11), previously known only from infrequent determinations of Ephemeris Time. There are in addition short-period fluctuations of doubtful significance, but the 1958 anomaly is too large to be dismissed.

When this feature was first noted early in 1958, it was interpreted as an anomalous decrease of the l.o.d. by something like ⅓ ms.

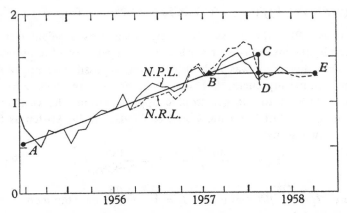

Fig. 9.10. The length of day derived from PZT observations (Washington and Richmond combined), referred to the caesium standards at the National Physical Laboratory, Teddington, and at the Naval Research Laboratory, Washington, D.C. Values for 1955, 1956 and 1957 are from Essen, Parry, Markowitz and Hall (1958); subsequent values have been kindly transmitted to us by Markowitz. Ordinate is milliseconds in excess over 86,400 seconds of Ephemeris Time (based on UT 2).

Accordingly, an effort was made to examine the meteorologic situation as a possible source of this decrease. Since then the persistent character of the deviation indicates rather a 'turning point' in the historical variation (§ 11.5), presumably of non-meteorologic origin. The following account of the meteorologic situations in spring 1958 is included only because it serves as an indication of the continuum spectrum to be expected from irregular atmospheric variations.*

Initially the situation looked favorable. Spring 1958 was characterized by a severe 'low zonal-index' situation. The upper-level westerlies migrated southward into the subtropics, where aloft the

---

* We are indebted to Professor Palmer for these calculations.

normal easterlies were replaced by the strongest westerlies on record. A weather anomaly of this type contains opposing elements. The shift of westerlies toward the equator would, by itself, increase the l.o.d. Weakening of the west winds and a poleward shift of air mass would decrease it.

In 'geostrophic latitudes' the spring 1958 anomaly is well represented by the function

$$p(\theta, P) = \mathscr{P}(P)f(\theta) = \mathscr{P}(P) \sin 5\theta \cos 1\cdot28\theta, \quad \theta \text{ from } 0° \text{ to } 72°,$$

where $p(\theta, P)$ is the pressure anomaly at some fixed height (represented by the mean pressure surface P) averaged over all longitudes. The fact that there was little 'tilt' makes it possible to apply the same latitude argument, $f(\theta)$, for all P. For spring 1958, $\gamma \approx 1\cdot5$ (9.4.3). Subject to the geostrophic approximation, the combined effects of the shift in air mass, and of winds, can be written as the sum of two terms:

$$\Psi_3 = - \frac{2\pi a^4 \mathscr{P}(P_0)}{gC} \left[I_1 + \frac{P_0 \gamma}{2a^2 \Omega^2 \rho_0} I_2 \right], \quad (9.9.1)$$

where $I_1 = \int_0^{72°} f(\theta) \sin^3 \theta \, \mathrm{d}\theta, \quad I_2 = \int_0^{72°} \frac{\mathrm{d} f(\theta)}{\mathrm{d}\theta} \sin \theta \tan \theta \, \mathrm{d}\theta.$

Sea-level pressure departures give $\mathscr{P}(P_0) = 10$ mb: it follows from (9.1.1) that

$$\Delta \text{ (l.o.d.)} = - 0\cdot056 + 0\cdot066 \text{ ms}$$

due to shifts in air mass and winds, respectively. Compared to the supposed anomaly of $- 0\cdot3$ ms the magnitude of both effects is small, and their sum is negligible and of the wrong sign.

The angular momentum of the stratosphere was not included in the preceding calculation. It turns out that a change by 50 ms^{-1} did occur in January 1958, and this would give Δ (l.o.d.) $= - 0\cdot2$ ms; but the effect did not differ from that of the preceding year. In the tropics, on the other hand, the integrated zonal winds showed a pronounced negative (easterly) anomaly in spring 1958 which was not found in spring 1957. The effect is of the sign and magnitude to account for the supposed anomaly in the l.o.d., thus indicating that the tropical atmosphere may play an active role in the continuum. Other factors may be important in the continuum; in particular,

Pacific sea-level in spring 1958 was anomalous from Mexico to Alaska. There is then nothing mysterious concerning irregular departures in the l.o.d. by a few parts in 10^9. Suppose a mean square variation by 10^{-18} has equal contributions from all frequencies between 1 c/year to 1 c/month. The frequency range is then from 0.3×10^{-7} c/s to 3.6×10^{-7} c/s, and the spectral density equals

$$\frac{10^{-18}}{3.3 \times 10^{-7} \text{ c/s}} = 3 \times 10^{-12} \text{ sec.} \qquad (9.9.2)$$

This lies above observational noise level (which is set largely by errors in sidereal time determination, fig. 8.1) for frequencies less than 10^{-7} c/s. The conclusion is that severe meteorological anomalies persisting over a season can be expected to cause detectable variations in the l.o.d.; to detect monthly anomalies the accuracy of sidereal time determination must be improved by an order of magnitude.

With regard to the situation in spring 1958, Markowitz has pointed out that European observations would favor a sharp decrease in d (l.o.d.)/dt late in 1957 ($ABDE$ rather than $ABCDE$, fig. 9.10) as the proper interpretation of the astronomical data. In that case the meteorologic anomaly is unrelated. The interpretation $ABCDE$ calls for a chance superposition of two geophysical events.*

* According to Markowitz there is a preliminary indication of a further decrease early in 1959, bringing the l.o.d. to approximately the same value (0.9 ms in fig. 9.10) as in the summer of 1955.

CHAPTER 10

THE CHANDLER WOBBLE

1. The astronomical evidence

The latitude component m_1 before and after removal of the seasonal terms is shown in fig. 7.4. The non-seasonal residue reveals 14-month oscillations in wave 'packets' consisting of something like 10 cycles. This is brought out more clearly from a plot (fig. 10.1)

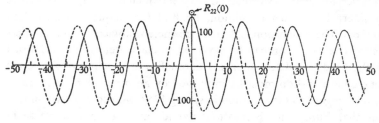

Fig. 10.1. The un-normalized autocorrelation $R_{11}(\tau)$ (solid) and cross-correlation $R_{12}(\tau)$ (dashed) for the unsmoothed ILS observations. The abscissa gives τ, the number of months m_2 is lagged. The ordinate is in units of $(0''01)^2$. $R_{22}(\tau)$ is similar to $R_{11}(\tau)$ but somewhat higher; only the point $R_{22}(0)$ is shown.

of the un-normalized autocorrelation $R_{11}(\tau)$ and cross-correlation $R_{12}(\tau)$, where

$$R_{uv}(\tau) = \langle m_u(t)m_v(t-\tau)\rangle, \qquad (10.1.1)$$

the average $\langle \rangle$ being taken over time t. The amplitude of the autocorrelation is reduced by approximately 10 per cent for a lag, τ, equal to one wave period, thus indicating a damping time of roughly a decade. From the cross-correlation we find that the oscillations are in phase if m_2 is advanced by one-fourth the wave period. This is the expected relation for positive (east-to-west) motion.

Fig. 10.2 shows the *smoothed* spectra (A.2.4, 6) of the *unsmoothed* ILS latitudes and of the latitude of Washington (fig. 7.4). Rudnick's unsmoothed spectrum is included for comparison. The analyses are not altogether equivalent. Rudnick's computation gives the spectra S^+ and S^- of positive and negative rotation.* Starting with

---

* Rudnick's published values refer to $(m^+)^2$ in units of $(0''01)^2$. In our convention, the power per harmonic equals the mean-square amplitude, $\frac{1}{2}(m^+)^2$, and the power density $S^+ = \frac{1}{2}T(m^+)^2$, where $T = 54.4$ years is the length of record. Accordingly, Rudnick's values are multiplied by 27·2 to give S^+ in $(0''01)^2/c/year$.

the relationships (6.7.6) it can be shown that

$$S^{\pm} = \tfrac{1}{4}[S_{11} + S_{22} \pm 2(-S'_{12})] \qquad (10.1.2)$$

We find that S_{11}, S_{22} and $-S'_{12}$ are nearly equal, so that

$$S^{+} \approx S_{11}, \quad S^{-} \ll S_{11}. \qquad (10.1.3)$$

If the data were consistent, then all spectra in fig. 10.2 should look nearly alike.

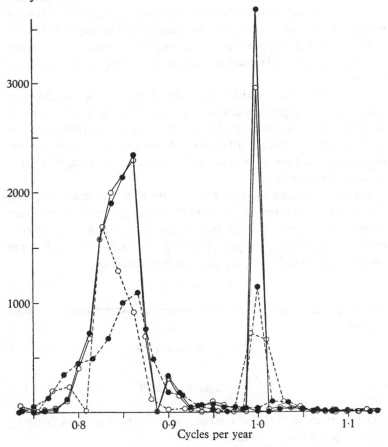

Fig. 10.2. Power spectra of latitude. Solid lines: $\overline{S_{11}}$ and $\overline{S_{22}}$, 1900 to 1954, from ILS. Dashed lines, open circles: $\tfrac{1}{2} S^{+}$, 1891 to 1945, from Kulikov's complication, according to Rudnick. Dashed lines, solid circles: Washington latitude for 1916–1952. The ordinate gives power density in units of $(0.''01)^2/c/\text{year}$.

Another difference is that Rudnick's spectrum is not smoothed. In the unsmoothed spectrum the energy associated with a single harmonic should be regarded as a random variable possessing a Rayleigh distribution and consequently a standard deviation as great at its expected value. The purpose of the smoothing procedure is to combine many adjoining harmonics into a single estimate with a smaller standard deviation. Resolution is traded for reliability. In the present application the amount of data is so limited that the resulting smoothing is negligible. The only reason for obtaining smoothed spectra is that the computing procedure was readily available. In the smoothed spectra a single harmonic of unit power is split into three adjoining harmonics containing $\frac{1}{4}$, $\frac{1}{2}$, $\frac{1}{4}$ units of power.

In view of these considerations the differences in the *shape* of spectra are no larger than is to be expected. Thus even half a century of observations is strikingly limited in its ability to determine the form of the spectrum. Differences in *area* under the curves are significant and may be due, at least in part, to the lack of overlap between the records.

We shall describe the spectral peak by three parameters: (1) its area, (2) its central frequency, and (3) some measure of its departure from a spectral line. The parameters depend somewhat on the type of curve to be fitted to the peak. Rudnick's procedure is to replace the frequency of nutation, σ_0, by the complex frequency

$$\sigma_0 = \sigma_0 + i\alpha, \tag{10.1.4}$$

where σ_0 and α are real positive constants. Accordingly, we must replace

$$\frac{d\mathbf{m}}{dt} - i\sigma_0\,\mathbf{m} = -i\sigma_0\mathbf{\Psi} \tag{6.3.6}$$

by*

$$\frac{d\mathbf{m}}{dt} - i\sigma_0\mathbf{m} = -i\sigma_0\mathbf{\Psi} \tag{10.1.5}$$

which contains an additional term, $\alpha\mathbf{m}$; this term leads to a damping of the free motion, according to

$$\mathbf{m} = e^{i\sigma_0 t} = e^{-\alpha t}\,(\cos \sigma_0 t + i \sin \sigma_0 t). \tag{10.1.6}$$

---

* Rudnick (1956) writes—$i\sigma_0\psi$ on the right side of the equation.

For circular positive polarization the power transmission (see 6.7.10) equals

$$I^2(f) = \frac{f_0^2}{(\alpha/2\pi)^2 + (f - f_0)^2} = \frac{S_0^+}{S_i^+} \qquad (10.1.7)$$

where $S_0^+(f)$ and $S_i^+(f)$ are the output and input spectra for positive motion. The power transmission is peaked at $f = f_0$ and attains half the peak value at $f = f_0 \pm \alpha/(2\pi)$. The area under the curve (including negative frequencies) is

$$2 \int_0^\infty \frac{f_0^2}{(\alpha/2\pi)^2 + (f - f_0)^2} \, df = \frac{2\pi^2 f_0^2}{\alpha}. \qquad (10.1.8)$$

The sharpness of the peak is conveniently portrayed by the dimensionless parameter

$$Q = \frac{\sigma_0}{2\alpha} = \frac{\pi f_0}{\alpha}. \qquad (10.1.9)$$

We may assume that the input spectrum is flat over the narrow range of frequencies contained in the Chandler wobble. The recorded spectrum is then of the form $S_i^+(f_0) \, I^2(f)$, and the total power in the Chandler wobble is given by

$$\frac{2\pi^2 f_0^2}{\alpha} S_i^+(f). \qquad (10.1.10)$$

The constant parameters f_0, α, S_i^+ can be determined by fitting (10.1.7) to the computed spectra. The fitting has been done by the method of maximum likelihood (§ A.2), following a suggestion by Jeffreys (personal communication). The results are given in table 10.1*; one of the fittings is shown in fig. 7.5.†

---

* The summations (A.2.8) extend over all harmonics in the Chandler peak. The indicated probability ranges in table 10.1 correspond to values for which φ has half its peak value. The ranges for f_0 and S_i^+ were obtained by varying the parameters f_0 and A separately: for α, A and B were varied concurrently so that A^2/B retained the value associated with maximum φ (total power is held constant).

For the smoothed spectra neighboring harmonics are not independent, but every other harmonic is nearly so, according to (A.2.6). For that reason separate analyses of even harmonics, and of odd harmonics, were conducted for the ILS observations. Furthermore, the smoothing lowers and spreads the spectra. If the unsmoothed spectra is of the form (A.2.7), the smoothed spectrum is

$$\overline{S_n} = \frac{A(1 + 2\epsilon^2)}{B(1 + 4\epsilon^2) + (f_n - f_0)^2}, \qquad (10.1.11)$$

where $\epsilon^{-1} = 2\tau_0\alpha/\pi$ is the number of harmonics (assumed large) between the half-power points. For the ILS values, $\epsilon^{-1} = 2\cdot2$ and $1\cdot2$ for even and odd harmonics, respectively, and no adjustment is possible. Values of $1/\alpha$ may then be regarded as lower limits. For the Washington latitudes, $\epsilon^{-1} = 2\cdot9$, and corrected values have been included in table 10.1.

† See also footnote on p. 174.

Table 10.1. Parameters of Chandler Wobble.

Interval	r	f_0^{-1} Years	α^{-1} Years	S_i^+ (0."01)²/c/year	Chandler Power (0."01)²	Q
ILS values smoothed by Kulikov; power spectrum according to Rudnick, 1956.*						
1891 to 1945	1 to 544	1·193 ± 0·011	11·4 (+ 5·3, − 4·2)	0·58 (+ 0·25, − 0·16)	96	30
Unsmoothed ILS values: smoothed spectrum for even and odd harmonics, respectively.						
even 1899 to 1954	1 to 480	1·183 ± 0·013	11·4 (+ 5·6, − 4·7)	0·86 (+ 0·45, − 0·27)	143	30
odd 1899 to 1954	1 to 480	1·178 ± 0·008	22·4 (+ 11·1, − 8·7)	0·61 (+ 0·31, − 0·18)	196	60
Washington latitudes: smoothed spectrum.						
1916 to 1952	1 to 360	1·183 ± 0·016	6·7 (+ 3·1, − 2·7)	1·02 (+ 0·41, − 0·27)	99	18
adjusted according to (10.1.11)			8·4 (+ 3·8, − 3·4)	0·76 (+ 0·31, − 0·20)		23
Smoothed ILS values: according to Jeffreys, 1952.						
1908 to 1921	0, 3, 6	1·202 ± 0·016	15·2 ± 1·6	—	—	40
Unsmoothed ILS values, according to Walker and Young, 1957.						
1899 to 1954	1, 2	1·287 ± 0·26	2·27 ± 0·49			6
	1 to 16	1·198				
	6 to 16	1·179				
1900 to 1934	1, 2	1·267 ± 0·041	2·81 ± 0·86			7
	1 to 16	1·202				
	6 to 16	1·186				
1900 to 1920	1, 2	1·238 ± 0·033	4·18 ± 1·94			11
	1 to 16	1·201				
	6 to 16	1·193				

* A noise level of 0·06 (0."01)² has been subtracted prior to fitting the spectrum.

The analysis by Jeffreys (1940) and by Walker and Young (1955, 1957) is closely related to the present analysis. Rudnick infers the power spectrum of the stationary series from the power spectrum of the 54-year sample. This inference consists essentially of smoothing the ragged spectrum with due regard to the large relative variance of any one spectral line.* Jeffreys, Walker and Young compute auto-correlation values and fit them by a smooth curve. The power spectrum is fitted by a curve of the type (10.1.7), the autocorrelation by (10.1.6). These are consistent choices, both following directly from the assumption of a complex frequency, σ_0, of the Chandler wobble.

If there were ample data, the two methods would converge to the same damping time α^{-1}. With a limited sample, Rudnick's (1953) work suggests the sampling error in the methods is comparable provided all possible lags are used in the autocorrelation. Using less than the complete autocorrelation leads to inferior results, according to Rudnick.

Table 10.1 summarizes results for various choices of lags r. Rudnick's and our results correspond to the use of all possible lags. Jeffreys's work is based essentially on a least-square fit to the auto-correlation $R(r)$ for $r = 0, 3, 6$ (unit lag is 0·1 years); this gave a period of $14\frac{1}{2}$ months and a damping time of 15 years. In their 1955 paper, Walker and Young used $R(r)$ for $r = 1, 2$ (unit lag is 1 month). They obtained periods of 15 to $15\frac{1}{2}$ months and damping times of 2·3 to 4·2 years. These results did not appear to be reasonable. In their 1957 paper the authors chose a least-square fit for $r = 1$ to 16, and $r = 6$ to 16. The results differ from one era to the next.

The consistent values are obtained for 1900 to 1920 using the largest values of r, and the authors consider 1·193 years to be their best estimate of the Chandler period. This is now in excellent accord with Jeffreys's value of 1·202 years and Rudnick's value of 1·193 years. With regard to the damping time the authors permit themselves

---

* Jeffreys has expressed some concern regarding the validity of harmonic analysis applied to latitude observations, and this has been amplified by Walker and Young. One must interpret Jeffreys's remarks concerning harmonic analysis as applying to the search for physically significant spectral lines (with stable phase). The Fourier line spectrum as a device for estimating the continuous power spectrum is without implication of the physical meaning of any one line.

only the conclusion that it is between 10 and 30 years; Jeffreys found 15 years and Rudnick 11 years.

Gutenberg (1956) has expressed some doubt that the damping times should be anywhere near as short as 10 periods. He suggests that it is not possible to discriminate the effect of damping from the beat phenomenon between the Chandler wobble and the annual term. But in the power spectrum the annual line falls outside the broadened peak of the Chandler wobble, so that the two spectral features are adequately resolved.

Melchior (1957, p. 234) has drawn attention to the fact that the autoregression schemes followed by Walker and Young 'imply the existence of a mathematical relation between the idea of elasticity (represented by the Chandlerian period) and that of viscosity (represented by the damping), and this relation is just Maxwell's law'. The curve-fitting procedure used by Walker and Young and by Rudnick implies only a complex frequency; this in turn implies only a damped linear system. The Maxwell body is just one of an infinite number of possible linear combinations of springs and dashpots. Moreover, the curve fitting is not sufficiently definitive to favor the linear models over other types; all that can be said is that the linear models cannot be excluded by the data. But even for non-linear models the parameters in table 10.1 retain much of their significance, with Q designating some appropriately defined width of the spectrum (§ 4.3).

2. The concept of an instantaneous Chandler period

The literature refers to two types of models: we shall call them the 'damped model' and the 'time-variable model'. In the preceding section we have been concerned with the damped model (Jeffreys, Walker and Young, Rudnick). The finite spectral width (or some equivalent feature) is associated with the damping of a tuned oscillator excited at random. Imperfections from elasticity in the Earth's mantle would give rise to such damping, as would the dissipation at the boundary between mantle and core if the core is viscous.

The majority of papers dealing with the Chandler wobble report that its period varies with time and that this implies corresponding variations of the pertinent physical parameters (rigidity, ellipticity, etc.). The authors include Chandler himself, Kimura, Berg, Backlund, Wahl, Stumpff, Witting, Ledersteger, Hattori, Nicolini and Melchior. The variable period is connected to a variable amplitude according to empirical laws formulated by Melchior (1957) and based largely on his work and that of Nicolini:

(1) The period of the Chandler motion fluctuates. The maximum departure from the mean value is approximately ± 4 per cent.

(2) Period and amplitude of the Chandler motion are proportional. The correlation coefficient, according to Nicolini, is + 0·88.

(3) A long Chandler period is correlated with a small amplitude of the annual motion.

The difference in the interpretation as characterized by the damped and time-variable models centers on the meaning of 'period' and the method whereby this is obtained from the record. Instantaneous period (or its inverse, the instantaneous frequency IF) can be determined from the time interval between alternate zero crossings in a record from which the annual term has been removed by suitable means. Similarly, instantaneous amplitude (IA) is the height of a peak above the record mean. The instantaneous values reported in the literature have been obtained by some such method.

Fourier frequencies (FF), on the other hand, are obtained by representing a record as a sum (or integral) of harmonic functions. The power spectrum of wobble (fig. 10.2) is an example of the application of FF. Here the power in the Chandler wobble is expressed as a sum of roughly nine values corresponding to various FF. It is in the nature of FF analysis that these values are fixed for any given record. Instantaneous values vary with time throughout the record.

Reasonably well-behaved oscillatory records can perfectly well be described in terms of IF and IA. FF likewise are a satisfactory means for describing wiggly records of stationary processes. IF and FF are not the same, but with some precaution they can be related. In general, the broader the peak (the smaller the Q) in an FF presentation, the larger will be the fluctuation of the IF throughout the record. The reported fluctuation of the Chandler frequency by as much as $\pm$ 4 per cent is roughly consistent with the Q of 25 derived from the FF.*

Melchior finds support for the time-variable model from the third empirical law. His argument involves the wobble equation (6.3.6). But this equation was derived with the explicit understanding that σ_0 does not vary in time. Melchior's hypothesis of a slightly damped time-variable model does permit him to apply the steady state results by replacing σ_0 with $\sigma_0(t)$, provided the time variation of σ_0 is slow compared to the damping time α^{-1}. The observed variations are rapid and therefore beyond the scope of his parametric treatment. Reducing the damping only increases the time required to reach equilibrium and does not help at all. Melchior is entitled to contemplate such rapid changes in the time-dependent parameters that might account for the fluctuations in IF, but he has not given the argument for the expected performance of his model.

The situation may be summarized as follows: The reported fluctuation in IF (empirical law 1) is in harmony with the results obtained from spectral analysis (or equivalent methods based on autocorrelation). The fluctuation in IF then does not favor the time-variable model over the linear damped oscillator. Empirical laws 2 and 3 have not been accounted for in terms of either model. Physical considerations speak against the time-variable model†: there has been no evidence of the remarkable physical changes that must accompany a change in the Chandler period by $\pm$ 4 per cent.

---

* The distribution of the interval, τ, between alternate zero crossings for a narrow spectrum formed by a Gaussian noise has been discussed by Longuet-Higgins (1958). Let

$$\mu_n = \int_{-\infty}^{\infty} f^n S(f) \, df$$

designate the nth moment of the power spectrum, $S(f)$. The mean interval, $\bar{\tau}$, is approximately μ_0/μ_1. Half the values of τ lie within the range

$$\bar{\tau}\left(1 - \frac{2}{\sqrt{3}}\delta\right) \quad \text{to} \quad \bar{\tau}\left(1 + \frac{2}{\sqrt{3}}\delta\right)$$

where

$$\delta^2 = \frac{\mu_0\mu_2 - \mu_1^2}{\mu_1^2}$$

is assumed small. For the spectral function (10.1.7) $\bar{\tau} = f_0^{-1}$, as expected. For this particular function, μ_2 and hence δ are infinite, and the approximations do not apply. But it will be noted that δ is essentially the root-mean-square width of the spectrum and can be expected to have a value near Q^{-1}, or 0·02 to 0·10. Departures in instantaneous period by 4 per cent, as reported, are not unexpected.

† See Newcomb's criticism of Chandler, § 7.1.

We prefer the linear damped model because it is simple and adequate and because we have found no compelling reasons for abandoning it in favor of the more complicated time-variable model.

3. Tidal-effective rigidity and viscosity

On the basis of table 10.1 we adopt

$$f_0^{-1} = 1 \cdot 195 \pm 0 \cdot 015 \text{ years} \qquad (10.3.1)$$

as an estimate for the period of the Chandler wobble. The tidal-effective Love number is then

$$k = 0 \cdot 29 \pm 0 \cdot 01 \qquad (10.3.2)$$

according to (6.2.5). A comparison of the value to Takeuchi's estimate is given in § 10.8. For the equivalent Earth model (§ 5.6) the tidal-effective rigidities (dimensionless and dimensional) then follow from (5.6.1, 2):

$$\mu = 2 \cdot 31 \pm 0 \cdot 10, \quad \tilde{\mu} = (8 \cdot 35 \pm 0 \cdot 36) \times 10^{11} \text{ dynes cm}^{-2}.$$
$$(10.3.3)$$

The rigidity of steel is 8 to 9 × 10^{11} dynes cm^{-2}.

The complex frequency, $\sigma_0 = \sigma_0 + i\alpha$, implies a complex Love number

$$\mathbf{k} = k \left(1 - i \frac{\mu}{2Q} \right) \qquad (10.3.4)$$

according to (6.2.5) and (5.6.2). For a Maxwell body (5.11.5)

$$\mathbf{k} = k \left(1 - i \frac{\mu}{1 + \mu} \frac{\tilde{\mu}}{\sigma_0 \tilde{\eta}_M} \right) \qquad (10.3.5)$$

provided $(\tilde{\mu}/\sigma_0\tilde{\eta}_M)$ is a small quantity. Equating imaginary parts and using (10.1.9) and (5.11.3) we obtain

$$\tilde{\eta}_M = \frac{2Q\tilde{\mu}}{(1 + \mu)\sigma_0}, \quad \alpha^{-1} = (1 + \mu)\tau \qquad (10.3.6)$$

where $\tau = \tilde{\eta}_M/\tilde{\mu}$ is the characteristic time. For $Q = 30$, the tidal-effective viscosity $\tilde{\eta}_M$ is 10^{20} g cm^{-1} sec^{-1}, and the damping time $\alpha^{-1} = 13$ years.

This interpretation of the damping cannot be reconciled with the time constant of isostatic adjustment. For a homogeneous Maxwell Earth the load operators are (5.8.4, 5.11.3, 5.7.2)

$$\hat{k'_n} = -\frac{1}{1 + N\hat{\mu}}, \quad \hat{\mu} = \frac{\mu\hat{D}}{\hat{D} + \tau^{-1}}, \quad N = \frac{2(2n^2 + 4n + 3)}{19n}.$$
$$\text{(5.7.2)}$$

A load suddenly applied at time 0 is reduced by deformation according to

$$(1 + \hat{k'_n})H(t) = \frac{N\mu}{1 + N\mu}\frac{\hat{D}}{\hat{D} + \gamma_n}H(t) = \frac{N\mu}{1 + N\mu}e^{-\gamma_n t}, \quad \text{(10.3.7)}$$

where
$$\gamma_n^{-1} = (1 + N\mu)\tau \quad \text{(10.3.8)}$$

is the compensation time. The operational solution follows the rules given by Jeffreys and Jeffreys (1950, 7·051). The factor $N\mu/(1 + N\mu)$ accounts for the immediate effect due to elastic deformation.

For a load of degree 2, $N = 1$ and

$$\gamma_2^{-1} = (1 + \mu)\tau = \alpha^{-1};$$

hence the compensation time equals the damping time of the Chandler wobble, 13 years. For higher order loads, the compensation times are somewhat longer: $\gamma_{10}^{-1} = 90$ years and $\gamma_{30}^{-1} = 300$ years. But gravity anomalies associated with ancient geologic structures of limited horizontal extent indicate adjustment times of millions of years! The foregoing application of the Maxwell body is intended for illustration only with no implication that it can serve as a meaningful model for the real Earth.

An equally absurd model is that of a Kelvin–Voigt body. From (5.11.3) we can write

$$\mathbf{k} = k\left(1 - i\frac{\mu}{1 + \mu}\frac{\sigma_0\,\tilde{\eta}_{K-V}}{\tilde{\mu}}\right) \quad \text{(10.3.9)}$$

where $\sigma_0\tilde{\eta}_{K-V}/\tilde{\mu}$ is a small quantity. Equating imaginary parts of (10.3.9) and (10.3.4)

$$\tilde{\eta}_{K-V} = \frac{(1 + \mu)\tilde{\mu}}{2Q\sigma_0}, \quad \alpha^{-1} = \frac{(1 + \mu)\tilde{\mu}}{\sigma_0^2\,\tilde{\eta}_{K-V}}. \quad \text{(10.3.10)}$$

For $Q = 30$, the viscosity is $2\cdot6 \times 10^{17}$ g cm^{-1} sec^{-1}. Knopoff and MacDonald (1958) show that a Kelvin–Voigt viscosity greater than 10 g cm^{-1} sec^{-1} is inconsistent with the observed damping of seismic waves. The Kelvin–Voigt model fails at high frequencies, the Maxwell model at low frequencies. The shortcomings of these two models have been discussed by Jeffreys (1917) and by Birch and Bancroft (1942).

4. Pole tide and Love numbers

'The sea would be set into vibration, one ocean up and another down. . . .' In these words Lord Kelvin* suggested that a wobble would induce an ocean tide. Kelvin overlooked the fact that such a tide would appreciably lengthen the wobble period leaving aside for the moment any effect of the Earth's elastic deformation; otherwise he would not have confined his search to the Eulerian period of ten months (§ 7.1).

The derivation of the *equilibrium* pole tide and its effect on wobble follows in a straightforward fashion from the equations already derived (see Federov, 1949; Haubrich and Munk, 1959). The equations of motion of the rotation pole are (6.3.5)

$$\frac{d\mathbf{m}}{dt} - i\sigma_e \mathbf{m} = -i\sigma_r \boldsymbol{\phi}, \qquad (10.4.1)$$

where $\qquad \boldsymbol{\phi} = \dfrac{c_{13} + ic_{23}}{C - A} = -\dfrac{a^4 \rho_w \int \xi \sin \theta \cos \theta \, e^{i\lambda} \, ds}{C - A} \qquad (10.4.2)$

is the excitation pole arising from the products of inertia of the pole tide, $\xi(\theta, \lambda; t)$, ρ_w is the density of sea water, and σ_e is the 'elastic wobble frequency' (as if the ocean were frozen). The solution to (10.4.1) will yield the observed frequency σ_0.

The equilibrium pole tide equals $(1 + k - h)U/g$, where

$$U = -\omega_1 \omega_3 x_1 x_3 - \omega_2 \omega_3 x_2 x_3$$
$$= -\Omega^2 a^2 \sin \theta \cos \theta \, (m_1 \cos \lambda + m_2 \sin \lambda)$$

is the potential produced by the wobble. To this we must add the Darwin correction (§ 9.2) in order to assure that ocean mass is conserved. When this is done, the resultant tide can be written

$$\xi = -\frac{1 + k - h}{g} \frac{\Omega^2 a^2}{2}$$
$$\times \left\{ m_1 \left(\sin 2\theta \cos \lambda - \frac{a_2^1}{5a_0^0} \right) + m_2 \left(\sin 2\theta \sin \lambda - \frac{b_2^1}{5a_0^0} \right) \right\} \quad (10.4.3)$$

* Presidential Address, 1876, Section of Mathematics and Physics, British Association for the Advancement of Science.

where a_n^m, b_n^m are the coefficients of $\mathscr{C}$ (oceans) (see § A.1). Combining (10.4.1–3), the equations of motion are put in the form

$$\left.\begin{array}{l} \dfrac{dm_1}{dt} + (\sigma_e - T_2\Omega)m_2 - R\Omega m_1 = 0 \\[2mm] \dfrac{dm_2}{dt} - (\sigma_e - T_1\Omega)m_1 + R\Omega m_2 = 0 \end{array}\right\} \qquad (10.4.4)$$

where

$$\frac{R}{1+k-h} = \frac{4\pi\Omega^2\rho_w a^6}{Ag}\left[\frac{b_2^2}{35} + \frac{b_4^2}{63} - \frac{a_2^1 b_2^1}{100a_0^0}\right] = -0.11 \times 10^{-4}$$

$$\frac{T_1}{1+k-h} = \frac{4\pi\Omega^2\rho_w a^6}{Ag}$$

$$\times\left[\left(\frac{a_0^0}{15} + \frac{a_2^0}{105} - \frac{4a_4^0}{315}\right) + \left(\frac{a_2^2}{35} + \frac{a_4^2}{63}\right) - \frac{(a_2^1)^2}{100a_0^0}\right] = 3\cdot36 \times 10^{-4}$$

$$\frac{T_2}{1+k-h} = \frac{4\pi\Omega^2\rho_w a^6}{Ag}$$

$$\times\left[\left(\frac{a_0^0}{15} + \frac{a_2^0}{105} - \frac{4a_4^0}{315}\right) + \left(\frac{a_2^2}{35} + \frac{a_4^2}{63}\right) - \frac{(b_2^1)^2}{100a_0^0}\right] = 2\cdot16 \times 10^{-4}$$

The solutions are

$$m_1 = M_1 \cos \sigma_0 t, \quad m_2 = M_2 \sin (\sigma_0 t + \beta) \qquad (10.4.5)$$

provided

$$\sigma_0^2 = \sigma_e^2 - (T_1 + T_2)\Omega\sigma_e - (R^2 - T_1 T_2)\Omega^2, \qquad (10.4.6)$$

$$\tan \beta = \frac{\Omega R}{\sigma_0}, \quad \frac{M_1^2}{M_2^2} = \frac{\sigma_e - T_2\Omega}{\sigma_e - T_1\Omega}. \qquad (10.4.7)$$

Thus the frequency has been diminished from σ_e to σ_0 according to (10.4.6). Inasmuch as $R \ll T_1$, T_2, this can be written

$$\sigma_0 \approx \sigma_e - \tfrac{1}{2}(T_1 + T_2)\Omega + 0(T_2\Omega/\sigma_e)^2. \qquad (10.4.8)$$

The observed frequency is $\sigma_0 = \Omega/437$. For the usually accepted values of Love numbers, $1 + k - h = 0\cdot68$, and this gives $\sigma_e = \Omega/404$ so that the oceans increase the period from 404 days to 437 days.* Without oceans, the annual term would be about half as far from resonance as it is now, and the annual wobble twice as large.

* The dominant coefficient in the expression for $T_1 + T_2$ is a_0^0. The decrease in frequency depends therefore to a first order on the fraction of the Earth's surface covered by oceans. This was noted by Larmor (1896) who first evaluated this effect.

But we are not concerned with the nutation frequency, σ_e, of an Earth without oceans but rather with the 'elastic tidal-effective Love number', k_e, that would be applicable under these conditions. In analogy with

$$\sigma_0 = \sigma_r \frac{k_f - k}{k_f} \qquad (6.2.5)$$

we write

$$\sigma_e = \sigma_r \frac{k_f - k_e}{k_f}. \qquad (10.4.9)$$

Equation (10.4.8) can then be written in either of the forms

$$\sigma_0 = \sigma_e - \frac{k_w}{k_f} \sigma_r, \quad k = k_e + k_w, \qquad (10.4.10)$$

provided we define k_w by

$$k_w = \tfrac{1}{2} k_f (T_1 + T_2) \frac{\Omega}{\sigma_r} = \frac{9}{8\pi} (1 + k - h) S \frac{\rho_w}{\bar{\rho}}, \qquad (10.4.11)$$

where $\bar{\rho}$ is the mean density of the Earth, and

$$S = \frac{8\pi}{15} \left(a_0^0 + \frac{1}{7} a_2^0 - \frac{4}{21} a_4^0 \right) - \frac{4\pi}{100 a_0^0} [(a_2^1)^2 + (b_2^1)^2] = 1 \cdot 20.$$

Thus k_w is that part of the Love number k which is due to the oceanic pole tide.

The problem is to evaluate k_e. We may regard $k = 0.29$ as known from the frequency of the Chandler wobble. There are two procedures: (1) we regard $h = 0.61$ as given, then it follows from (10.4.10) and (10.4.11) that

$$(\Omega/\sigma_e) = 404, \quad h = 0.610, \quad k_e = 0.235, \quad k_w = 0.055; \qquad (10.4.12)$$

(2) for an 'equivalent' Earth (§ 5.6) the ratio h_e/k_e depends only on the density distribution, not on the rigidity. Then

$$\frac{h_e}{k_e} = \frac{h_f}{k_f} = 2 \cdot 05, \qquad (10.4.13)$$

where $h_e U/g$ is the elastic deformation of the *solid* Earth, and $k_e U/g$ the potential of this deformation. For the case of no load $h = h_e$. We now obtain

$$(\Omega/\sigma_e) = 396, \quad h = 0.457, \quad k_e = 0.223, \quad k_w = 0.069, \qquad (10.4.14)$$

which differs from (10.4.12) chiefly with regard to h.

The principal result is that the tidal-effective Love number is reduced from 0·29 to approximately 0·22 after allowing for an equilibrium response of the oceans. The results are not appreciably altered by a first-order correction for load deformation (Haubrich and Munk, 1959). A complete treatment along the lines of (§ 5.12) has not been attempted.

The asymptotic behavior is as expected. For $\rho_w = 0$ we have $k_w = 0$ and $\sigma_0 = \sigma_e$, which is the appropriate frequency for an Earth without oceans. For a rigid Earth ($k_e = 0$, $h = 0$) completely surrounded by oceans, $S = 8\pi/15$, and

$$k_w = k = \frac{1}{\dfrac{5}{3} \dfrac{\bar{\rho}}{\rho_w} - 1}, \qquad \sigma_0 = \sigma_r \left(1 - \frac{k}{k_f}\right).$$

For a homogeneous Earth, $k_f = 3/2$ (§ 5.4). Thus as the ocean density approaches that of the Earth, $\rho_w \rightarrow \bar{\rho}$, we find that $\sigma_0 \rightarrow 0$: the rotational axis becomes unstable. For the actual density distribution $k_f = 0.96$ so that instability is reached when $\rho_w = 4.5$ g cm^{-3}.

The foregoing calculations emphasize the large effect of a pole tide on the Chandler wobble, provided the tide is given by the equilibrium theory. But there is a question whether the equilibrium theory applies. Even if the equilibrium theory is adequate for p_2^0 tides of relatively short period, such as the fortnightly, monthly, semi-annual and annual tides, this does not necessarily constitute an argument for the equilibrium response to the p_2^1 potential of the pole tide. Indeed, the work of Jeffreys and Vicente on the dynamics of a fluid core shows that the equilibrium theory is not applicable to the core (even though these authors use an equilibrium correction for the ocean).

Studies of the diurnal tide (also due to a p_2^1 potential) come closest, but these are usually restricted at the start to the assumption of a diurnal frequency. The problem is closely connected to the possibility of steady (zero frequency) motion of the p_2^1 types. Laplace shows that for an ocean of constant depth covering the whole Earth there is no diurnal tide (Lamb, 1932, p. 342). Under these conditions, the pole tide would be insignificant compared to the equilibrium response unless the boundary layer associated with the viscous flow along the sea bottom has a thickness comparable to the depth of the ocean.

Under actual conditions, the pole tide is superimposed upon other motion, vastly larger, and this increases the frictional coupling between oceans and earth. Let u designate the water velocity associated with the pole tide, U the maximum speed of a short-period tidal current, and h the water depth. The frictional force per unit mass can be written γu, where

$$\gamma = \frac{4}{\pi}(0\cdot0025)\frac{U}{h}$$

according to Bowden (1953). Proudman (personal communication) has suggested that the equilibrium assumption depends on the smallness of the ratio f_0/γ (see Proudman, 1916). Setting $h = 4$ km, $U = 6$ cm sec^{-1} gives $2\pi/\gamma = 1540$ days which is not small compared to $f_0^{-1} = 435$ days. But in regions with very large tidal currents equilibrium conditions might be approached.

For oceans of constant depth bounded by parallels of latitude (including a polar ocean), Goldsbrough (1913, 1914) demonstrates the possibility of resonance at periods of the same order as that of the Chandler wobble. In the case of meridional boundaries such long-period resonances are not possible (Colborn, 1931), provided the boundaries are vertical. Proudman's (1913) work indicates that these results would not necessarily apply to the case of an ocean shelving gradually.

Thus theoretical considerations have not yet settled the applicability of equilibrium theory to the pole tide. Can observations provide a clue?

Tide records are read hourly to an accuracy of perhaps ± 1 cm. A typical amplitude of the equilibrium pole tide is $\frac{1}{2}$ cm.* Searching for the pole tide is then a marginal undertaking, even if we allow for the fact that the analysis is based on monthly averages of hourly readings. There have been a number of attempts: Christie (1900);

* This would be the value at $\theta = 45°$ for a wobble of $0\overset{''}{\cdot}14$ amplitude, according to (10.4.3). A rough estimate can be made from the following elementary considerations. The polar radius is shorter by 20 km than the equatorial radius, thus giving an average 'slope' of 20 km in 10,000 km, or 1:500. If the Earth were rigid and the pole moved towards a fixed observer by $0\overset{''}{\cdot}14$ (= 14 ft = 430 cm), then the equilibrium sea-level would *sink* by $430/500 = 0\cdot86$ cm. To allow for self-attraction and elastic deformation of the Earth, multiply by $1 + k - h$.

Bakhuyzen (1913) uncovered records of Amsterdam sea-level going back to 1700; Przbyllok (1919); Baussan (1951); Maximov (1956). The procedure has been to divide the data into as many 7-year series as possible and to derive for each of these by Fourier analysis the amplitude and phase of the sixth harmonic, corresponding exactly to a 14-month period. Haubrich and Munk (1959) have determined the power spectrum of sea-level for all tide stations that were in operation at the turn of the century. This involves no preconceived assumption as to the frequency of the pole tide and, furthermore, furnishes an estimate of the noise level at adjoining frequencies. This noise level is very high, so much so that previous investigators would have found the amplitude at 15 months or 13 months not much less than at 14 months.

Fig. 10.3 shows the average spectrum of 11 records (six Netherlands stations are combined into a single record). Seasonal terms were subtracted from the records before analysis. The most prominent peak between zero frequency and the annual frequency occurs at 0·84 c/year corresponding to a period of 1·19 years. This is in excellent agreement with the estimate 1·195 ± 0·015 based on latitude observations. For each station the equilibrium tide level was computed month by month alongside the recorded level, using the unsmoothed latitude observations (fig. 7.4) and allowing for latitude and longitude according to (10.4.3). The co-spectrum (or in-phase component) shows a positive maximum at the pertinent frequencies, whereas the quadrature spectrum (or out-of-phase component) has none. This is the expected result for an equilibrium tide, or an equilibrium tide multiplied by a positive constant. The features are weak, and the significance marginal.

But there are some unexpected results. Comparing areas under the spectral peak associated with the equilibrium tide, and that of the recorded tide (after subtracting a noise level of 10 cm^2/c/year), the power of the recorded tide is found to be four times as large; hence its amplitude is twice the equilibrium amplitude. On further examination it is found that Swinemünde, the Netherlands group, Marseilles and Bombay contribute most of the power of this peak: furthermore, that the early Swinemünde tide record (1811 to 1906) does not reveal a corresponding peak. Perhaps the precision of automatically recording tide gauges is essential. Measurements in the nineteenth century were conducted largely with tide poles, and these might be inadequate to measure pole tides.

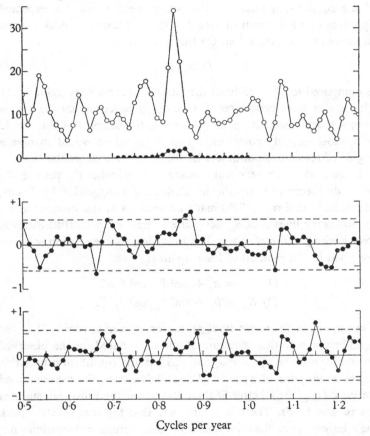

Cycles per year

Fig. 10.3. *Upper diagram:* The power spectra of sea-level (open circles) and of the equilibrium pole tide (solid circles), in cm²/c/year. This is based on an average of 11 stations (six Netherland stations have been combined into one record) for the years 1900 to 1954. Lower curves show the normalized co-spectrum and quadrature spectrum. Dashed lines indicate the 95 % significance level.

The equilibrium spectrum is based on the value $1 + k - h = 0.68$. For the equivalent Earth model (10.4.13) we obtain $1 + k - h = 0.84$, and the discrepancy is then considerably reduced. But the essential thing is that the foregoing comparison is based on the equilibrium argument, and there is no justification for this assumption. The pole tide observations do not support it.

The spectral peak of the recorded tide is far narrower than that of the equilibrium tide. The three-point peak is just the expected signature of a line spectrum (see A.2.5), or at least of a peak narrow compared to the resolution. On this basis we find

$$Q > 100 \qquad (10.4.15)$$

as compared to $Q \approx 35$ from the latitude observations (table 10.1). This is a most surprising result. One might conceive that the oceans have a sharp resonance peak at just this frequency. But this would be a most unlikely coincidence and furthermore would involve a degree of resonance not otherwise found in oceanic oscillations.

There is always a very real concern as to whether the peak in the pole tide spectrum is significant. Following a suggestion by Tukey, Munk and Haubrich (1958) made an analysis of the variance along the following lines: Let S_{fg} designate computed spectral densities at frequencies f and stations g. The following two regression schemes are fitted by the method of least squares:

$$(1) \;\; S_{fg} = a_g + \sin^2 \theta_g \cos^2 \theta_g S_f'',$$
$$(2) \;\; S_{fg}' = b_g + \sin^2 \theta_g \cos^2 \theta_g S_f'''.$$

Here a_g, b_g is the noise level at station g; $\sin^2 \theta_g \cos^2 \theta_g S_f''$ is the equilibrium pole tide spectrum at station g based on the observed latitude spectrum S_f''; $\sin^2 \theta_g \cos^2 \theta_g S_f'''$ is the equilibrium pole tide on the assumption of a spectral line at 0·84 c/year. For the combined stations, (1) gives a better fit than (2) but for the European stations the reverse holds. There is a hint here that the true spectral peak may be narrower than indicated by the latitude observations but much further work needs to be done to clarify the situation.

5. Ellipticity of the Chandler wobble

Larmor (1896) has pointed out that the pole tide imposes a slight ellipticity on the path described by the pole of rotation. The ellipticity equals

$$\varepsilon = 1 - \frac{M_1}{M_2} = \frac{1}{2}\frac{\Omega}{\sigma_e}(T_1 - T_2) = 0\cdot017 \qquad (10.5.1)$$

with a major axis pointing toward λ_0 east longitude, where

$$\tan 2\lambda_0 = \frac{2R}{T_1 - T_2} = -0.20, \quad \lambda_0 = -6°. \quad (10.5.2)$$

For comparison, the ratio of amplitudes of m_1 and m_2 was taken from the analysis of the unsmoothed latitude data, 1899–1954. The result is

$$\varepsilon = 0.01 \pm 0.05,$$

and this is not inconsistent with the computed ellipticity. Fedorov (1949) has computed the ellipticity by a similar method and obtains $\varepsilon = 0.01$. He states that this value does not agree with the observed ellipticity. Lambert (1922) attempted to evaluate the ellipticity from a number of six-year series of latitude observations, 1900–1917. His values of ε range from 0.02 to 0.20, and the direction of the major axis from 59° W to 116° W.

An ellipticity of the pole path could be caused also by a difference in the equatorial moments of inertia. From geodetic measurements Helmert (1915) reported a difference of 230 m between the equatorial semi-axes. Let A, B, C denote the principal moments of inertia. From Helmert's measurements Schweydar (1916) obtained

$$\frac{B - A}{C + \frac{1}{2}(A + B)} = \frac{1}{46},$$

which gives an ellipticity $\varepsilon = 0.016$. Modern measurements have neither confirmed nor dismissed the degree of triaxiality proposed by Helmert.

6. Generation

An excitation, $\Psi(t)$, which is not purely harmonic will excite resonance response. Any mobile part of the Earth is an obvious suspect. As usual it is easier to rule out causes than to confirm them.

There have been numerous attempts to link the variations in the Chandler wobble to earthquakes and volcanic eruptions (Cecchini, 1928, pp. 91–92). In particular, the increase in activity around 1907 (fig. 7.4) has been ascribed to the great San Francisco earthquake of 1906. Quantitative considerations rule out this possibility. Consider a block 100 km by 100 km in area and 30 km high, thus extending from the bottom of the Earth's crust to the surface. The total mass, m, is then of the order of 10^{21} g. The block is assumed to be under compression and to expand during the earthquake, so that the surface is lifted by an amount $h = 1$ meter. The excitation pole is then of the order (omitting latitude and longitude factors)

$$\frac{m(a+h)^2 - ma^2}{C-A} = \frac{2mah}{C-A} = 4\cdot6 \times 10^{-11} = 0\rlap{.}''00001$$

and the resulting wobble is of the same order. The observed wobble is of the order $0\rlap{.}''1$. For horizontal (strike-slip fault) displacements the same formula applies, provided we interpret h as the representative horizontal slip. Conceivably this might be 10 m and the dimensions of the block 1000 km by 100 km; even so the effect is negligible compared to observed values.

The atmosphere provides a possible mechanism, as pointed out by Volterra (1895) and Jeffreys (1940). According to Jeffreys 'the distribution of air over the Earth's surface, though mainly periodic, is not strictly so; it does not accurately repeat itself every year. It is therefore possible that the free variation of latitude is maintained by the irregular variation of the products of inertia, superimposed on the mean annual variation.' The required amount of irregular variation turns out to be of the right order of magnitude, as pointed out by Rudnick (1956, p. 142).

For a first estimate we consider spike-like departures of $\Psi(t)$ from the normal seasonal variation. Suppose these have a duration τ short compared to 14-month. The solution (6.7.3) is applicable. For definiteness, set $\tau = \sigma_0^{-1} = 2\cdot2$ months. The induced free wobble then has an amplitude of $\sigma_0\tau|\Psi| = |\Psi|$. The observed increase in the free motion between 1906 and 1908 could have been generated by one such spike of amplitude $|\Psi| \approx 0\rlap{.}''1$. This is six times the amplitude of the annual excitation function! Spikes of this order of magnitude occurring at random intervals of a few decades would maintain the Chandler wobble at its observed level.

But spikes are an inefficient means of exciting wobble. For a quantitative estimate we suppose the annual excitation function to be modulated so that its amplitude is sometimes larger, sometimes smaller. The modulation is assumed to contain all frequencies less than 0·5 c/year; for definiteness we choose a triangular modulation spectrum as shown by the dashed line in fig. 10.4 (*top*). The sum and difference frequencies between the annual line and the modulation produce an input spectrum consisting of the annual line plus triangular side bands between $1 \pm 0·5$ c/year. According to table 10.1 the input spectrum has a power density of roughly $0·6(0''01)^2$/c/year at the Chandler frequency. With this value the area of the triangle

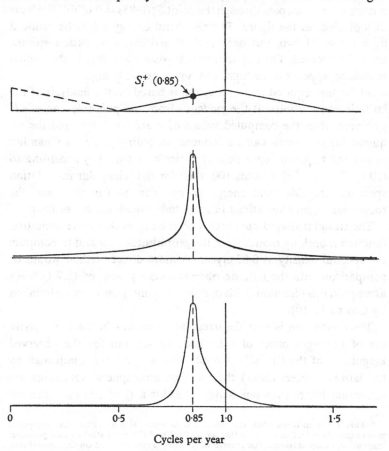

Fig. 10.4. *Top*: The input spectrum is schematically presented by an annual line plus triangular side bands, resulting from a modulation spectrum indicated by the dashed line. *Middle*: The power transmission of the Earth, peaked at the Chandler frequency, 0·85 c/year. *Bottom*: The power spectrum of the pole of rotation; this curve is the product of the upper two. Adapted from Rudnick, 1956.

equals $0 \cdot 43(0''01)^2$. This is the power in the side bands. The power of the annual excitation function for positive motion is $\frac{1}{2}(0''0168)^2 = 1 \cdot 4(0''01)^2$, according to Jeffreys's values for the modified excitation (table 9.2). The side-band power is then 30 per cent of the annual term, and the root-mean-square variations in amplitude are of the order of 50 per cent. The numerical values can be varied by changing the shape of the modulation spectrum, with the restraint that the input spectrum must pass through the point $S_i^+(0 \cdot 85) = 0 \cdot 6(0''01)^2/\text{c/year}$ as indicated on the figure. The side-band energy might be reduced by a factor of two, but not more, if artificial modulation spectra are to be avoided. The required year-to-year variability in the annual excitation appears to be high, but not impossibly high.

So far the required input spectrum is based on the analysis of the latitude observations. If the evidence from the pole-tide analysis is pertinent, then the computed values of α are too large, and the required input spectra can be reduced accordingly. For a Chandler power the required input density varies as α (or Q^{-1}) according to (10.1, 12, 13); if Q exceeds 100, then for the triangular modulation spectrum the side-band energy is less than 10 per cent, and the root-mean-square variations in amplitude are about 30 per cent.

The straightforward approach would be to evaluate the excitation function month by month from the geophysical data and to compute the spectral density at $0 \cdot 85$ c/year. A more direct method would be comparison with the latitude observations by way of (6.7.1).* An attempt in this direction (with indifferent results) has been undertaken by Cowan (1950).

The conclusion is that the irregular variations in the atmosphere are of the right order of magnitude to account for the observed magnitude of the Chandler wobble. For a Q of 30 (as indicated by the latitude observations) the required atmospheric variations are somewhat larger than one would like; for a Q of 100 (as indicated

---

* This has been done while this book was in press. Hassan (1960) has computed mean monthly values of $\psi_i(t)$ for the period 1873 to 1950 using station-level pressures from all available stations. The spectra of ψ_i show a rise centered on the annual line, as expected, but nevertheless the spectral density near $0 \cdot 85$ c/year fails to account for the observed wobble by one to two orders of magnitude (Munk and Hassan, 1961). Nor do the cross-spectra between m_i and ψ_j indicate the expected phases. Apparently the atmosphere plays a minor roll in the generation of the Chandler wobble, and our discussion must be modified accordingly.

by the pole tide) the values are more acceptable. Earthquakes are not a possible cause. Irregular motion in the core cannot be excluded (§ 11.12).

In closing we can estimate the amount of wobble if the resonance period were to be made coincident with the period of one year. As Lambert (1931, p. 97) has pointed out, the resonance frequency $\sigma_0 = [(C - A)/A]\Omega$ would have been higher by the required amount some hundreds of million years ago if the present rate of tidal retardation in Ω can be extrapolated into the past (§ 11.6). At peak amplification the wobble is Q times 0″0168 as compared to an amplification $7 \times 0″0168$ under present circumstances. Apparently the annual displacement of the pole will not exceed a few hundred feet under any circumstances.

7. Dissipation

The spectral analysis of latitude and tide observations lead to values of Q for the Chandler wobble. The values of Q give information regarding the dissipation of energy in the Earth at the Chandler frequency provided the apparent damping is not due to exciting impulses of a non-random kind. The identification of the energy sink is important to the problem of polar wandering. If the energy dissipates in the mantle, the resulting anelasticity may fix the time-scale for secular shifts of the pole. If the oceans or core are responsible for the damping, no such statement is possible.

First we need to estimate the rate of energy dissipation. The kinetic energy of a rigid body freely wobbling about a mean axis, x_3, is

$$K = \tfrac{1}{2}(A\omega_1^2 + A\omega_2^2 + C\omega_3^2);$$

the kinetic energy of the rigid body in steady rotation, having the same angular momentum is

$$K_0 = \frac{1}{2}\frac{1}{C}[A^2(\omega_1^2 + \omega_2^2) + C^2\omega_3^2].$$

The time-varying part of the kinetic energy is then

$$\Delta K = K - K_0 = \tfrac{1}{2}HA\Omega^2(m_1^2 + m_2^2). \tag{10.7.1}$$

Since the Earth is not rigid, any calculation of the total energy associated with the Chandler wobble must include the elastic energy of deformation plus the potential energy resulting from the redistribution of mass. If the Earth were perfectly elastic, then the mean kinetic energy equals the mean of the sum of the elastic and potential energies. The energy is not exactly equipartitioned if dissipation is present in the system. The energy dissipated per cycle is $2\pi/Q$ times the peak energy. To terms in order Q^{-1}, the kinetic energy of the wobble equals the potential and elastic energies,

$$\Delta K = \Delta P + \Delta U + 0(\Delta K/Q),$$

and to a similar degree of approximation the total energy involved in the wobble is twice the mean kinetic energy.

For a root-mean-square amplitude of $0''14$, we obtain

$$2\Delta K = 8 \times 10^{21}\ \text{ergs} = 4 \times 10^{-15}K \qquad (10.7.2)$$

for the mean energy of the Chandler wobble, and

$$2\Delta K\,\sigma_0 Q^{-1} = 10^{15}Q^{-1}\ \text{ergs sec}^{-1} \qquad (10.7.3)$$

for the mean rate of energy dissipation. This is very small compared to 3×10^{19} ergs sec^{-1} dissipated by tides (§ 11.6), or to $2\cdot5 \times 10^{20}$ ergs sec^{-1} flowing out from the Earth due largely to radioactivity.

The interpretation of the damping of the Chandler wobble has had a long and complex history. Jeffreys (1920) at first attributed the damping to tidal friction in the oceans. In a later paper, Jeffreys (1949) thought tidal friction to be inadequate and suggested elastic afterworking of the mantle as a suitable energy sink. This raised some problems in the interpretation of seismic data. Jeffreys in the third edition of *The Earth* added the core as a possible energy sink. Bondi and Gold (1955) dismiss the core and the oceans since in their model neither has a sufficient moment of inertia to affect the mantle. Jeffreys (1956) accepts the argument by Bondi and Gold and refers the damping back to the mantle. The problem is apparently closed but we wish to reopen it.

The wobble will tend to set up motions in the ocean and in the core, and indeed the ocean tide has been observed. Any friction at the bottom of the ocean or at the core-mantle boundary will oppose the motion of the solid Earth shifting in response to the wobble of the instantaneous axis of rotation. For the case of an excitation due entirely to a torque L_i, the equations of motion (6.1.2, 3) are

$$\frac{\dot{m}_1}{\sigma_r} + m_2 = \frac{L_1}{\Omega^2(C-A)}, \quad \frac{\dot{m}_2}{\sigma_r} - m_1 = \frac{L_2}{\Omega^2(C-A)}, \quad \dot{m}_3 = \frac{L_3}{C\Omega}.$$

The approach followed by Bondi and Gold consists essentially of setting $L_i \sim - m_i$: the frictional torque is taken about the instantaneous rotation axis. But this is not the only choice. We shall set $L_i \sim \varepsilon_{ijk}m_j(\mathrm{d}m_k/\mathrm{d}t)$: the frictional torque acts about an axis perpendicular to the plane defined by the instantaneous rotation axis and the vector $\mathrm{d}m_k/\mathrm{d}t$, and directly opposes the motion of the solid Earth shifting in response to the wobble of the instantaneous axis. The frictional torque is then about an axis perpendicular to that taken by Bondi and Gold. The model explicitly recognizes that the wobble induces meridional motion, whereas Bondi and Gold's torque can only result from zonal motion about the instantaneous rotation axis.*

The wobble equations can now be written

$$\frac{\dot{m}_1}{\sigma_r} + m_2 = \lambda(m_2\dot{m}_3 - \dot{m}_2) \approx - \lambda\dot{m}_2, \quad \frac{\dot{m}_2}{\sigma_r} - m_1 \approx \lambda\dot{m}_1, \quad (10.7.4)$$

where λ is a constant characterizing the frictional interaction. The solution

$$\mathbf{m} = \mathbf{M} \exp\left[(-\lambda + i)\sigma_r t\right]$$

is a typically damped oscillation. For vanishingly small frictional interaction $\dot{m}_1 \approx - \sigma_r m_2$, $\dot{m}_2 \approx \sigma_r m_1$, and the frictional terms approach $-\lambda\sigma_r m_i$ in accordance with the Bondi-Gold model. But for extreme damping the two models behave altogether differently.

Core.—A detailed calculation is needed to interpret γ. In the case of damping by the core, an upper limit can be obtained using the Bondi-Gold method. Consider a shell with moments of inertia A, A and C enclosing a sphere having a moment A'. The two bodies are set in rotation, and we examine the motion assuming that the coupling torque is given by $\Omega^{-1}\gamma\varepsilon_{ijk}\omega_j(\dot{\omega}_k - \dot{\omega}_k')$ where ω_k is the angular

* The discussion by Klein and Sommerfeld (1903, pp. 588–9, 727) is pertinent. Their solution of the damping of a top by air resistance is somewhat analogous to the Bondi-Gold model, provided the core had infinite inertia. With respect to the damping due to a pole tide "kann man sich vorstellen dass die Reibung hier der *Änderung der Rotationsaxe* entgegenwirkt". Their solution is similar to ours.

velocity of the shell and ω_k' is the angular velocity of the inner body. γ is an interaction constant. The normal modes for the system include the damped wobble. This mode shows zero damping for $\gamma = 0$, $\gamma = \infty$; in the latter case the bodies are rigidly joined. The damping is a maximum for

$$\gamma = \gamma_0 = \frac{A}{C - A},$$

and for this special case the wobble is attenuated according to

$$\exp\left(- \tfrac{1}{2}\gamma_0^{-1}\Omega t\right) = \exp\left(- \tfrac{1}{2}\sigma_s t\right) \qquad (10.7.5)$$

where σ_s is the frequency of wobble of the rigid shell. The maximum damping rate is independent of the core's moment of inertia. The corresponding solution by Bondi and Gold contains a time factor

$$\exp\left[- \frac{1}{2}\gamma_0^{-1}\Omega t \frac{A'}{A + A'} \right]$$

which depends on the moment of inertia, A', of the core and has a damping time $(1 + A/A')$ times that in (10.7.5).

According to (10.7.5) the amplitude decays to e^{-1} of its initial value in a time $\pi^{-1}(2\pi/\sigma_s)$; for $\sigma_s = \sigma_0$ this equals 100 days. Thus the *maximum* rate of damping is more than sufficient to account for the observed damping time of at least 10 years. The question is whether the *actual* dissipation in the core is sufficient to account for the damping time.

We may examine the question by application of the boundary layer theory to a viscous core. The thickness of the boundary layer is* $(2\eta/\Omega)^{\frac{1}{2}}$ and the rate of dissipation per unit area is of the order $\rho V^2(\eta\Omega)^{\frac{1}{2}}$, where V is the differential velocity between core and mantle at the boundary. Jeffreys and Vicente (1957a, b) have shown that to a first approximation the core remains motionless, and under this assumption we obtain

$$V \approx \frac{2\pi|\mathbf{m}|a_{\text{core}}}{(2\pi/\sigma_s)} \approx \frac{2\pi \times 250 \text{ cm}}{4 \times 10^7 \text{ sec}} \approx 4 \times 10^{-5} \text{ cm sec}^{-1}$$

---

* Jeffreys has called our attention to the fact that for periods long compared to a day the boundary layer argument applies provided we replace σ by Ω. We are concerned with nearly horizontal motion, $\mathbf{u} = u_x + iu_y$. For simple harmonic oscillations the equations of motion are

$$i\sigma\mathbf{u} + if_C\mathbf{u} = \eta(\partial^2 u/\partial z^2)$$

and the Coriolis frequency f_C swamps the oscillation frequency σ. The result is implicit in the work of Bondi and Lyttleton (1948). The flow in the boundary layer is an Ekman spiral.

for an upper limit of the differential speed. The dissipation for the entire Earth is at the rate

$$4\pi a_{\text{core}}^2 \, \rho V^2 (\eta \Omega)^{\frac{1}{2}} \approx 2 \times 10^8 \eta^{\frac{1}{2}} \text{ ergs sec}^{-1}. \qquad (10.7.6)$$

Jeffreys's (1956) discussion of the viscosity of the core would indicate 10^8 cm^2 sec^{-1} as an upper limit for η. The corresponding thickness of the boundary layer is 15 km, and the rate of dissipation according to (10.7.6) is 2×10^{12} ergs sec^{-1}, as compared to the 'observed' rate of $10^{15} Q^{-1}$ ergs sec^{-1} (10.7.3).

Mantle.—An alternative energy sink is in the mantle. Laboratory and seismic data show that the Q of inorganic non-ferromagnetic solids is independent of frequency over a frequency range 10^7 to 10^{-1} c/s (Knopoff and MacDonald, 1958). For any model involving linear friction Q is frequency-dependent. Knopoff and MacDonald have therefore accounted by non-linear friction for damping in this frequency range, and they obtain values between 100 and 250 for the Q of the mantle. If this behavior can be extrapolated to 10^{-7} c/s, as perhaps indicated by the Q of the pole tide, then the conclusion is that the same non-linear processes responsible for the damping of earthquake waves are responsible for damping the Chandler wobble. A lower Q (as indicated by the latitude observations) requires a mechanism other than solid friction if the motion is to be damped in the mantle. The usual resort is to a Maxwell body with a kinematic viscosity of $2 \cdot 5 \times 10^{19}$ cm^2 sec^{-1} (§ 10.3). Jeffreys (1956) states that such a viscosity in the lower mantle coupled with material having 300 bars strength in the upper 700 km could account both for the damping and gravity anomalies. Later work by Jeffreys (1958a) emphasizes a model based on Lomnitz' (1956, 1957) experiments on the finite strain of rocks. It is not obvious that Lomnitz' results are applicable to the present problem.

Oceans.—In our model the maximum damping rate is independent of the inertia of the core, and this suggests that, at least in principle, the damping could result also from the motion of the oceans relative to the sea bottom. Jeffreys (1952, p. 245) has examined this possibility and finds it to be ineffective. His argument consists of three steps: (1) the rate of dissipation for the semi-diurnal tide is of the order 10^{19} ergs sec^{-1}. On the supposition that the dissipation rates are proportional to the square of the equilibrium amplitudes, the pole tide dissipates roughly

$$\left(\frac{\frac{1}{2}\text{ cm}}{30\text{ cm}}\right)^2 10^{19} = 3 \times 10^{15} \text{ ergs sec}^{-1}.$$

(2) But only the *departure* from equilibrium response is associated with dissipation. For a uniform Earth covered by oceans Jeffreys estimates this departure to be 2·5 per cent for the pole tide. For the semi-diurnal tide he sets the departure at 100 per cent, and obtains

$$\left(\frac{2\cdot5}{100}\right)^2 3 \times 10^{15} = 2 \times 10^{12} \text{ ergs sec}^{-1}.$$

(3) But the dissipation by 10^{19} ergs sec^{-1} of the semi-diurnal tides takes place largely in shallow seas, where the magnification due to local resonances may be important. No such magnification is to be expected for the pole tide. This leads to a further reduction 'by some hundreds', so that the final value is of the order of 10^{10} ergs sec^{-1} as compared to the 'observed' value of $10^{15} Q^{-1}$ ergs sec^{-1}.

(1) is based on a comparison of a p_2^1 and p_2^0 tide, and this requires justification; (2) follows from a theory for a global ocean, but in accordance with the remarks in (§ 10.4) the presence of shelving continents may entirely alter the results; with regard to (3) there is now some doubt as to whether the dissipation takes place in shallow seas (§ 11.8) and some evidence that the pole tide is in fact amplified (§ 10.4). It would appear that the oceans cannot be ruled out as a possible source of damping.

The conclusions are then as follows:

(1) If the Q is between 100 and 200, as vaguely suggested by the pole-tide analysis, then solid friction in the mantle can account for the damping.

For a Q of 30 to 50, as indicated by the latitude analysis, there are many possibilities:

(2) The damping can be in the oceans.

(3) The lower mantle is a possible energy sink; a model involving a Maxwell viscosity is, however, unsupported by laboratory and seismic evidence.

(4) Damping by viscosity in the core appears to be ruled out; electromagnetic damping is still a possibility (Jeffreys, 1956).

(5) Impulses of a non-random kind (originating in the core, oceans or atmosphere) can absorb as well as excite the wobble. The computed Q is then not due to damping, but associated with the interaction between these loosely-coupled components. The situation is appallingly uncertain.

8. The Love number k

Takeuchi (1950) calculated

$$k_e = 0.29$$

for the Earth on the basis of Bullen's distribution of density and elasticity. The close agreement with the observed value of the tidal-effective rigidity

$$k = 0.29 \pm 0.01 \qquad (10.3.2)$$

is misleading. In the first place, the value of k_e depends critically on the surface density, ρ. Takeuchi (1951) obtains

$$k_e = 0.281 \quad \text{for} \quad \rho = 3.0 \text{ g cm}^{-3},$$
$$k_e = 0.256 \quad \text{for} \quad \rho = 2.7 \text{ g cm}^{-3},$$

for the same model of the Earth. Furthermore, the parameters k_e and k are not immediately comparable; k_e is defined for a static response of the core and does not take into account the effect of oceans on the wobble. For a comparison we must first allow for the response of core and oceans, and this is due primarily to the frictional interaction between the shell and the fluid. Unlike classical tides, gravitational effects are of second order.

The oscillations excited by the wobble are only in part communicated to a liquid core so that the inertia of the core plays only a minor role in determining the period. Indeed, if the core were a perfect fluid occupying a spherical cavity within a rigid shell, the core would not partake in the wobble at all. The fraction of the core involved in the wobble depends on the ellipticity, density distribution and viscosity of the core and on the electromagnetic forces acting on core and mantle. Only the ellipticity and density distribution as modified by the elasticity of the shell have been investigated (Jeffreys and Vicente, 1957a, b). The subtraction of the inertia of the core then reduces the period of the dynamic model, as compared to the static model, in the ratio of the inertia of the core to the inertia of core plus shell. For a rigid outer Earth, the ratio is one-tenth, and the reduction in period is 30 days.

In quite the same way an inviscid ocean covering the whole Earth would *shorten* the period, though by a very small amount since the ocean's moment of inertia is minute. If, on the other hand, the ocean remained always aligned with respect to the wobbling axis, then the ocean would *lengthen* the period. The lengthening due to such an equilibrium tide is 33 days.

If, without any real evidence, we assume an equilibrium pole tide, then by a remarkable coincidence the adjustments in the Love number k due to core and oceans appear to cancel. The equilibrium pole tide reduces the value from 0·29 to 0·23. Jeffreys and Vicente (1957) demonstrate that the dynamic effects of the core increase k by about 0·08. With this cancellation we obtain a close agreement between the theoretical tidal-effective Love number

$$k = k_e - 0·06 + 0·08 = 0·31$$

and the observed value, 0·29. However, the precision of the theoretical value of k is far smaller than that of k_e because of the two large and as yet uncertain corrections.

9. Summary

The statistical properties of the latitude time series are those associated with a damped oscillator excited at random. Peak response is at a period of 1·20 years; the 'Q' is poorly determined but appears to lie between 30 and 40.* From the peak period one determines $k = 0·29$ for the tidal-effective Love number, a value consistent with Takeuchi's estimate corrected for the (uncertain) response of core and oceans. The Q can be interpreted in terms of a viscosity of 10^{20} g cm^{-1} sec^{-1} for a Maxwell Earth, or 3×10^{17} g cm^{-1} sec^{-1} for a Kelvin–Voigt Earth. Both models have severe short-comings, and the problem of the dissipation is not solved. Irregular variations of the atmosphere are the most likely cause of the wobble.

---

* *Note added in press.* A very complete analysis by Peter Fellgett based on the method of maximum likelihood is now in the press (*Monthly Notices Royal Astronomical Society*). His results are: $F_0 = 1·180 \pm 0·012$ c/year, $\alpha^{-1} = 12·4$ years, Chandler power $= 259·2$ $(0''·01)^2$. According to Fellgett the value of α^{-1} is 'uncertain by a factor of at least 10, and damping times as short as $2\frac{1}{2}$ years or as long as several hundred years are not excluded'. Fellgett has substantiated his uncertainty estimates with artificial time series of random numbers.

CHAPTER 11

HISTORICAL VARIATIONS

1. Astronomical evidence concerning wobble

We are concerned with the drift of the pole of rotation after the annual term and Chandler wobble have been removed. The rejection of these two terms is usually accomplished by taking running means over six years, an interval equal to five Chandler and six annual periods. The elimination would be complete if the Chandler spectrum consisted of a line at 1 cycle in 1·2 years, but in fact the spectrum shows a band structure (§ 7.5).

A glance at fig. 7.4 establishes the absence of any obvious drift. The annual and Chandler motions combined contain 93 per cent of the record energy, according to Rudnick. The residual is small and of doubtful significance. Nevertheless, there have been some quantitative estimates of polar drift. These should be viewed in the context of the overwhelming pressure exerted by geophysicists for any kind of numbers, coupled with an *a priori* plausibility of a polar drift. Evidence from latitude observations is then largely confined to establishing an upper limit to polar drift. Even so, this has geophysical application as a restraint on certain speculations.

A detailed discussion of the first 15 years of latitude observations was given by Lambert (1922). His results, together with those given by Wanach a few years later, are as follows (in 0″·01 per annum):

Source	Interval	dm_1/dt	dm_2/dt
Lambert, 1922	1900–1911	− 0·02 ± 0·10	− 0·63 ± 0·09
Lambert, 1922	1900–1914	+ 0·05 ± 0·04	− 0·62 ± 0·04
Wanach, 1925	1900–1926	+ 0·35 ± 0·06	− 0·31 ± 0·08

Lambert's calculations indicate a polar displacement of 9 ft towards Chicago during 1900–1914; Wanach obtains 12 ft towards Greenland during 1900–1925. There is also some indication of a polar shift towards North America in an apparent decrease of European latitudes during the nineteenth century, accompanied by an increase in American latitudes. In spite of the agreement concerning at least the general direction of the polar drift, the results were not considered significant by the astronomer Schlesinger who concluded that the apparent changes in latitude were due to erroneous proper-motion of

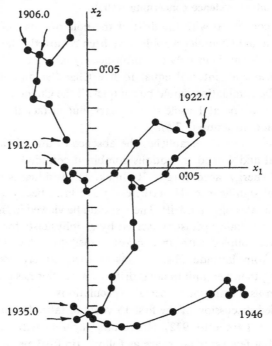

Fig. 11.1. The drift of the pole according to Sekiguchi (1954). Each point corresponds to an annual mean position. The location for the years 1906·0, 1912·0, 1922·7, 1935·0, 1946·0 has been labeled. In the original paper the ILS coordinates, $x = x_1$, $y = -x_2$, are used.

the stars plus some residual *proper-motions of the observatories* (a nice simplification of geologic complexity by astronomical terminology). We refer to Lambert's paper and to Markowitz (1945) for further details.

One might have thought that the subsequent 30 years of latitude observations would have settled the dispute concerning the reality of the apparent drift. But this is not the case. Fig. 11.1 shows the secular motion of the pole according to Sekiguchi (1954) as amended by Melchior (1955). Discontinuities in the direction of drift occur in 1912, 1923, 1935, the years when changes in the catalogue occurred. Sekiguchi considers the changes to be real, but it seems unlikely, as pointed out by Melchior (1957), that these perturbations should occur just when the catalogue is changed. Melchior's conclusion is that the changes merely show the proper motions of the catalogue stars. The analysis by Cecchini (1950) emphasizes rather the dependence of the result on the particular stations selected for the analysis. A uniform reduction of all observations is now under way at the Royal Observatory of Belgium, under the direction of Melchior and Becq.

At the moment the best over-all estimates based on the latitude observations appear to be the following:

Cecchini (1952): $\delta m_1 = 0\overset{''}{.}072$, $\delta m_2 = -0\overset{''}{.}056$
Comstock (1954): $0\overset{''}{.}075$ $-0\overset{''}{.}195$

The foregoing values refer to the displacement from the mean position during 1900–1905 to the mean position in 1949–1950. This amounts to a displacement by about 10 ft towards Greenland during the first half of the twentieth century.

The observations at Greenwich and Washington do not necessarily contradict these values, nor do they offer any support. Spencer Jones (1939a, b) concludes that for the 25 years 1912–1937 '. . . the mean latitude of Greenwich has not varied by as much as $0\overset{''}{.}10$, . . .', and Markowitz (1942) found that in the interval 1916–1940 '. . . the mean latitude of Washington is constant within the errors of observation and that the uncertainty of the change is about $0\overset{''}{.}001$ per year'. Thus the observations at Greenwich and Washington are remarkable for the absence of any observable secular drift.

2. Modern observations concerning the l.o.d.

On the basis of gravitational theory, the orbital motion of the planets, including the Earth, about the Sun, and of satellites about their primaries, can be calculated as a function of time. The time that satisfies the equations of motion is termed Newtonian Time, or Ephemeris Time ($ET_\odot$, $ET_{\mathbb{C}}$, . . .). The time kept by the rotating Earth is Universal Time. UT is defined in terms of the mean solar day, that is, the interval between two successive meridian transits of the 'mean Sun' over a point on the Earth's surface. In practice what is observed are meridian transits of zenith stars.

Tabulated positions of Sun, Moon and planets in accordance with Newtonian mechanics are called *ephemerides*. The independent argument of the ephemerides is $ET_\odot$. If the rotation of the Earth were uniform, and certain other conditions fulfilled, then the observed positions and ephemerides would be in agreement. In fact certain discrepancies have been discovered. For navigational purposes the attempt is made to allow for these discrepancies by including empirical terms in the *Improved Ephemerides*. For our purposes these empirical corrections have to be removed from the tables in order that a comparison can be made between the observed positions and the (unimproved) ephemerides.

The discrepancies can arise from various sources:

(1) The gravitational theory used to compute the ephemeris is incomplete. Dicke (1957) has proposed a theory with a varying gravitational constant, and this introduces discrepancies between $ET_\odot$ and UT (§ 9.8).

(2) Moon and planets are assumed to be rigid bodies. The deformable nature of the real planets results in processes (such as tidal friction) not included in classical gravitational theory. In general these processes will modify the mean motion of the planets.

(3) Numerical approximations in the development of the theory may be insufficient as compared to the accuracy of the observations. Woolard (1953), using an electronic computer, has shown that errors in Brown's lunar tables are sufficiently large to produce observable discrepancies. The errors result from counting twice the effect, upon several terms, of long-period variations in the lunar orbit. Though these errors are observable, they appear to be small compared to other uncertainties. We will attribute all discrepancies to causes other than deficiencies in the gravitational theory and numerical calculations, though future work may establish their importance.

(4) The rotation of the Earth is variable. As a consequence the position of the Sun, Moon and planets observed at a given instant of UT does not agree with the ephemeris position. If there is a secular change in the rate of rotation, then the observed positions depart

from their ephemerides by continually increasing amounts. The principal effect is to produce discrepancies in the mean longitudes, though long-term variations will result also in measurable discrepancies in the perihelia and nodes.

Let $\Delta L_\odot = a_\odot + b_\odot T + f_\odot(T)$, T is $\text{ET}_\odot$ (11.2.1)

denote the difference between the observed and theoretical longitudes of the Sun. T is $\text{ET}_\odot$ measured in Julian centuries of 36525 mean solar days counted from 1900 January 0 Greenwich mean noon. The Sun's tropical mean longitude increases at a rate of 1 sec of arc in 24·349 sec of time. The discrepancy in time is then related to the discrepancy in longitude according to

$$\Delta t = \text{ET}_\odot - \text{UT} = 24 \cdot 349\ \Delta L_\odot.$$ (11.2.2)

The constants $a_\odot$ and $b_\odot$ are arbitrary; they may be chosen to achieve a particular origin and rate of Ephemeris Time. They are chosen so as to make $\Delta L_\odot$ as small as possible over the period of observations discussed by Newcomb; $\Delta L_\odot$ does not vanish at 1900·0.

The Moon has the most rapid motion in celestial longitude (0″55 per sec) and is therefore the most favorable object for determining Ephemeris Time. In its eastward motion among the stars it passes in front of certain bright stars having no mean motion or parallax. The occultation of fixed stars fixes the celestial longitude of the Moon. The Moon's longitude can also be determined from observations of meridian transits with a transit circle. The two methods are in agreement with regard to any of the major features in the lunar discrepancy (fig. 11.8), but there are some differences larger than can be accounted for by accidental errors (Brouwer and Watts, 1942). The source of the differences may be in the limb irregularities. Occultations have generally been considered to provide the more accurate method, particularly in the older observations.

The observed minus the theoretical mean longitude of the Moon can be written

$$\Delta L_\mathbb{C} = a_\mathbb{C} + b_\mathbb{C} T + f_\mathbb{C}(T)$$ (11.2.3)

where $a_\mathbb{C}$ and $b_\mathbb{C}$ are again empirically determined constants. The numerical values are specified by setting

$$\text{WDD} = 0,\quad \frac{d}{dt}(\text{WDD}) = 0$$

180   THE ROTATION OF THE EARTH

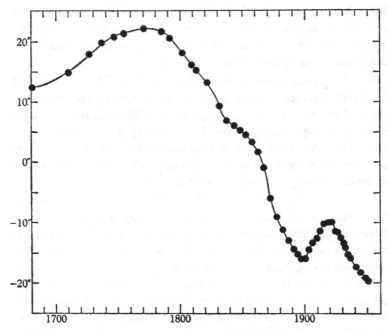

Fig. 11.2. The discrepancy, $f_{\mathbb{C}}(T)$, of the Moon's longitude, based on occultations.

at 1900·0, where WDD is the 'weighted discrepancy difference' to be defined in (11.2.4). Fig. 11.2 shows a plot of the lunar discrepancy curve, $f_{\mathbb{C}}(T)$, based on occultations.* In interpreting this figure it should be remembered that yearly mean longitudes are shown and

* Numerical values were obtained from the formula
$$f_{\mathbb{C}}(T) = \text{Occ} - (\text{Th} - 5\farcs22\,T^2),$$
where Occ − Th are given by Spencer Jones (1932, table 5) for the period 1680 to 1908 and by Brouwer (1952b, table VIIIa) for 1908 to 1950 (see also Spencer Jones, 1939a, table I). The theoretical values Th were derived from Brown's table, which includes the 'Great Empirical Term'
$$\text{GET} = 10\farcs71 \sin (140\fdg0\,T + 240\fdg7)$$
to allow for the hump between 1680 and 1900. Spencer Jones and Brouwer define
$$\text{Th} = \text{Brown's table} - \text{GET} - 4\farcs65 + 12\farcs96\,T + 5\farcs22\,T^2$$
so that the Great Empirical Term has been removed, but an empirical secular variation, $5\farcs22\,T^2$ (based on de Sitter's discussion of ancient observations), has been put in. By writing Th − $5\farcs22\,T^2$ instead of Th, all empirical terms (except for the arbitrary constants $a_{\mathbb{C}}$, $b_{\mathbb{C}}$) have been removed from the theory and are included in the discrepancy curve, $f_{\mathbb{C}}(T)$.

that the data have been smoothed in various ways. Three features are apparent: (1) a 'secular' decrease; (2) a smooth hump between 1680 and 1850, which represents Newcomb's 'Great Empirical Term'; and (3) a relatively high-frequency wiggle for 1900 to 1950. This structure during the last 20 per cent of the record differs markedly from the remaining record, and one is tempted to ascribe this to improved observational techniques. However, the same structure shows up in the lunar meridian observations, and, as we shall see, in the discrepancy curves for Sun and Mercury. Furthermore, no major change in the observational program took place at 1900.

Fig. 11.3 shows the discrepancy for Sun and Mercury,* multiplied by the ratios

$$\frac{n_{\mathbb{C}}}{n_{\odot}} = 13\cdot37, \quad \frac{n_{\mathbb{C}}}{n_{\mathrm{\mathstrut \ddot y}}} = 3\cdot22$$

of the mean motion of the Moon to the mean motion of the Sun and Mercury, respectively. If the variable rotation of the Earth were the sole cause of the discrepancies, then a discrepancy, Δt, between UT and $ET_{\odot}$ will produce a discrepancy in longitude, ΔL, which is proportional to the mean motion (as observed from the Earth) of the bodies. Under this assumption the discrepancies of Moon, Sun and Mercury, as plotted, should be the same. There is, in fact, a close resemblance. This was first noted by Glauert (1915). Spencer Jones (1939a, b) clearly established this relationship, and Clemence's (1943) exhaustive study of Mercury provided further confirmation.

But the resemblance is not complete. There is a long-term drift between the lunar discrepancies, on the one hand, and those of the

---

* The plotted values are

$$\frac{n_{\mathbb{C}}}{n_{\odot}} f_{\odot}(L) = (\mathrm{obs} - \mathrm{Newcomb's\ Tables} + a_{\odot} + b_{\odot}\, T) = B_{\odot} + 16\farcs71\, T^2$$

$$\frac{n_{\mathbb{C}}}{n_{\mathstrut \ddot y}} f_{\ddot y}(L) = (\mathrm{obs} - \mathrm{Newcomb's\ Tables} + a_{\ddot y} + b_{\ddot y}\, T) = B_{\ddot y} + 16\farcs39\, T^2$$

where $B_{\odot}$ and $B_{\ddot y}$ are the fluctuations listed in tables II and IV, respectively, of Spencer Jones (1939). The quadratic terms remove the secular variation, $1\farcs23^2\, T$ (based on de Sitter's discussion of ancient observations), which Spencer Jones added to the theoretical values of Newcomb. Accordingly, the secular drift is retained in the figure.

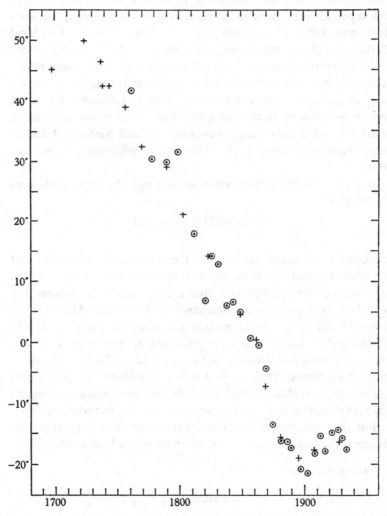

Fig. 11.3. ⊙ designates values of $(n_{\mathbb{C}}/n_{\odot})\,f_{\odot}(T)$; + designates values of $(n_{\mathbb{C}}/n_{\text{\textup{\textmercury}}})\,f_{\text{\textup{\textmercury}}}(T)$. These are the discrepancies for Sun and Mercury (November transits), weighted for their mean motion. There are no observations of solar longitudes for the period 1680 to 1740. The reduction of the observations of Mercury for this period depend on the extrapolated longitude of the Sun.

Sun and Mercury on the other. The 'weighted discrepancy difference'

$$\text{WDD} = f_{\mathbb{C}}(T) - \frac{n_{\mathbb{C}}}{n_{\odot}} f_{\odot}(T) \qquad (11.2.4)$$

makes this drift explicit. Suppose we write

$$\ddot{f}_{\mathbb{C}}(T) = \dot{n}_{\mathbb{C}}(T) - (n_{\mathbb{C}}/\Omega)\, \dot{\Omega}(T), \quad (n_{\mathbb{C}}/\Omega) = 0\cdot 0366,$$
$$\ddot{f}_{\odot}(T) = \dot{n}_{\odot}(T) - (n_{\odot}/\Omega)\, \dot{\Omega}(T), \quad (n_{\odot}/\Omega) = 0\cdot 00274, \qquad (11.2.5)$$

for the second derivatives of the longitude discrepancies of Moon and Sun, allowance having been made for the calibration terms $a + bT$. The effects arising from a variable rate of rotation of the Earth are contained entirely in the second terms. It follows that

$$\frac{d^2(\text{WDD})}{dt^2} = \dot{n}_{\mathbb{C}}(T) - \frac{n_{\mathbb{C}}}{n_{\odot}}\, \dot{n}_{\odot}(T) \qquad (11.2.6)$$

depends only on the weighted difference in the orbital accelerations, the effect of the Earth's variable rotation having been removed. If, furthermore, the orbital accelerations are assumed constant, then

$$\text{WDD} = -cT^2, \quad -2c = \dot{n}_{\mathbb{C}} - (n_{\mathbb{C}}/n_{\odot})\dot{n}_{\odot}. \qquad (11.2.7)$$

The important thing is that for both Sun and Mercury the weighted discrepancy difference can be represented reasonably well by the parabola* (fig. 11.4)

$$\text{WDD} = -11''2T^2, \qquad (11.2.8)$$

which bears out the assumption of constant orbital accelerations and determines the value of c. Lunar tidal friction is apparently responsible (§ 11.6). The WDD curve shows no sign of the Great Empirical Term and of the twentieth-century wiggle, so that the Moon has shared these fluctuations with Sun and Mercury. Accordingly these arise from variations in the Earth's rate of rotation. The time-scale of the GET is of the order of centuries, that of the 1900 to 1950 structure of the order of decades.

* The numerical coefficient, $-11''2$, is based on a least-square fit by Spencer Jones (1939a, b). Combining (11.2.5) with de Sitter's value for the ancient eclipses, $f_{\mathbb{C}} = 5''22\, T^2$, he obtains on elimination
$$f_{\odot}\,(T) = 1''23\, T^2$$
for the secular acceleration of the Sun in modern times. But it should be noted that the foregoing value of $-11''2\, T^2$ for the discrepancy difference is independent of the ancient observations.

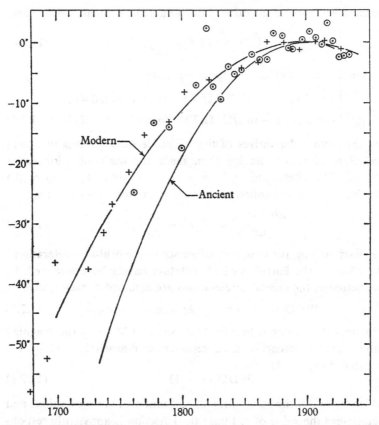

Fig. 11.4. Weighted discrepancy differences: ⊙ for the Sun, + for Mercury (prior to 1740 these depend on the extrapolated longitude of the Sun). The curve labeled "Modern" corresponds to WDD $= -11''\!.22T^2$, "Ancient" to $-18''\!.85T^2$ (see § 11.3).

If we assume that, aside from a *known* quadratic term, the discrepancy in longitude of the Sun is precisely equal to that of the Moon times the ratio of their mean motions, then we can translate lunar occultation measurements into differences between Ephemeris and Universal time,

$$\Delta t = \mathrm{ET}_\odot - \mathrm{UT} = 24 \cdot 349 \left[a_\odot + b_\odot T + \frac{n_\odot}{n_{\mathbb{C}}} cT^2 + \frac{n_\odot}{n_{\mathbb{C}}} f_{\mathbb{C}}(T) \right],$$

(11.2.9)

where T is $\mathrm{ET}_\odot$. This equation permits us to solve for $\mathrm{ET}_\odot$ using

$f_{\mathbb{C}}(T)$ determined by lunar occultations. This method, which is the one used in practice, depends on the value c and the assumption that c is constant. Furthermore any fluctuation of the Moon's angular velocity not due to gravitational effects is assumed small compared

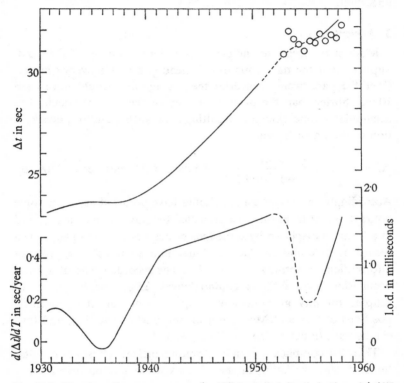

Fig. 11.5. The time discrepancy curve, $\Delta t\ (T)$ (*top*) and its derivative, d $\Delta t/dT$ (*bottom*). The curve for 1930 to 1950·5 is based on Ephemeris Time, using the smoothed values of Brouwer (1952*b*). The curve from 1955·5 to 1958 is based on atomic time (Essen, Parry, Markowitz, and Hall, 1958). The circles are based on the unsmoothed observations with the Moon camera (Markowitz, 1959).

with observational accuracy. These assumptions are forced on us by the practical limitations on the observation of the Sun's longitude.

Fig. 11.5 is based on Markowitz's (1959) attempt to join the results from lunar occultations, the moon camera, and the atomic frequency standard. The discrepancy curve, $\Delta t(T)$, was derived from lunar occultations in accordance with (11.2.9); the derivative then corresponds to the excess or deficiency of the l.o.d. as compared to the mean solar day of epoch 1900 January 0. The assumed linear increase in the l.o.d. from 1955·5 to 1958 corresponds to the straight line in fig. 9.10. The joining of ephemeris and atomic time is uncertain.

3. Ancient eclipses

Modern observations indicate a long-term increase of the l.o.d. Suppose that the day is increasing steadily by 1 ms every century. Over the past twenty centuries the average day would have been 10 ms shorter than the mean solar day of the present epoch. The accumulated time difference resulting from such a uniform deceleration of the Earth is then

$$\Delta t = 10 \times 10^{-3} \frac{\text{sec}}{\text{day century}} \times 365 \text{ days} \times 20 \text{ centuries} \approx 7300 \text{ sec.}$$

Accordingly an ancient eclipse should have occurred at a time some hours different from the time predicted by gravitational theory, and at a location displaced by some tens of degrees of longitude. Thus a record of time and/or place of lunar and solar eclipses provides information concerning $\Delta L_{\mathrm{C}} - \Delta L_{\odot}$. The recorded *time* of a lunar occultation gives ΔL_{C}, of equinox gives $\Delta L_{\odot}$. The *magnitude* of eclipses for any given location depends only on the geocentric positions of Sun and Moon, not on the Earth's rotation, and thus can be used to determine $\Delta L_{\mathrm{C}} - (n_{\mathrm{C}}/n_{\odot})\Delta L_{\odot}$.

There are complications in interpreting ancient eclipses or occultations. A slow variation in the Moon's mean motion arises solely from gravitational effects. The development of a theory of this secular change is a fascinating study in the obtuseness of scientists. Halley in 1695 compared observations of the Moon made in his time with positions derived from ancient eclipses and concluded that the Moon was being accelerated. Astronomers working in the first half of the eighteenth century confirmed Halley's discovery. Euler and Lagrange attempted in vain to account for the acceleration in terms of Newtonian mechanics. Finally, in 1787 Laplace announced that he had discovered the explanation: perturbations due to the planets cause a secular decrease in the eccentricity of the Earth's orbit. This effects the disturbance of the Moon by the Sun and leads to a secular increase in the Moon's orbital velocity. The final result appears as a mathematical series. On evaluating the first term Laplace obtained $f_{\mathbb{C}} = 10''18T^2$, as compared to the observed value of about $10''T^2$. Damoiseau, Plana and Hansen developed further terms in the series and confirmed Laplace's result. The agreement was hailed as one of the major scientific accomplishments of the eighteenth century.

In 1853 Adams announced that he had reworked the theory and had found that neglected terms mounted up; he obtained $5''70T^2$, about half the observed acceleration. Delaunay confirmed Adams's conclusion, but the new results were not generally accepted by the astronomical world, including LeVerrier and Hansen. Even some early twentieth century texts on celestial mechanics refer only to Laplace. The principal reason for the rejection of Adams's work appears to be that it destroyed a widely acclaimed theory. Acceptance came only after it was found that oceanic tidal friction could lengthen the day.

Many ancient observations have been analyzed. It is somewhat surprising to find Greek and Babylonian observations predominating. Only a single record of an observation by ancient Egyptians exists, and this is questionable (Neugebauer, 1957). The Chinese observations are generally discounted (Cowell, 1905); whether for lack of ancient Chinese observational technique or modern Chinese scholars is not apparent. One discrepancy value might possibly be obtained from a solar eclipse observed in Loyang, in Eastern China, on 16 December, A.D. 65 (Ginzel, 1906, p. 460). The Mayan and other Central Americans developed a sophisticated astronomy, yet we know of no discussion of these observations relevant to the problem at hand.

Since the study of ancient observations requires competency in both astronomy and antiquities, the field has never been over-crowded. Prior to 1906, the few professional astronomers engaged in this research found little to agree upon and often expressed their misgivings of other work in bitter phrases. During the period 1906 to 1926, J. K. Fotheringham, a classicist turned astronomer, evaluated many of the older observations and is largely responsible for the values of ΔL currently in the literature. According to Neugebauer (1957) much progress has since been made in the study of ancient astronomy, but this newly gained knowledge has not yet been applied to the problem of determining ΔL.

Tables of ΔL and of their standard deviations capture neither the flavor nor the uncertainties of ancient observations. The problems faced by a classicist-astronomer are best shown by example. Given the statement as taken from a Babylonian Tablet 'On the twenty-sixth day of the month Sivan in the seventh year the day was turned to night and fire in the midst of heaven . . .', identify the eclipse referred to (if it truly is an eclipse) and locate the place and/or date of observation. Fotheringham (1920) discusses this particular passage in detail. There is an uncertainty as to the actual date referred to and even to the symbol used in the date. The reference to 'fire in the heavens' has caused great difficulty. Does this mean there was a corona, or was the event a thunderstorm or a fall of meteors? Some would refer the whole passage to a divinity scholar rather than to an astronomer. Fotheringham (1920 p. 107) concludes that on 1062 B.C., July 31, the eclipse was total at Babylon.

Both Fotheringham and de Sitter attach great weight to the solar eclipse which was presumably observed by Hipparchus in 128 B.C. The pertinent quotation is 'Once it (the Sun) was observed to be totally eclipsed at Hellespont, when at Alexandria it was eclipsed with the exception of one fifth of its own diameter . . .' No clue is given by Ptolemy as to the identification of the eclipse except for the name of Hipparchus and the magnitudes at Hellespont and Alexandria. Only two eclipses could satisfy the conditions stated, one in 309 B.C. and the other at 128 B.C. It is generally assigned to 128 B.C. on the assumption that Hipparchus would be more likely to compare two contemporary observations than an earlier set of observations. Fotheringham (1909) obtains three possible belts of totality for the eclipse (fig. 11.6). On this basis de Sitter (1927) calculated $f_{\mathbb{C}}$ and $f_{\odot}$ as shown in fig. 11.6. The limits of uncertainty reflect the dependence of the values on the three belts of totality.

Fotheringham (1920) analyzed a number of other solar eclipses. The range of values is very large as can be seen in fig. 11.6. Fotheringham considers the triangle defined by Hipparchus, Plutarch and

Eponym Canon eclipses as the best limits. However, the latter two are most uncertain. Plutarch's is based on the dialogue De Facie, the beginning of which is lost. The quotation is 'Grant me that no one of the phenomena relating to the sun is so like another as an eclipse to a sunset, that recent concurrence of sun and moon, which, beginning just after noon, showed us plainly many stars in all parts of the

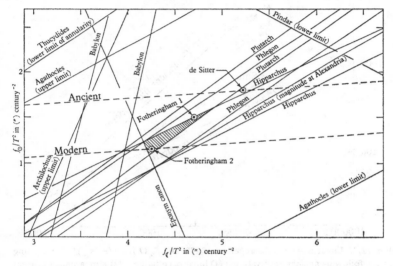

Fig. 11.6. Fotheringham's (1920) summary of consistent values of $f_{\mathbb{C}}/T^2$ and $f_{\odot}/T^2$, for various ancient solar eclipses. In the original paper the $f_{\mathbb{C}}/T^2$ scale includes the gravitational term, $6''1$ century^{-2}, and this has been subtracted. The dashed lines correspond to WDD $= - 11''22 \; T^2$ for the modern observations, and WDD $= - 18''85 \; T^2$ for the ancient observations.

heavens, and produced a chill in the temperature like that of twilight. If you have forgotten it . . .' Unfortunately, the narrator is unknown and so are the homes of the addressed company.

Fotheringham (1918) analyzed the twenty equinox observations of Hipparchus listed by Ptolemy in the *Almagest*. The method of observation is unknown, but Hipparchus presumably interpolated between observed meridian transits. Times are given to the nearest quarter of a day, and this is perhaps the largest source of uncertainty. The value obtained by Fotheringham is shown in fig. 11.7 together with the standard deviation.

The occultations listed in the *Almagest* have been analyzed by Fotheringham and Miss Gertrude Longbottom (1915). The descriptions of the occultations are remarkably precise regarding the occulted star though the position of the Moon is not always clear. Newcomb had earlier dismissed the occultations principally because Ptolemy

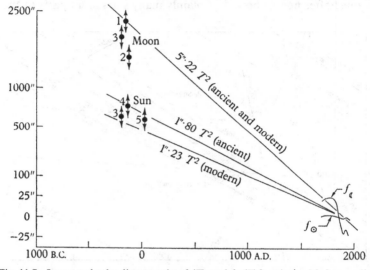

Fig. 11.7. Lunar and solar discrepancies, $f_{\mathbb{C}}(T)$, and $f_{\odot}(T)$ [not $(n_{\mathbb{C}}/n_{\odot})f_{\odot}$], according to the following ancient observations: (1) lunar occultations; (2) Hipparchus's eclipse; (3) solar eclipses (including that of Hipparchus); (4) equinox observations reported by Hipparchus; (5) lunar eclipses. The vertical axis is linear in $|f|^{\frac{1}{2}}$ so that the secular accelerations are plotted as straight lines. de Sitter's interpretation of the ancient observations is $f_{\mathbb{C}} = 5''22\ T^2$, $f_{\odot} = 1''80\ T^2$, as shown. Spencer Jones's result for the modern observations, $f_{\odot} = 1''23\ T^2$, is included for comparison. The observed discrepancies since 1680 are sketched.

used them to prove Hipparchus's erroneous value for the constant of precession. It appears that the errors introduced are small.

Fotheringham's final choice is*

$$f_{\mathbb{C}} = 4''7T^2, \quad f_{\odot} = 1''5T^2.$$

de Sitter reviews Fotheringham's results and obtains the values

$$f_{\mathbb{C}} = (5''22 \pm 0''30)T^2, \quad f_{\odot} = (1''80 \pm 0''16)T^2.$$

These lie above the Fotheringham mean values because de Sitter

* After subtraction of the gravitational term $f_{\mathbb{C}} = 6''1\ T^2$ (Spencer Jones, 1956).

heavily weights a lunar eclipse in Babylon of 424 B.C. For comparison, Spencer Jones's interpretation of the modern observations gave

$$f_{\mathbb{C}} = (5''22 \pm 0''30)T^2, \quad f_{\odot} = (1''23 \pm 0''04)T^2.$$

Curves corresponding to de Sitter's and Spencer Jones's sets of values are included in figs. 11.4, 11.6 and 11.7.

4. Spectra of modern observations

Figures 11.8 to 11.10 show various power spectra of the function $f_{\mathbb{C}}(t) - 5''22T^2$, i.e., the lunar discrepancy curve minus the 'secular' term. The material covers roughly two centuries of yearly values, and these have received various degrees of smoothing. Early observations

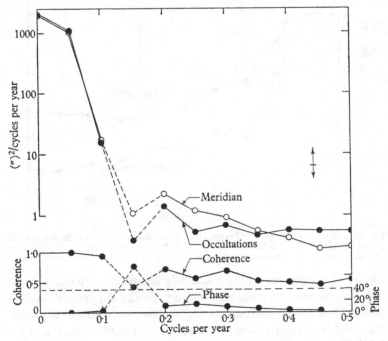

Fig. 11.8. Spectra of the Moon's discrepancy derived from occultations (solid circles) and from meridian observations (open circles) for the period 1751·5 to 1908·5. The 95 % uncertainty limits are indicated by the vertical arrow. The lower diagram gives the coherence between the records (see § A.2), the horizontal dashed line the approximate significance level of coherence at $2\nu^{-\frac{1}{2}} = 0.36$ for $\nu = 32$ degrees of freedom, and the phase lag of the meridian observation relative to the occultation. Negative side bands from power at very low frequency is responsible for the drop at 0·15 c/year, and the values at this frequency are not significant.

are published at longer intervals, and from these we have taken yearly values by straight-line interpolation. For such a short and inhomogeneous time series the usefulness of a statistical treatment is clearly limited; a glance at fig. 11.2 shows that the assumption of a stationary time series is untenable.

Lunar discrepancy curves have been obtained from two sources: occultations and meridian observations. According to fig. 11.8 the

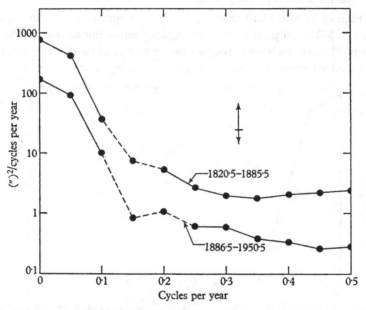

Fig. 11.9. Spectra of the Moon's longitude from occultations (fig. 11.2) for the periods 1820·5 to 1885·5 and 1886·5 to 1950·5. The 95% uncertainty limit is indicated by vertical arrow. The low point at 0·15 c/year is due to side bands, and not significant.

spectra of the two curves are practically identical for frequencies up to 0·1 c/year, the coherence being almost one and the phase difference zero. This is strong evidence for the significance at low frequencies. But even at higher frequencies the spectra are not far apart, and the coherence is significant.

Separate spectra for the early and late half of the observations are shown in fig. 11.9. The latter record has lower spectral densities at all frequencies. This immediately raises doubt concerning the significance of the observations, for such a change is the result expected from an improvement in observational technique (Blackman and Tukey, 1958, p. 560–562). At high frequencies observational error is undoubtedly an important factor; for an error of $1''$ per lunation, the expected spectral level of observational error is (8.2.3)

$$1.9 \times 10^7 \sec^3 = \frac{0.2('')^2}{c/year}$$

in agreement with the 1886·5 to 1950·5 results. But at frequencies of less than 0·1 c/year the excellent coherence between the occultation and meridian observations would suggest that the decrease is real and related to the twentieth century wiggle (fig. 11.2).

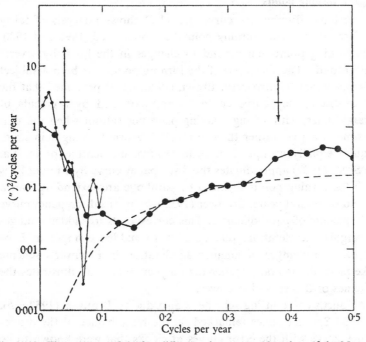

Fig. 11.10. The spectrum of the first differences between yearly values of the Moon's longitude discrepancy, based on the entire length of occultations record, 1751·5 to 1950·5. This is roughly proportional to the spectrum of the l.o.d. A year-to-year difference by $1''$ corresponds to a difference of $1\overset{s}{.}82$ per year (the Moon moves $1''$ in $1\overset{s}{.}82$) or $1\overset{s}{.}82/3\cdot16 \times 10^7$ per second. The above spectrum in units of $(1\overset{s}{.}82/3\overset{s}{.}16 \times 10^7)^2 = 3\cdot3 \times 10^{-15}$ gives the contribution, per unit frequency band (c/year), towards the mean square *fractional* variation in the l.o.d. A higher resolution was used in the low-frequency portion at the expense of spectral reliability, as shown by the 95% uncertainty arrows. The broken curve is the assumed spectrum of observational error.

194   THE ROTATION OF THE EARTH

Figure 11.10 shows the spectrum of the first differences,*

$$\delta[f_{\mathcal{C}}(T) - 5''22T^2]$$

of the annual occultation discrepancies for the entire period of observations, 1751·5 to 1950·5. According to (11.2.9) this is closely related to the derivative of the time discrepancy, Δt, and the spectrum may be regarded as the spectrum of the l.o.d. The characteristic feature is the minimum at roughly 0·15 c/year. Towards the high frequencies the rise is due primarily to experimental error. The low-frequency rise is believed to be genuine.

5. The turning points

The lunar discrepancy curve (fig. 11.2) shows intervals of fairly uniform slope, with 'turning points' at about 1785, 1898 and 1920. The turning points correspond to changes in the l.o.d. by several milliseconds. The sharpness of the turning points has been a subject of considerable controversy. Brown (1926, fig. 1) presumed that the lunar discrepancy curve could be approximated by segments of straight lines, with 'changes taking place per saltum when the unit of time is a year rather than gradually'. Abrupt changes in l.o.d. require equally abrupt changes in relative momentum or inertia. de Sitter (1927) approximates the discrepancy curve by straight lines with the turning points replaced by parabolic arcs extending over a period of several years. Brouwer (1952a, b) fits the discrepancy curve by segments of parabolic arcs. This corresponds to sudden changes in angular acceleration, due possibly to sudden changes in L or dC/dt. The changes of angular acceleration in Brouwer's scheme take place in intervals smaller than a year. Fig. 11.11 illustrates the schemes of Brown and Brouwer.

A fourth curve-fitting scheme is also due to Brouwer (1952a, b). In 1950 Spencer Jones remarked on the resemblance of the discrepancy curve with the error curves of clocks that were known to be affected by frequent small erratic changes in the clock rate. Brouwer and van Woerkom have examined the properties of functions representing the accumulation of random numbers. Let $\delta^2(\Delta t)$ designate

* This could have been obtained from the discrepancy spectrum by multiplying with $\sin^2(\frac{1}{2}\pi f/f_N)$, where $f_N = 0·5$ c/year is half the sampling frequency (A.2.5). Because of the steep rise of the discrepancy spectrum at the origin it is safer to compute first differences and then form the spectrum of that time series.

second differences between successive yearly values of the discrepancy, Δt. Brouwer and van Woerkom performed experiments with random numbers for δ^2 and were able to produce synthetic Δt curves which resemble the observed discrepancy curve to a remarkable degree. Brouwer's conclusion is then that the discrepancy curve can be adequately presented by segments of parabolic arcs extending over periods from 4 to 15 years, but that accumulation of many small random events may also fit the observations.

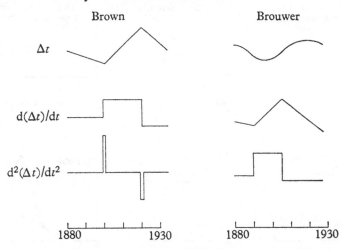

Fig. 11.11. The amount of time, Δt, by which the Earth is slow, and the first two derivatives of this function. The first derivative is proportional to the l.o.d., the second derivative to angular deceleration. The two columns represent schematically the interpretations by Brown and Brouwer, respectively, of the variation during 1880 to 1930.

The curve-fitting problem is to match the observed spectrum of (fig. 11.10) to a spectrum proportional to an error spectrum plus that of one of the curve-fitting schemes. The simplest assumption regarding the errors in Δt is that they are independent, thus neglecting any smoothing. The spectrum of Δt is then white and that of $\delta^2(\Delta t)$ is proportional to (§ A.2)

$$\sin^4\left(\tfrac{1}{2}\pi f/f_N\right),$$

where $f_N = 0 \cdot 5$ c/year is half the sampling frequency. This spectrum is shown in fig. 11.12, a.

The two Brouwer schemes deal with correlated and uncorrelated fluctuation in $\delta^2(\Delta t)$, as has been made explicit by Blackman and Tukey (1958, p. 249–252, 555–563). For the random case, successive values of $\delta^2(\Delta t)$ are uncorrelated, and accordingly the spectrum of δ^2 should be flat (fig. 11.12, c). The non-random model is characterized by blocks of constant values of δ^2, with the values assigned to different blocks being independent. The power spectrum for constant-by-five blocks has been evaluated by Blackman and Tukey;

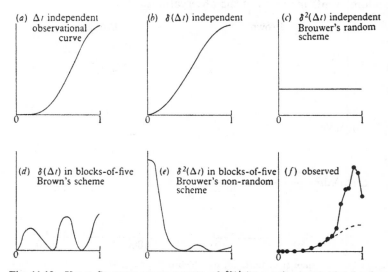

Fig. 11.12. *Upper diagrams*: power spectra of $\delta^2(\Delta t)$ assuming uncorrelated values of Δt, $\delta(\Delta t)$ and $\delta^2(\Delta t)$. *Lower left diagrams*: power spectra of $\delta^2(\Delta t)$ for $\delta(\Delta t)$ and $\delta^2(\Delta t)$ in blocks-of-five, but with independence between blocks. (f) shows the observed spectrum (solid) and the assumed spectrum of observational error (dashed).

it consists of a sharp low-frequency peak and some ripples (fig. 11.12e). If the scheme had consisted of constant-by-eight blocks with independence between blocks, then the low-frequency peak would have been even sharper and the ripples higher. For a random distribution in the length of blocks, but still with independence between them, the resulting spectrum is the corresponding average of the fixed-block spectrum.

In a similar way we can construct spectra corresponding to fluctuations in the first difference, $\delta(\Delta t)$. For the uncorrelated case the spectrum of δ is white and that of δ^2 proportional to $\sin^2{(\tfrac{1}{2}\pi\,f/f_N)}$;

for the correlated case we obtain $\sin^{-2}(\frac{1}{2}\pi f/f_N)$ times Brouwer's non-random spectrum (see fig. 11.12, b, d). The last case corresponds to the Brown scheme where the discrepancy curve is approximated by straight lines and sharp corners.

Fig. 11.12f shows the observed spectrum of $\delta^2[f_{\mathbb{C}}(T) - 5\overset{''}{.}22T^2]$ or, roughly, of $\delta^2(\Delta t)$. This is to be made up of a spectrum proportional to the error spectrum (as represented by the dashed curve) plus one of the others. It will be noted that the residual vanishes at low frequencies. This would seem to rule out Brouwer's non-random scheme, but it is also inconsistent with Brouwer's random case. The residual is most closely approximated by correlated fluctuations in $\delta(\Delta t)$, provided we smooth out the ripples by assuming blocks of variable length. In other words, the observed spectrum can be fitted in terms of independent errors in Δt, plus rather sudden variations in $\delta(\Delta t)$, that is, in l.o.d., the variations taking place in an interval small compared to the length of the blocks. This is just about the interpretation given by Brown and de Sitter, and we are not far from where we started by an inspection of the record.

The statistical discussion has been given in some detail because it forms such an important part of the recent literature on the subject. The emphasis on curve fitting (stemming probably from the analogy with clock errors) would appear, in retrospect, not to have been very fruitful. The spectrum in the l.o.d. (fig. 11.10) is perhaps as satisfactory a statistical summary as any. The assumed error spectrum is again represented by a dashed curve. It is drawn according to $\sin^2(\frac{1}{2}\pi f/f_N)$, which is the expected spectrum in $\delta(\Delta t)$ from independent errors in Δt. The shape of the error spectrum is then determined, but its vertical position in the figure is arbitrary. In any event the residual spectrum has a pronounced maximum at low frequencies, and a weak maximum at high frequencies.

The high time-resolution provided by the caesium frequency standard gives hope that in the near future the problem of the turning points can be settled. As far as the evidence goes (fig. 9.10), the interpretation $ABDE$ favors a Brouwer-type discontinuity, whereas $ABCDE$ would call for a combination of Brown- and Brouwer-type discontinuities.

198 THE ROTATION OF THE EARTH

6. Dissipation of tidal energy

The observed discrepancies in the motion of Moon and Sun have fascinated philosophers and scientists alike. The role of tidal friction was proposed in 1754 by Kant in an essay 'Untersuchung der Frage, ob die Erde in ihrer Umdrehung um die Achse, wodurch sie die Abwechselung des Tages und der Nacht hervorbringt, eine Veränderung seit den ersten Zeiten ihres Ursprunges erlitten habe, und woraus man sich ihrer versichern könne.' Laplace rejected this suggestion because it implied a secular acceleration in the motion of the Sun and planets, as well as of the Moon, and this had not yet been observed. In 1865 Delaunay revived the hypothesis to account for the secular acceleration of the Moon, but not until 1905 did Cowell discover the prerequisite acceleration of the Sun. The first discussion of the role of oceanic tides appears in the *Treatise on Natural Philosophy* (Thomson and Tait, 1879, p. 191). In 1881 Frederick Engels* criticized Thomson and Tait, claiming that only bodily tides can account for the lunar discrepancy. Engels's critique is based on a misconception concerning the energetics of the situation.

It is remarkable how far one can go in the interpretation of the discrepancy curves without introducing geophysical observations. The key lies in Murray's (1957) procedure for separating that part of the discrepancy due to orbital acceleration from that due to the angular acceleration of the Earth. The WDD (11.2.7) depends only on the orbital accelerations. First we shall show that $(n_{\mathbb{C}}/n_\odot)\dot{n}_\odot$ is negligible compared with $\dot{n}_{\mathbb{C}}$, so that the WDD determines $n_{\mathbb{C}}(T)$. Then we subtract the first term from the lunar discrepancy,

$$f_{\mathbb{C}}(T) = \int n_{\mathbb{C}}(T)\,dT + (n_{\mathbb{C}}/\Omega)\int \Omega(T)\,dT,$$

leaving only the effect of the variable rotation of the Earth, $\Omega(T)$. But knowing $n_{\mathbb{C}}(T)$ we can compute that part, $\Omega|_{\mathbb{C}}$, depending on lunar tidal torque, and subtract its contribution to $f_{\mathbb{C}}(T)$. For a given ratio of lunar to solar tidal torque we can subtract from the remaining lunar discrepancy the part, $\Omega|_\odot$, due to solar tidal torque. Finally we subtract the effect of $\Omega'|_\odot$ due to a known solar torque on the atmosphere. In this manner we subtract from the observed

---

* In an essay that forms part of *Dialectics of Nature* (Engels, 1954).

lunar discrepancy one tidal effect after another, and are left with a residual lunar discrepancy from which all tidal effects have been removed.

If the ocean has any viscosity or if the response of the solid Earth is not perfectly elastic, then the lunar and solar tides will be shifted

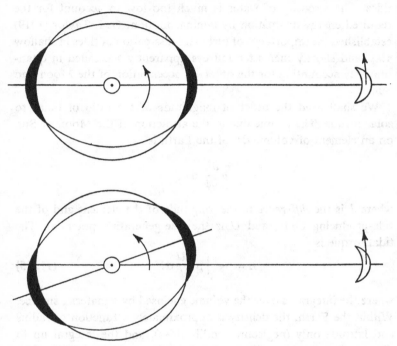

Fig. 11.13. The upper diagram shows the tidal bulge if there is no friction. In the case of friction there is a delay in the time of high tide, and the resulting distortion of the tidal bulge leads to a deceleration in the Earth's rate of rotation, and to an acceleration in the Moon's orbital motion.

in phase (see fig. 11.13). There is a delay in the time of maximum elevation or depression of a tide at a given place relative to lunar (or solar) transit times. The gravitational attraction on the bulges is assymetric to the line of centers and gives rise to a torque tending to decelerate the rotational motion of the Earth. Angular momentum is conserved, and the momentum lost by the Earth must appear in the orbital motion.

Tidal friction is qualitatively capable of producing a decrease in the Earth's angular velocity and at the same time accelerating the Sun and Moon. Darwin ascribed the observed discrepancies to bodily tides, but the required degree of fluidity or plasticity is ruled out by modern seismology (§ 11.7). Darwin was aware of these difficulties and in several places hints at the importance of ocean tides. The viscosity of water is much too low to account for the required energy dissipation by laminar motion. G. I. Taylor (1919) established the importance of turbulent dissipation of tides in shallow seas, and shortly thereafter Jeffreys apparently succeeded in quantitatively accounting for the observed acceleration of the Moon and Sun.

We shall need the order of magnitude of the ratio of lunar to solar torque. The couple due to the attraction of the Moon or Sun on an element of volume dV of the Earth is

$$\rho \frac{\partial U}{\partial \phi} dV,$$

where ϕ is the *difference* in the longitude of the element and of the tide-producing body, and U is the tide-generating potential. The tidal torque is

$$L = - \int \rho \frac{\partial U}{\partial \phi} dV, \qquad (11.6.5)$$

where the integral is over the volume enclosed by a material surface. Within the Earth, the density is approximately a function of radius and latitude only (neglecting bodily tides), and the integral up to the mean sea-level contributes nothing to the torque. Denoting the height of the ocean tide by ξ, the integral up to the actual ocean surface is

$$L = - \int \rho \mathscr{C}(\theta, \lambda) \xi \frac{\partial U}{\partial \phi} dS, \qquad (11.6.6)$$

where $\mathscr{C}$ is the ocean function (§ A.1). But the tide-generating potential U and the equilibrium height ξ of the tide are both proportional to M_{d}/r^3. The periods of the M_2 and S_2 tides are about equal. Provided the laws governing the tides are linear in ξ, then the tidal amplitudes should be in ratio of the disturbing potentials. Observation of tides on oceanic islands suggest that the empirical ratio approximates the ratio given by equilibrium theory. The ratio of solar to lunar torques due to the ocean tide is then

$$\frac{L_\odot}{L_{\mathbb{C}}} = \left(\frac{M_\odot\, r_{\mathbb{C}}^3}{M_{\mathbb{C}}\, r_\odot^3}\right)^2 = \frac{1}{5 \cdot 1}. \tag{11.6.7}$$

A similar argument can be made for the ratio of equilibrium bodily tides. The ratio (11.6.7) is somewhat increased if lunisolar terms are taken into account (Groves and Munk, 1959). Jeffreys (1952) shows that the presence of non-linear friction in the oceans increases the ratio to

$$L_\odot/L_{\mathbb{C}} = 1/3 \cdot 4. \tag{11.6.8}$$

Next we estimate the ratio of the orbital accelerations of Moon and Sun. Let $r_{\mathbb{C}}$ designate the geocentric distance of the Moon,

and $\qquad r'_{\mathbb{C}} = \dfrac{M_\oplus}{M_\oplus + M_{\mathbb{C}}} r_{\mathbb{C}}, \quad r'_\oplus = \dfrac{M_{\mathbb{C}}}{M_\oplus + M_{\mathbb{C}}} r_{\mathbb{C}} = r_{\mathbb{C}} - r'_{\mathbb{C}}$

the distances of Moon and Earth from their center of mass. The joint angular momentum is

$$(M_{\mathbb{C}} r'^2_{\mathbb{C}} + M_\oplus r'^2_\oplus)n_{\mathbb{C}} = \frac{M_\oplus M_{\mathbb{C}}}{M_\oplus + M_{\mathbb{C}}} r_{\mathbb{C}}^2 n_{\mathbb{C}}. \tag{11.6.9}$$

The tidal bulge raised by the lunar tide exerts a torque L on this system. An equal and opposite torque diminishes the angular momentum $C\Omega$ of the Earth's rotations so that the angular momentum of orbital motion plus rotation is conserved:

$$\frac{M_\oplus M_{\mathbb{C}}}{M_\oplus + M_{\mathbb{C}}} \frac{\mathrm{d}(r^2_{\mathbb{C}} n_{\mathbb{C}})}{\mathrm{d}t} = L = -\frac{\mathrm{d}(C\Omega)}{\mathrm{d}t}. \tag{11.6.10}$$

Combining the equations of motion and Kepler's third law

$$r_{\mathbb{C}}^3 n_{\mathbb{C}}^2 = G(M_\oplus + M_{\mathbb{C}}) = \text{constant} \tag{11.6.11}$$

we have $\dot n_{\mathbb{C}} = \dfrac{3(M_\oplus + M_{\mathbb{C}})}{r_{\mathbb{C}}^2 M_\oplus M_{\mathbb{C}}} \dfrac{\mathrm{d}(C\Omega)}{\mathrm{d}t} = -\dfrac{3(M_\oplus + M_{\mathbb{C}})}{r_{\mathbb{C}}^2 M_\oplus M_{\mathbb{C}}} L$ (11.6.12)

and a similar expression for $\dot n_\oplus$. Allowing for the fact that $M_{\mathbb{C}} \ll M_\oplus \ll M_\odot$, we obtain, approximately,

$$\frac{\dot n_\odot}{\dot n_{\mathbb{C}}} = \frac{M_{\mathbb{C}}}{M_\oplus} \frac{r_{\mathbb{C}}^2}{r_\odot^2} \frac{L_\odot}{L_{\mathbb{C}}} \approx 10^{-8}.$$

The ratio $\dot n_\xi/\dot n_{\mathbb{C}}$ is even smaller. The orbital acceleration of the Sun and Mercury due to tidal forces are then completely negligible

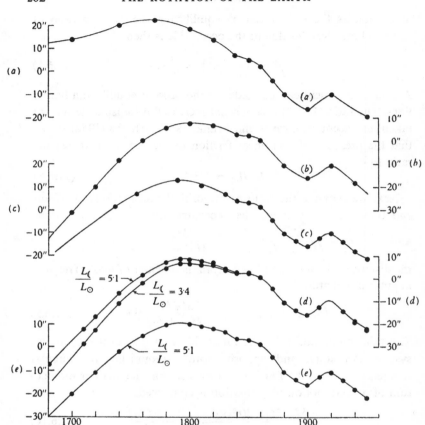

Fig. 11.14. (a) The observed smoothed lunar discrepancy curve, $f_{\mathbb{C}}(T)$. The remaining curves show the discrepancy residuals after allowing for various tidal effects (see table 11.1).

compared with the lunar acceleration. This is consistent with the observational result that the discrepancies of Sun and Mercury are the same (fig. 11.7). The discrepancy difference between Moon and Sun (11.6.2, 3) is then due entirely to the orbital acceleration of the Moon:

$$\frac{d^2(\text{WWD})}{dt^2} = \dot{n}_{\mathbb{C}} = -22''4 \text{ century}^{-2} = -1.09 \times 10^{-23} \text{ rad sec}^{-2}.$$

(11.6.13)

By Kepler's third law this corresponds to an increase in the Moon's orbital radius by 1.0×10^{-7} cm sec^{-1}. The torque exerted by the tidal bulge on the Moon is found from (11.6.10) to equal

$$L_{\mathbb{C}} = 3.9 \times 10^{23} \text{ dyn cm} \qquad (11.6.14)$$

and the rate at which this torque does work equals

$$- \dot{E} = L_{\mathbb{C}}(\Omega - n_{\mathbb{C}}) = 2.7 \times 10^{19} \text{ ergs sec}^{-1}. \qquad (11.6.15)$$

If the tidal friction is to account for the advance in longitude of the Moon, then on the average 2.7×10^{19} ergs must be dissipated every second within the earth or oceans. This value is independent of the observations and interpretations of ancient eclipses but is determined solely from observations since 1680 concerning the celestial positions of the Sun, Moon and Mercury.*

Fig. 11.14 illustrates the various components of the lunar discrepancy curve. Fig. 11.14a shows $f_{\mathbb{C}}(T)$ as copied from fig. 11.2; 11.14b gives $f_{\mathbb{C}}(T) - 11''2T^2$, that is, the lunar discrepancy after removal of the effect of lunar tidal friction on the Moon's orbital motion. We have not yet subtracted from the discrepancy the effect of the Earth's deceleration by lunar tidal friction. This is found from (11.6.10):

$$\dot{\Omega}|_{\mathbb{C}} = - L_{\mathbb{C}}/C = - 9''86 \times 10^2 \text{ century}^{-2}$$
$$= - 4.81 \times 10^{-22} \text{ rad sec}^{-2}, \qquad (11.6.16)$$

and this corresponds to an increase in the l.o.d. by 1.8 ms per century. The discrepancy due to the combined effect of the orbital acceleration of the Moon and the angular deceleration of the Earth is given by

$$\ddot{f}_{\mathbb{C}}(T) = - 22''4 - \frac{n_{\mathbb{C}}}{\Omega}(- 986'') = 13''7 \text{ century}^{-2}$$

and accordingly $f_{\mathbb{C}}(T) - 6''85T^2$ gives the lunar discrepancy resulting solely from extra-lunar effects (fig. 11.14c). All values are summarized in table 11.1.

The Earth's angular velocity is diminished also by the solar tidal torque, and this deceleration further contributes to the lunar discrepancy. Using the ratios $L_{\odot}/L_{\mathbb{C}} = 1/5.1$ and $1/3.4$ corresponding

* Jeffreys (1952, p. 227) uses $\dot{E} = - 1.4 \times 10^{19}$ ergs sec^{-1}. This value is based on ancient observations. His method for separating lunar and solar torques is different from the one used here.

to Jeffreys's models for linear and non-linear friction, respectively, we obtain the values in table 11.1. Fig. 11.14d shows

$$f_{\mathbb{C}} - 10''40T^2, \quad f_{\mathbb{C}} - 12''15T^2.$$

These are the lunar discrepancies after removal of the retarding effects of lunar and solar tides for the two ratios $L_{\odot}/L_{\mathbb{C}}$. This figure is not as close to the observations as the curve in fig. 11.14c. The lunar retarding torque is determined directly by astronomical

Table

	Symbols	Units	Fig.
1. Lunar discrepancy	$f_{\mathbb{C}}/T^2$	$('')$ century^{-2}	11.2
2. Solar discrepancy	$f_{\odot}/T^2$	$('')$ century^{-2}	11.3
3. Weighted discrepancy difference	$(f_{\mathbb{C}} - \frac{n_{\mathbb{C}}}{n_{\odot}} f_{\odot})/T^2$	$('')$ century^{-2}	11.4
4. Lunar tidal torque	$L_{\mathbb{C}}$	10^{23} dyn cm	
5. Resulting dissipation	$- \, dE/dt$	10^{19} ergs sec^{-1}	
6. Resulting deceleration of Earth	$\dot{\Omega}\|_{\mathbb{C}}$	$('')$ century^{-2}	
7. Lunar discrepancy after allowing for lunar orbital acceleration		$('')$ century^{-2}	11.14b
8. Lunar discrepancy after removal of all lunar tidal effects		$('')$ century^{-2}	11.14c
9. Assumed torque ratio	$L_{\mathbb{C}}/L_{\odot}$		
10. Solar oceanic tidal torque	$L_{\odot}$	10^{23} dyn cm	
11. Resulting dissipation	$- \, dE/dt$	10^{19} ergs sec^{-1}	
12. Resulting deceleration of Earth	$\dot{\Omega}\|_{\odot}$	$('')$ century^{-2}	
13. Lunar discrepancy after removal of lunar and solar oceanic tidal effects		$('')$ century^{-2}	11.14d
14. Solar atmospheric torque	$L'_{\odot}$	10^{23} dyn cm	
15. Resulting dissipation	$- \, dE/dt$	10^{19} ergs sec^{-1}	
16. Resulting deceleration of Earth	$\dot{\Omega}'\|_{\odot}$	$('')$ century^{-2}	
17. Lunar discrepancy after removal of all lunar and solar effects		$('')$ century^{-2}	11.14e
18. Tidal energy dissipation in ocean and earth, (5) + (11)	$- \, dE/dt$	10^{19} ergs sec^{-1}	
19. Total tidal acceleration, (6) + (12) + (16)	$\dot{\Omega}$	10^{-22} rad sec^{-2}	
20. Non-tidal acceleration	$\dot{\Omega}$	10^{-22} rad sec^{-2}	11.19
21. Proportional change in the Earth's angular velocity after removal of all known tidal effects, since time of observation ($T \approx - 19$)	$\dfrac{\delta\Omega}{\Omega}$	10^{-7}	

observation; the solar torque does depend on assumptions regarding the ratio of amplitudes of lunar and solar tides.

For later reference (§ 11.9) we have included in table 11.1 the discrepancies from an accelerating torque exerted by the Sun on the Earth's atmosphere. Fig. 11.14e shows the lunar discrepancy after removal of all known lunar and solar effects.

The interpretation of the modern observations is based on the assumption that both the tidal-retarding and accelerating torques

11.1

Modern		Ancient					
		de Sitter		Fotheringham 1		Fotheringham 2	
		5·22		4·7		4·25	
		1·80		1·5		1·18	
−11·2		−18·85		−15·2		−11·2	
3·9		6·6		5·3		3·9	
2·7		4·6		3·7		2·7	
−986		−1670		−1340		−986	
$(f_{\mathbb{C}}/T^2) - 11·2$		−13·63		−10·5		−7·0	
$(f_{\mathbb{C}}/T^2) - 6·85$		−6·4		−4·6		−2·6	
5·1	3·4	5·1	3·4	5·1	3·4	5·1	3·4
0·76	1·10	1·29	1·94	1·04	1·56	0·76	1·10
0·55	0·80	0·93	1·41	0·75	1·13	0·55	0·80
−193	−290	−326	−511	−264	−412	−193	−290
$(f_{\mathbb{C}}/T^2) - 10·4$	$(f_{\mathbb{C}}/T^2) - 12·15$	−12·4	−15·8	−9·4	−12·1	−6·25	−7·9
−0·30		−0·30		−0·30		−0·30	
−0·22		−0·22		−0·22		−0·22	
76		76		76		76	
$(f_{\mathbb{C}}/T^2) - 8·9$	$(f_{\mathbb{C}}/T^2) - 10·7$	−11·4	−14·3	−7·9	−10·6	−4·7	−6·5
3·2	3·5	5·5	6·0	4·5	4·8	3·2	3·5
−5·3	−5·8	−9·4	−10·1	−7·4	−8·1	−5·3	−5·8
		2·8	3·0	2·2	2·4	1·6	1·7
		2·6	3·3	1·8	2·4	1·0	1·4

have remained constant over the 270 years of observations. Accordingly the discrepancy difference in fig. 11.8 has been fitted by a parabola; observations do not rule out small higher order terms in T. In § 11.9 and 11.11 geophysical arguments are presented for constant torques. The ancient eclipses, however, appear to require drastic changes in the torques during the past 2000 years (Murray, 1957). de Sitter's values for the discrepancies of Moon and Sun, as derived from ancient observations, yield (11.6.12)

$$L_{\mathbb{C}} = - \frac{r_{\mathbb{C}}^2 M_{\oplus} M_{\mathbb{C}}}{3(M_{\oplus} + M_{\mathbb{C}})} \left(5''22 - \frac{n_{\mathbb{C}}}{n_{\odot}} 1''8 \right) = 6 \cdot 6 \times 10^{23} \text{ dyn cm}$$

for the average lunar retarding torque over the past 2000 years; Fotheringham's (1920) values lead to

$$5 \cdot 3 \times 10^{23} \text{ dyn cm}$$

as compared to $3 \cdot 9 \times 10^{23}$ dyn cm (11.6.14) from the modern observations.

Curves corresponding to the modern observations and to de Sitter's interpretation of the ancient observations have been drawn in fig. 11.3, 11.6 and 11.7. de Sitter's values certainly do not fit the modern observations. The discussion in § 11.4 points out the uncertainties in the interpretation of ancient observations. The difference between modern observations and Fotheringham's values is no greater than that between de Sitter's and Fotheringham's interpretation of essentially the same data. In fact the point marked "Fotheringham 2" on fig. 11.6 is consistent with the modern value $L_{\mathbb{C}} = 3 \cdot 9 \times 10^{23}$ dyn cm and at the same time falls into the preferred triangle of discrepancies. The astronomic evidence suggests a change in tidal torque, but the evidence is by no means overwhelming.

Table 11.1 summarizes the interpretations of modern and ancient observations. The last columns (Fotheringham 2) are then an interpretation of the ancient observations subject to the restraint that the tidal torque has remained constant. This suggests a non-tidal increase of Ω by $1 \cdot 0$ or $1 \cdot 4$ parts in 10^7 during the last 2000 years, depending on whether $L_{\mathbb{C}}/L_{\odot}$ is $5 \cdot 1$ or $3 \cdot 4$.

The facts that need to be geophysically interpreted are (1) the dissipation of oceanic tidal energy by approximately 3×10^{19} ergs sec^{-1}, with a hint of larger values in ancient times; (2) an apparent increase in the relative angular velocity (a decrease in the l.o.d.) by 1 to 2 parts in 10^7 since ancient times by factors other than tidal friction; (3) the residual discrepancy curve (fig. 11.14, *e*) showing the non-tidal fluctuations since 1680.

7. Bodily tides

If the Earth and Moon were perfectly elastic, then the work done by the Moon on the Earth over a tidal cycle exactly equals the work done by the tides on the Moon. The time average of the work vanishes as does the average rate, dE/dt, at which mechanical energy is lost by the Earth–Moon system:

$$\left\langle \frac{dW}{dt} \right\rangle = 0, \quad \left\langle \frac{dE}{dt} \right\rangle = 0.$$

If the response of the solid Earth is anelastic, there is a flow of mechanical energy from the orbital motion of the Moon and from the rotational motion of the Earth into thermal energy within the Earth. The Moon does work on the Earth, $\langle dW/dt \rangle$ is positive, and mechanical energy is lost by the Earth–Moon system,

$$\langle dE/dt \rangle = - \langle dW/dt \rangle.$$

The rate at which a body force, F_i, derivable from a potential U, does work is

$$\frac{dW}{dt} = \int \rho u_i \frac{\partial U}{\partial x_i} \, dV = \int \rho u_i n_i U \, dS + \int \frac{\partial \rho}{\partial t} U \, dV. \quad (11.7.1)$$

The volume integral on the right side vanishes for a homogeneous incompressible Earth.

For a rough estimate of dissipation, we assume that the bodily tide follows an equilibrium theory except for a phase lag by 2ϕ (ϕ is called the 'luni-tidal interval'). Then approximately (table 7.1)

$$U_{\mathbb{C}} = \tfrac{1}{2}(1 + k)g K_{\mathbb{C}} b_{\mathbb{C}} \sin^2 \theta \cos 2\sigma_{\mathbb{C}} t,$$

$$\xi = \tfrac{1}{2} h K_{\mathbb{C}} b_{\mathbb{C}} \sin^2 \theta \cos 2(\sigma_{\mathbb{C}} t - \phi), \quad (11.7.2)$$

where $2\sigma_{\mathbb{C}} = 2(\Omega - n_{\mathbb{C}}) = 2\pi/12^{\text{h}}42$ is the frequency. The average rate of work per unit area done by the Moon on an homogeneous Earth is then

$$\langle \rho u_i n_i U \rangle = \langle \rho (d\xi/dt) U \rangle$$

$$= - \tfrac{1}{2}\rho g h(1 + k)(K_{\mathbb{C}}b_{\mathbb{C}})^2 \sigma_{\mathbb{C}} \sin^4 \theta \langle \sin 2(\sigma_{\mathbb{C}}t - \phi) \cos 2\sigma_{\mathbb{C}}t \rangle$$

$$= \tfrac{1}{4}\rho g h(1 + k)(K_{\mathbb{C}}b_{\mathbb{C}})^2 \sigma_{\mathbb{C}} \sin^4 \theta \sin 2\phi.$$

Upon integration over the Earth's surface we obtain

$$-\left\langle \frac{dE}{dt} \right\rangle = + \frac{8\pi}{15} \rho g h(1 + k)(K_{\mathbb{C}}b_{\mathbb{C}})^2 \sigma_{\mathbb{C}} a^2 \sin 2\phi$$

$$= 2 \cdot 39 \times 10^{20} \sin 2\phi \text{ ergs sec}^{-1} \qquad (11.7.3)$$

for the energy dissipated into heat; numerical values are $\rho = 2 \cdot 75$ g cm^{-3} for the surface density, $h = 0 \cdot 59$, $k = 0 \cdot 29$, $K_{\mathbb{C}} = 53 \cdot 7$ cm, $b_{\mathbb{C}} = 0 \cdot 908$, $\sigma_{\mathbb{C}} = 7 \cdot 02 \times 10^{-5}$ sec^{-1}. Evaluating $\int \rho u_i n_i U \, dS$ for all interior layers and allowing for compressibility might conceivably double the computed dissipation.

For large Q we have $\tan 2\phi \approx 1/Q$. Note the following numerical examples:

$$Q = \quad 9, \quad 2\phi = 3°2, \quad -\left\langle \frac{dE}{dt} \right\rangle = 2 \cdot 7 \quad \times 10^{19} \text{ ergs sec}^{-1}$$

$$40 \qquad\qquad 0°7, \qquad\qquad\qquad 0 \cdot 60 \times 10^{19}$$

$$100 \qquad\qquad 0°3, \qquad\qquad\qquad 0 \cdot 25 \times 10^{19}$$

The first case corresponds to the 'observed' rate of dissipation (11.6.15) due to lunar tides. The required value of Q is too small. From seismic and polar tide studies $Q = 100$–200, whereas the latitude analysis of the 14-month wobble gave $Q = 30$ to 40. The semi-diurnal tides are intermediate in frequency between the seismic and wobble frequency, and the Q derived from these sources should bracket the Q applicable to tides even if the dissipation mechanism is frequency dependent. The conclusion is that bodily tides can account from 10 per cent to 25 per cent of the observed dissipation (the evaluation of the second term in (11.7.1) might conceivably double these values). Had we used Jeffreys's estimate of $1 \cdot 4 \times 10^{19}$ ergs sec^{-1} for the dissipation, the corresponding values are 18 per cent to 43 per cent. A rate of dissipation of $0 \cdot 6 \times 10^{19}$ erg sec^{-1} corresponds to a steady heat flow of 4×10^{-8} cal cm^{-2} sec^{-1}. The observed flow is of the order 10^{-6} cal cm^{-2} sec^{-1}.

The value of energy loss obtained here is a factor of 100 greater

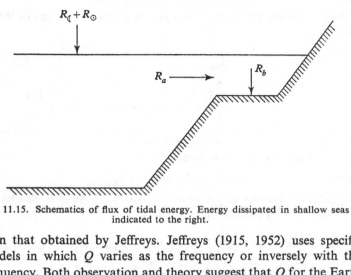

Fig. 11.15. Schematics of flux of tidal energy. Energy dissipated in shallow seas is indicated to the right.

than that obtained by Jeffreys. Jeffreys (1915, 1952) uses specific models in which Q varies as the frequency or inversely with the frequency. Both observation and theory suggest that Q for the Earth is remarkably frequency independent and that the damping of wobble, seismic damping and tidal dissipation are consistent with a Q varying by at most a factor of four in the frequency range 1 c/s to 3×10^{-8} c/s.

8. Ocean tides

Fig. 11.15 illustrates the flux of tidal energy in the oceans. $R_{\mathbb{C}}$ and $R_{\odot}$ are the mean rates, per unit surface area, at which the Moon's and Sun's gravitational attraction do work on the water. It is usually assumed that the dissipation within the volume of the ocean can be neglected. If this is so, then under steady conditions the incoming flux must equal the work done by tidal currents on the sea bottom (R_b ergs cm^{-2} sec^{-1}). But over nearly all the area of the oceans the bottom tidal currents are weak and the dissipation negligible, as will be shown. Only in a few shallow seas is the dissipation of consequence. The area of these shallow seas is small compared to the total ocean area, and the work done by Sun and Moon on these limited areas is correspondingly small. Nearly all of the energy dissipated in the shallow seas is convected into the regions by tidal currents. R_a designates the flux across a unit vertical surface comprising the entrance(s) into the shallow sea.

The energy dissipation can then be estimated by any one of three methods:

$$-\frac{dE}{dt} = \int_{\text{total sea surface}} (R_{\text{(}} + R_{\odot}) \, dS \qquad (11.8.1, a)$$

$$= \int_{\text{entrance(s) into shallow seas}} R_a \, dS \qquad (11.8.1, b)$$

$$= \int_{\text{area of shallow seas}} R_b \, dS \qquad (11.8.1, c)$$

All three methods have been used.

Let $\qquad \xi_{\text{(}} = A_{\text{(}} \cos 2(\sigma_{\text{(}}t - \phi_{\text{(}})$

designate the observed elevation of the M_2-tide above mean sea-level, in the notation of (11.7.2). Then

$$R_{\text{(}} = \left\langle \rho U_{\text{(}} \frac{d\xi_{\text{(}}}{dt} \right\rangle = \tfrac{1}{2}\rho g K_{\text{(}} b_{\text{(}} \sigma_{\text{(}} A_{\text{(}} \sin^2 \theta \sin 2\phi_{\text{(}} \quad (11.8.2)$$

$$R_{\odot} = \tfrac{1}{2}\rho g K_{\text{(}} b_{\odot} \sigma_{\odot} A_{\odot} \sin^2 \theta \sin 2\phi_{\odot},$$

where $2\pi/2\sigma_{\text{(}} = 2\pi/12^{\text{h}}42$, $b_{\text{(}} = 0.908$; $2\pi/2\sigma_{\odot} = 2\pi/12^{\text{h}}$, $b_{\odot} = 0.423$, corresponding to the M_2- and S_2-tides, respectively.

The rate of dissipation of energy by friction equals the frictional stress multiplied by the relative velocity of the surfaces between which the friction acts. Writing $\gamma\rho u^2$ for the stress on the bottom, the energy dissipation per unit area per unit time is

$$R_b = \gamma\rho \left\langle |u^3| \right\rangle = \frac{4}{3\pi} \gamma\rho u_0^3, \qquad (11.8.3)$$

where $u = u_0 \cos \sigma t$ and $(4/3\pi)$ is the average of $|\cos^3 \sigma t|$. The empirical formula for the stress, $\gamma\rho u^2$ with $\gamma = 0.002$, is widely applicable. It appears to apply to wind stress on ground and the stress of rivers on their bed. With regard to tidal currents, the most pertinent investigations are those by G. I. Taylor (1919) to which we shall make reference, and the measurements by Bowden and Fairbairn (1956) of turbulence associated with tidal currents. From simultaneous measurements of the horizontal and vertical components of turbulent velocity at a distance 75 cm above the sea bottom, they computed the Reynolds stress and found it to be consistent with a drag coefficient ranging from $\gamma = 0.0020$ to 0.0025.

Next we consider the flux of energy by tidal currents. Let $u(t)$,

$\Delta p(t)$ designate the tidal velocity and pressure departure, respectively. The flux of energy across a vertical face of unit area oriented normal to u is $\Delta p\, u$. For a *progressive* tide wave

$$\Delta p = \rho g \xi, \quad u = C\xi/h, \tag{11.8.4}$$

where h is depth, and C phase velocity. The energy flux then equals, on the average,

$$\langle \Delta p\, u \rangle\, h = \rho g \langle \xi^2 \rangle\, C = \tfrac{1}{2}\rho g A^2 C = EC, \tag{11.8.5}$$

where A is surface amplitude, and E the wave energy per unit area. For long waves, group and phase velocities are equal, and we have obtained the well-known result that energy flux equals energy times group velocity.

In the foregoing example of a progressive wave the maximum current (in the direction of wave motion) occurs beneath the crest. This is the situation for ocean swell traveling toward the beach. It would apply to tide waves traveling into a bay if all of their energy were absorbed. If all energy were reflected, then in the resulting *standing* wave pattern maximum inward current would occur at a time midway between low and high tide. In actual cases it is usually found that the maximum current occurs more nearly at mid-tide, and this corresponds more nearly to a standing wave than a progressive wave. To treat the case of partial reflexion we set

$$\Delta p = \rho g A \cos 2\sigma t, \quad u = (CA/h) \sin 2(\sigma t + \phi)$$
$$\langle \Delta p\, u \rangle\, h = \tfrac{1}{2}\rho g A^2 C \sin 2\phi = EC \sin 2\phi. \tag{11.8.6}$$

We may regard the situation as one where the incident wave carries EC ergs sec^{-1} cm^{-1} into the bay, the reflected waves $r^2 EC$ out of the bay. The net gain is $EC(1 - r^2)$, so that

$$\sin 2\phi = 1 - r^2. \tag{11.8.7}$$

For perfect reflexion $r = 1$ and $2\phi = 0$; for perfected absorption $r = 0$ and $2\phi = 90°$.

Let S designate the surface area of a bay or channel. Then

$$R_a = (1/S)(\tfrac{1}{2}\rho g C)\int A^2 \sin 2\phi \; \mathrm{d}x \tag{11.8.8}$$

designates the average dissipation, with the integral being across the entrance(s). In most instances the shallow water formula $C = \sqrt{gh}$ is an adequate approximation, and all that needs to be known is the tidal amplitude and the relative phase of height and current at the entrance(s). Because of friction and the effect of the Earth's rotation, the shallow water formula may not be adequate. In that case one requires magnitude and phase of both height and current.

It remains to be shown that the dissipation against the deep sea bottom is negligible, as has been assumed. The velocity C is of the order $\sigma a = 5 \times 10^4$ cm sec^{-1}, $A = 25$ cm, and $u = CA/h = 3$ cm sec^{-1} at most. Swallow (1955) has measured deep sea velocities by tracking neutrally buoyant floats, and finds tidal components of the order of 1 cm sec^{-1}. R_b is then of the order of 0·002 ergs cm^{-2} sec^{-1}, and $- dE/dt$ of the order 10^{16} ergs sec^{-1}, which is entirely negligible compared to the 'observed' dissipation, 3×10^{19} ergs sec^{-1}.

Continental shelves (ocean depths less than 200 m) occupy 5·5 per cent of the Earth's surface, or $2·8 \times 10^{17}$ cm^2. If all of the observed dissipation took place on the shelf, this would amount to 114 ergs cm^{-2} sec^{-1} on the average. When this is equated to 0·002 ρu^3, we obtain 38 cm sec^{-1} for the required mean-cubed current over the shelf, or 3/4 of a knot. Peak current during one cycle is $3\pi/4$ times this amount. The required currents are of the order of magnitude of the observed tidal velocities, though they seem high. In any event the dissipation in shallow water deserves careful consideration.

In a remarkable paper Taylor (1919) discussed tidal dissipation in the Irish Sea. He obtained $6·4 \times 10^{17}$ and $5·1 \times 10^{17}$ ergs sec^{-1} using the flux and bottom friction method, respectively. The work done by the Moon on the Irish Sea is $0·4 \times 10^{17}$ ergs sec^{-1} and thus relatively small, as expected. Prior to Taylor's work, Street (1917) had estimated the dissipation assuming a laminar boundary layer, and had obtained less than (1/150)th of the foregoing values. The agreement between the flux and friction methods established the validity of Taylor's procedure, and the numerical value (approximately 2 per cent of the total tidal dissipation) suggests the important role played by ocean tides in the secular acceleration of the Moon.

Within one year of the publication of Taylor's estimate of tidal dissipation in the Irish Sea, Jeffreys (1920) and Heiskanen (1921) had extended the estimates to all oceans. Jeffreys obtained $1·1 \times 10^{19}$

ergs sec^{-1}, 80 per cent of what he required from his interpretation of the astronomical observations (only 34 per cent according to our estimate); Heiskanen's estimate (as revised by Lambert, 1928) based on about the same data is 1.9×10^{19} ergs sec^{-1}. Both of these values depend chiefly on the method (11.8.1c) involving bottom friction in shallow seas. Heiskanen also evaluated the work done by the Moon and Sun (11.8.1, a) and obtained 1.0×10^{19} ergs sec^{-1} (as revised by Lambert). Because of a misunderstanding Heiskanen added his two results, whereas they are, in fact, equivalent estimates of the same process, as pointed out by Lambert. In all events, there is sufficient accord between the geophysical and astronomical estimates that the problem has been considered as solved for 40 years. We wish to reopen it.

In table 11.2 we have summarized the results of Jeffreys and Heiskanen. Published tidal velocities usually refer to maximum currents during spring tide. There is then a need for two reductions. We must multiply by $(4/3\pi)$ to obtain a mean dissipation during the tidal cycle, in accordance with (11.8.3). This reduction has been incorporated in the table. But during spring tides the tidal currents are at a maximum, and this requires a further reduction. Jeffreys (1952, p. 230) derives a factor 0.51, and this has been applied to the totals. Heiskanen overlooked the need for this further correction, and accordingly his published values are about twice too large. One-third to one-half of Heiskanen's total dissipation takes place along open coast lines of continents, whereas Jeffreys ignores this possibility. We believe Heiskanen's estimates to be generally too large. For example, from the mouth of the Gulf of California to Vancouver Island he assumes a tide current of 1.5 knots. Measurements of *total* current on the shelf rarely exceed 0.5 knots (Shepard, Revelle and Dietz, 1939). Systematic measurements off Los Angeles (Stevenson, Tibby and Gorsline, 1956; San Diego Geologic Diving Consultants, 1956) gave maximum tidal velocities of 0.1 to 0.25 knots during spring tides in 100 ft of water. For depths up to 150 ft the most reliable source of information is the experience of divers, and on the basis of their many reports, tidal velocities in excess of 0.5 knots can be ruled out as far as the open California coast line is concerned.

Table 11.2. Energy dissipation in units of 10^{19} ergs sec^{-1}.

	Jeffreys	Heiskanen
European Waters		
Irish Sea	0·06	0·04
English Channel	0·11	0·23
North Sea	0·07	
Other seas		0·16
	0·24	0·43
Asiatic Waters		
Yellow Sea	0·11	
Malacca Strait	0·11	0·18
Other seas	0·01	0·73
	0·23	0·91
North American Waters		
Northwest Passage	0·16	
Bay of Fundy*	0·04	0·05
Other seas		0·30
	0·20	0·35
South American Waters		0·40
Australian Waters		0·34
African Waters		0·08
Arctic Waters		0·13
Bering Sea	1·50	0·96
Total, Spring Tide	2·17	3·60
Times factor 0·51	1·1	1·9

* The largest known tides are in the Bay of Fundy, but the dissipation is relatively small. A new calculation by McLellan (1958) gives $0·027 \times 10^{19}$ ergs sec^{-1} by the flux method, 0·029 by the bottom friction method (using $0·002 \, \rho u^3$), as compared to Jeffreys's 0·04 and Heiskanen's 0·05.

Patagonia.—Along the coast of Patagonia a shelf roughly 500 km wide extends for nearly 2000 km. Tidal amplitudes decrease from 12 ft at 50° S to 1 ft at 37° S. The formula (11.8.4) suggests maximum velocities somewhat less than 10^2 cm sec^{-1} over the shelf. The result of current measurements taken by the Hydrographic Service of Argentina at nine offshore stations has been kindly placed at our disposal by Luis Capurro, Capitan de Fragata. Maximum values vary from $2\frac{1}{2}$ knots in the Falkland passage to 1 knot off Rio de la Plata, and this confirms our estimate of 10^2 cm sec^{-1}. The resultant dissipation is roughly $0·2 \times 10^{19}$ ergs sec^{-1} after allowing for the reduction factors. This is in accord with Heiskanen's estimate for this region. We can make another estimate by the flux method.

Alfred Redfield (personal communication) suggests as a possible interpretation of the tidal observations that a tide wave enters from the Antarctic Ocean through the passage between Falkland and Staten Islands and is attenuated in its northward travel over the shelf. Set $A = 1 \cdot 5$ m, $h = 50$ m, and an entrance of 500 km. The inflow of energy equals

$$\tfrac{1}{2}\rho g A^2 \sqrt{gh} \times 500 \text{ km} \approx 10^{18} \text{ ergs sec}^{-1}$$

and is of the same order of magnitude as the previous estimate.

Bering Sea.—In the eastern Bering Sea there is a large shoal area, measuring roughly 1000 km by 1000 km, with depths generally less than 60 m. Both Jeffreys and Heiskanen are in agreement with regard to the overriding importance of this shelf as a sink of tidal energy. Three-fourths of the global dissipation takes place here according to Jeffreys. The theory then rests on the selection of a representative value for u_0 on the shelf. According to Jeffreys '. . . it is stated that the maximum rate of water, where clear of the passes between the Aleutian Islands, is usually about $2\tfrac{1}{2}$ knots when the depth is less than 100 fathoms'. Both authors take into account reported velocities of $2\tfrac{1}{2}$ knots at St. Matthew Island and the Pribilof Islands. Jeffreys sets u_0 at $2\tfrac{1}{2}$ knots, Heiskanen at 2 knots.

The Coast and Geodetic Survey *Coast Pilot* for Alaska and the *Current Tables, Pacific Ocean*, though they are not clear on this point, do not call for a vast system of tidal currents over the shelf. The reported velocities refer to points near land, a feature that both Jeffreys and Heiskanen attempted to take into account. Forty years have passed, and still no adequate observations of off-shore tidal currents have been published. The drift of the ice-breaker *Northwind* (U.S. Hydrographic Office, 1958) does not indicate strong tidal currents over the northern part of the shelf, east of St. Lawrence Island. In the Bering Strait itself a steady northward current of about 1 knot has been observed, but the superposed tidal oscillation is less than $\tfrac{1}{4}$ knot. To the south there have been numerous oceanographic anchor stations, but no measurements of current.* A ship at anchor in a two-knot current takes a noticeable

* In July 1960 William Revelle made two measurements from a ship anchored in the eastern Bering Sea, and obtained $\cdot 9$ and $\cdot 1$ knots, respectively.

strain on the anchor line. There is general agreement among expedition members that tidal currents in excess of 1 knot can be ruled out on the basis that no such strain was evident. We have been permitted to report that on the submerged poleward journey of the *Nautilus* during August 1958, any two-knot tidal current over the Bering shelf would have led to obvious discrepancies between the steered course and the course made good, and no such discrepancies were observed. We believe that 1 knot might be an upper limit for u_0; because of its dependence on u_0^3 the dissipation is then reduced by an order of magnitude.

Of course, the fact that anchor lines were not taut and the *Nautilus* not set off its course hardly constitutes the type of evidence that is considered desirable. But there are considerations involving the advection method (11.8.1, *b*) which further indicate that the dissipation is less by an order of magnitude than the values in table 11.2. From tide tables it is found that high water and maximum northward current occur at about the same time along the portion of the Aleutian chain which forms the southernmost boundary of the shelf. This is consistent with the case of a progressive tide wave traveling northward onto the shelf where it is eventually absorbed. Semidiurnal amplitudes are of the order of 25 cm at the Aleutians and decrease northward (except for Bristol Bay). At the Strait the amplitudes are of the order of 10 cm. The possibility of a large standing wave with nodes at the southern boundary of the shelf is ruled out. Setting $C = \sqrt{gh}$, $h = 60$ m, $A = 25$ cm, gives (11.8.8)

$$-\frac{dE}{dt} = R_aS = \tfrac{1}{2}\rho g C \int_0^{1000 \text{ km}} A^2 \, dx = 0.8 \times 10^{17} \text{ ergs sec}^{-1},$$

or 1 per cent of the desired amount. In this calculation the application of the shallow water formula for C implies an orbital velocity of $u_0 = CA/h = 20$ cm sec^{-1} = 0.4 knots according to (11.8.3). This is, of course, much lower than the value chosen by Jeffreys and Heiskanen. But even if we assume that all of the Bering Sea rises synchronously, the required inflow (neglecting island obstructions) can hardly exceed

$$u_0 = \sigma_{\mathbb{C}} A L^2 / hL,$$

where L^2 is the surface area and h the depth just north of the Aleutians. If we take $L = 1000$ km, this gives $u_0 = 60$ cm sec^{-1} and $- dE/dt = 2.4 \times 10^{17}$ ergs sec^{-1}.

Torque method.—The work done by the Moon and Sun furnishes an independent method for estimating $- dE/dt$. Amplitudes $A(\lambda, \theta)$ and phase $\phi(\lambda, \theta)$ can be read of co-tidal charts,* $R(\lambda, \theta)$ evaluated according to (11.8.2), and the dissipation obtained upon integration over all oceans (11.8.1). The difficulty with this method is illustrated by fig. 11.17: a large variability in phase associated with amphidromic points. Positive and negative contributions to $- dE/dt$ are not far different, and the sum is most uncertain. Furthermore phase and amplitude over wide expanses of the open sea are only vaguely known. All this is in marked contrast with the atmospheric tide (fig. 11.17) where the phase lies within the limits $- 60°$ to $- 75°$ over a large part of the Earth.

A global calculation was undertaken by Heiskanen based on the co-tidal charts published by Sterneck in 1920. Groves and Munk (1959) have made a new calculation based on Dietrich's (1944) compilation from approximately three times the number of tide stations available to Sterneck. The improvement is rather less than these numbers suggest since there is no proportional increase of stations in the southern oceans where they are most needed. Island stations in the southern hemisphere established during the IGY should give worthwhile information.

The result of the calculations is summarized in table 11.3. Positive and negative values are itemized to indicate the degree of uncertainty; we estimate that the totals may be far smaller than we have obtained, or twice as large. It is, perhaps, encouraging that the totals are positive for each ocean (i.e., energy is being dissipated). For the lunar terms, $- dE/dt$ adds up to 3.2×10^{19} ergs sec^{-1}, as compared to the astronomically determined value 2.7×10^{19} ergs sec^{-1} (table 11.1);

* Actually co-tidal charts show lines of equal phase, $2\alpha = 0, 30°, \ldots, 330°$ relative to the Moon's transit over the Greenwich meridian. Thus the line $2\alpha = 30°$ connects all points having high tide

$$\frac{30°}{360°} 12^h42 = 1.035 \text{ solar hours} = 1 \text{ lunar hour}$$

after the Moon's Greenwich transit. At east longitude λ the Moon transits $2\lambda°$ (or $2\lambda°/30°$ lunar hours) *earlier* than at Greenwich, and the phase lag relative to the Moon's local transit is

$$2\psi_{\mathcal{C}}(\lambda, \theta) = 2\alpha(\lambda, \theta) - 2(-\lambda) = 2(\alpha + \lambda).$$

Similar remarks apply to the S_2-tide.

Table 11.3. Work done by the tidal forces of Moon and Sun on the oceans, in units of 10^{19} ergs sec^{-1}.

	Heiskanen's Lunar Semid.	Lunar			Solar			Total
		Semid.	Diurnal	Total	Semid.	Diurnal	Total	
Pacific								
+	3·8	2·8	0·6	3·4	0·7	0·2	0·9	4·3
−	− 2·5	− 1·9	− 0·0	− 1·9	− 0·5	− 0·0	− 0·5	− 2·4
total	1·3	0·9	0·6	1·5	0·2	0·2	0·4	1·9
Atlantic								
+	1·8	2·1	0·1	2·2	0·5	0·0	0·5	2·7
−	− 1·2	− 1·1	− 0·1	− 1·2	− 0·2	− 0·0	− 0·2	− 1·4
total	0·6	1·0	0·0	1·0	0·3	0·0	0·3	1·3
Indian								
+	1·8	1·5	0·2	1·7	0·4	0·1	0·5	2·2
−	− 1·6	− 0·9	− 0·1	− 1·0	− 0·2	− 0·0	− 0·2	− 1·2
total	0·2	0·6	0·1	0·7	0·2	0·1	0·3	1·0
Total	2·1	2·5	0·7	3·2	0·7	0·3	1·0	4·2

for the solar terms the corresponding values are 1·0 and 0·6 × 10^{19} ergs sec^{-1}. The agreement is better than we had any right to expect.

The Q of the oceans.—The rate of energy dissipation, as indicated by the astronomical evidence, enters as an important factor in problems concerning theory and prediction of ocean tides. Aside from local resonances, the total energy contained in the ocean tides at any moment cannot differ from that given by equilibrium considerations by more than a factor Q^{-1}. The equilibrium energy of the lunar tide would be

$$\tfrac{1}{2}\rho g \int (U_{\mathbb{C}}/g)^2 \, dS = \rho g K_{\mathbb{C}}^2 b_{\mathbb{C}}^2 \tfrac{1}{2} 2\pi a^2 \int_0^\pi (\tfrac{1}{2}\sin^2 \theta)^2 \sin \theta \, d\theta$$

$$= \frac{4\pi}{30} \rho g a^2 K_{\mathbb{C}}^2 b_{\mathbb{C}}^2 = 4 \times 10^{23} \text{ ergs}$$

if the Earth were covered by oceans. For the actual oceans we take 2·5 × 10^{23} ergs. From astronomical data we then infer

$$\frac{2\pi}{Q} = 2\pi \frac{dE/dt}{2\sigma_{\mathbb{C}} E} = 2\pi \frac{2\cdot7 \times 10^{19}}{(1\cdot4 \times 10^{-4})(2\cdot5 \times 10^{23})} = \frac{2\pi}{1.3} = 4\cdot8$$

for the relative dissipation per cycle. Once every 5 hours all of the tidal energy is dissipated! Taking Jeffreys's estimate of 0·51 × 1·5 × 10^{19} = 0·77 × 10^{19} ergs sec^{-1} for the dissipation on the Bering

Sea shelf, then once every 9 hours all of the global tidal energy finds its way into the Bering Sea. The rate of energy dissipation is remarkable, but the concentration of so large an amount in one area would be even more remarkable. Jeffreys himself has been concerned about this and in a recent review (Jeffreys, 1958b) he wrote that '. . . it was always difficult to see how so much energy got into the shallow seas at all'.*

If the ocean tide approximated an equilibrium configuration (which it does not, see fig. 11.17) then the maximum dissipation could be found by setting $2\phi = 90°$ in (11.8.2); the result is $11·2 \times 10^{19}$ ergs sec^{-1}, and this exceeds the amount $2·7 \times 10^{19}$ ergs sec^{-1} required by astronomical considerations. An equilibrium amplitude and a phase lag $2\phi = 14°$ would yield the required dissipation. On this basis the ocean dissipation cannot be ruled out.

Independent evidence concerning the high rate of tidal dissipation in the oceans is desirable. A related initial value problem is afforded by tsunamis (tidal waves). A large tsunami, such as the one following the Kamtschatka Earthquake of 4 November 1952, releases energy into the Pacific of the same order as that associated at any moment with Pacific tides. Tsunamis are waves long compared to the ocean depth, and the bottom currents associated with them should be comparable with tidal currents provided the energies are commensurate. It turns out that most of the tsunami energy is dissipated in one day, although the activity can remain above background for one week. This evidence, as far as it goes, does not conflict with the indicated rate of tidal dissipation.

Discussion.—The astronomic observations call for a tidal dissipation of 3×10^{19} ergs sec^{-1}, and the values in table 11.3 indicate no great difficulty in getting this much energy into the oceans. Concerning the location and mechanism of the dissipation we are left with a dilemma. If our estimates concerning the Bering Sea are correct, then the dissipation in shallow seas would amount at most to 10^{19} ergs sec^{-1}. There is always the possibility that we have overlooked regions of dissipation. But any region of concentrated dissipation is unlikely for the reasons discussed above. What is required is a process leading to a more evenly distributed dissipation. The only possibility that

* See also *The Observatory*, 78: 93–95, 1958.

has occurred to us is that the bulk of the dissipation is associated with internal (or baroclinic) tides. Serial observations of temperature in the oceans almost invariably show that the depth of a given isotherm fluctuates with tidal frequencies. Typical amplitudes for M_2 are of the order of 10 m. This is from one to two orders of magnitude larger than equilibrium values. On the other hand the density contrast is of the order 10^{-3} times that at the surface, and the work done directly by the Moon and Sun on internal tides is correspondingly small.

There have been no satisfactory theories of how internal tides of such magnitude are generated. The phase velocity of free internal waves, neglecting the Earth's rotation, is roughly 20 km per hour in the open sea, as compared to a velocity of 1500 km per hour for the equilibrium tidal bulge. Defant (1950) has suggested that this 'mismatch' can be removed by allowing for the effect of rotation on the phase velocity and that the resulting coupling could lead to a large internal response to the tide-generating forces. Recent observations do not bear out this suggestion. Simultaneous measurements at two points separated by 100 km (Reid 1956) show no obvious phase relations; if the internal tides were due to the tide-generating forces, then the phases at the two stations should have been virtually identical.

Charles Cox (personal communication) has suggested that in regions of variable depth the internal (baroclinic) and external (barotropic) modes are not independent and that a flux of energy must take place from each mode to all others. More specifically, the degree of coupling depends on the extent to which a spectrum of the sea bottom topography contains power at the local wave length of the tides. The results obtained by Reid (1956) are consistent with the hypothesis that internal tides are generated all along the coast line, with the degree of conversion depending on local bottom topography. Cox estimates that the conversion of energy from surface to internal modes may amount to 5 ergs cm^{-2} sec^{-1} for the North Atlantic deep waters. There may be a sizeable conversion back into surface modes so that 5 ergs cm^{-2} sec^{-1} represent some sort of upper limit. At this rate the global conversion amounts to $1 \cdot 5 \times 10^{19}$ ergs sec^{-1}. Anyway the order of magnitude here presented does not rule out the possibility that the energy of the surface tides is effectively converted into internal wave motion and then dissipated within the ocean volume. Our reference to the dissipation of tsunamis is then not pertinent as it stands.

Finally, we wish to consider the variation of tidal dissipation over the past. de Sitter and Murray have suggested that in Egyptian times the dissipation had been twice what it is now, though the evidence on this point is not compelling, as we have shown. Inasmuch as we have no understanding of the dissipation today a discussion of past dissipation can hardly be definite. Suppose the dissipation is dependent, by one mechanism or another, on the dimensions and extent of the continental shelf. An examination of the frequency distribution of elevations (Meinardus, 1942) reveals a maximum for elevations within a few hundred meters of the present sea-level, so that the total area of dissipation is relatively sensitive to changes in sea-level. Of the Earth's total surface area, about 7 per cent lie between 0 and 200 m above sea-level, 5 per cent within the corresponding interval beneath sea-level, 1 per cent between the depths of 200 m to 400 m. A rise in sea-level by 200 m would lead to an increase by 40 per cent of areas covered by less than 200 m of water, assuming, of course, that the topography remains unchanged. A drop by 200 m would reduce this area by 5:1. The change per meter is then of the order of 1 per cent. Since Egyptian times the sea-level may have varied by a few meters. It is difficult to see how such a change could lead to a substantial alteration in the dissipation.

9. Atmospheric tides

The variation of atmospheric pressure in the tropics is remarkable for its regularity. Aside from an occasional tropical storm, a barometer records a monotonous semi-diurnal variation. The maxima occur at approximately 10 a.m. and 10 p.m. local time and the minima at 4 a.m. and 4 p.m. (see fig. 11.16). The amplitude near the equator is 1·2 millibars. A similar variation but of smaller amplitude is observed at middle latitudes during periods of settled weather. The passage of fronts occasions much greater variations in atmospheric pressure, but the semi-diurnal oscillation can still be detected through analysis of long runs of observations. At high latitudes the

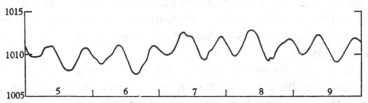

Fig. 11.16. Five days of barometric record (millibars) at a tropical station (Batavia).

situation is different. Maxima and minima occur at a fixed Greenwich time, rather than at 10 and 4 local time.

An explanation was offered by Schmidt in 1890: the total oscillation is made up of two principal components, a p_2^2 traveling wave which follows the Sun around the Earth and corresponds to the S_2-tide, and a p_2^0 standing wave. Detailed analyses (Simpson 1918, Haurwitz 1956) have confirmed Schmidt's suggestion. In the tropics the p_2^2 traveling wave is strikingly exhibited by the constancy in phase (fig. 11.17). The p_2^0 standing wave is apparent from the variable phase at latitudes greater than 60°.

The standing wave has no cumulative effect on the Earth's rate of rotation, but the traveling wave gives rise to a torque in the same way as bodily tides and ocean tides. We shall need phase and amplitude. According to Simpson (1918) and Haurwitz (1956) the S_2-tide can be represented by the formulae

$$\Delta p = 1\cdot25 \sin^3 \theta \cos 2(\sigma_\odot t + 32°)\,\text{mb}$$
$$\Delta p = 1\cdot16 \sin^3 \theta \cos 2(\sigma_\odot t + 34°)\,\text{mb} \qquad (11.9.1)$$

respectively, where t is the time after the (mean) Sun has passed over the local meridian (12 noon local), and $\sigma_\odot = 2\pi/12^\text{h}$. Haurwitz's values are based on data from 269 stations. The numerical constants are well established.

The remarkable feature is that the atmospheric S_2-tide exhibits a phase *lead*, with maximum pressure at any point on the Earth occurring two hours in advance of the Sun or anti-Sun. Maximum pressure corresponds to an excess in mass, and accordingly (Kelvin, 1882) the Sun exerts an *accelerating* torque on the Earth's atmosphere (in contrast to the situation in fig. 11.13). The rate at which this

Fig. 11.17. *Top*: The atmospheric semi-diurnal oscillation S_2 (after Haurwitz, 1956). Solid lines: phase lag 2φ relative to local transit of Sun ($2\varphi = 0°$ designates maximum pressure at 12 a.m. and 12 p.m. local time, $2\varphi = -30°$ at 11 a.m. and 11 p.m., etc.). Dashed lines: amplitude in millibars. *Bottom*: The oceanic semi-diurnal oscillation M_2. $2\varphi = 30°$ designates maximum tide one lunar hour ($1\cdot035$ solar hours) following the transit of Moon and Anti-Moon across the local (not Greenwich) meridian. Adapted from Dietrich (1944).

torque increases the mechanical energy of the Earth-Sun system is obtained from (11.8.1), setting $\rho g A = \Delta p$, and using (11.9.1):

$$\frac{dE}{dt} = + 2 \cdot 2 \times 10^{18} \text{ ergs sec}^{-1}. \tag{11.9.2}$$

The rate is fixed by observation and in no way depends on the theory of the atmospheric tide.

Holmberg (1952) has presented an intriguing theory concerning the present value of the l.o.d., based on the supposition that on the average the accelerating atmospheric torque is balanced by the decelerating oceanic torque. The theory has received considerable attention (for example Hoyle, 1955). Central to Holmberg's theory is the concept of a sharp atmospheric resonance at 12 hours. There is now considerable doubt concerning the existence of this resonance peak. A discussion of this point requires an excursion into the jungle of theories relating to atmospheric oscillations.

The pressure departure at a fixed level due to an equilibrium gravitational tide is $\Delta p = \rho U$, where ρ is the density of air. The computed and observed pressure amplitudes at the equator and the observed phase lag are as follows:

	period	equilibrium	observed	phase lag
S_2	12^h00	0·013 mb	1·16 mb	− 68°
M_2	12^h42	0·028 mb	0·08 mb	+ 18°

The observed lunar semi-diurnal tide has about three times the equilibrium amplitude, but only 7 per cent of the observed amplitude of the solar semi-diurnal tide.* To account for this discrepancy a sharp resonance peak very near the semi-diurnal frequency has been proposed. This is illustrated by curve a, fig. 11.18. The response curve for a linear damped oscillator has been fitted to the appropriate magnifications at 12^h00 and 12^h42. The resultant curve has a peak amplification of 100 ($Q = 100$) at a period rather poorly determined by the data but lying within a few minutes of 12 hours.

There are a number of difficulties with the resonance theory:

* The lunar atmospheric tide exhibits a phase lag and thus is responsible for a decelerating torque on the Earth. This torque is about 1/20 of the accelerating solar torque. The phase of M_2 shows a puzzling annual variation. In both hemispheres the phase is retarded by about one hour in December as compared with June. These are the months of maximum and minimum angular momentum of the atmosphere, as exhibited by the annual variation in the l.o.d. (§ 9.6).

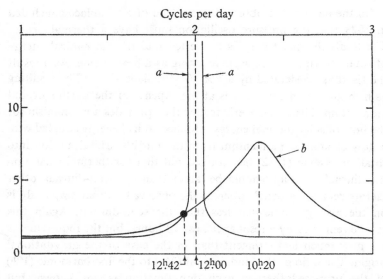

Fig. 11.18. Response of the atmosphere to external excitation. a is the required response if the atmospheric S_2-tide is of gravitational origin; peak amplification is 100:1. b is derived from observed magnification and phase of M_2-tide.

(1) G. I. Taylor (1929) has shown that the free period of the atmosphere can be obtained from the phase velocity of long air waves. The Krakatoa eruption of 1883 and the Great Siberian Meteor of 1908 gave velocities consistent with a free period of about $10\frac{1}{2}$ hours. The free period very near 12 hours required by the magnification of the S_2-tide is not in accord with these observations. However, Pekeris (1937, 1939) has suggested that the atmosphere has two resonance periods, one at $10\frac{1}{2}$ hours and one at 12 hours (corresponding to zero and one vertical nodal planes, respectively), and he found some evidence (though marginal) of a second arrival on the Krakatoa records which would be consistent with the 12-hour peak. But numerous Japanese reports of pressure waves from atomic blasts do not support this evidence for a second arrival.

(2) The gravitationally excited oscillations are expected to have a phase lag (at least for the barotropic mode), not a phase lead as observed for S_2.

(3) M_2 does in fact show a phase lag, as expected, but the lag is inconsistent with a linear oscillator of such high Q.

On the basis of the observed phase lead of S_2, Laplace concluded in 1823 that the pressure oscillation must have a thermal origin. The Sun's thermal energy is converted into the mechanical energy of rotation with the atmosphere acting as a heat engine. As a result the Earth is accelerated by $3 \cdot 7 \times 10^{-22}$ radians sec^{-2}. The resulting gain in angular momentum is at the expense of the Earth's orbital momentum. The atmospheric torque then provides a mechanism for the flow of solar thermal energy into mechanical energy coupled with a flow of angular momentum from the Earth's orbital motion into rotation. Kelvin (1882) has pointed out that for thermal excitation the diurnal variation should be larger than the semi-diurnal, other factors remaining equal, whereas the observed diurnal amplitude is on the average somewhat less than the semi-diurnal. Again this can be remedied by a resonance near 12 hours, but the requirements are now much less stringent than in the case of the gravitational origin. The ratio of the fundamental (24^h) to the first harmonic (12^h) of the proposed thermal excitation function is not known, but ratios of the order 3:1 to 10:1 seem reasonable. The required amplification at 12^h relative to that at 24^h should be within these limits.

The resonance curve for a linear damped oscillator having the observed magnification *and phase* of the M_2-tide is shown by Curve *b*, fig. 11.18. The difficulty (3) is then removed by hypothesis. The curve is peaked at 10^{h}2, not far from Taylor's value. The phase and amplitude of S_2 are now referred to an unknown excitation function and provide no immediate difficulty. The magnification at 12 hours is by a factor 3·3, and this implies that thermal excitation must exceed gravitational excitation by 30:1. Siebert (1954) claims this to be the case and finds no difficulty in reconciling amplitude and phase of S_2 with processes of radiative excitation.

We now return to Holmberg's hypothesis. He suggests that the rotation of the Earth was progressively slowed by oceanic friction until half the l.o.d. approached the period of atmospheric resonance. A further retardation was accompanied by an increase in the S_2-tide and a corresponding increase in the accelerating couple associated with this tide until a balance was attained between the two couples. In this way the close coincidence of half the l.o.d. with the atmospheric resonance peak near 12 hours (if there is one) is explained. There is no need for the couples to be in exact balance at all times. At the present time, for example, the l.o.d. is somewhat too long for balance, and the oceanic decelerating couple exceeds the atmospheric accelerating couple by 10:1. At other times in the geologic past the reverse situation applies. According to this statistical interpretation the present amplification of the atmospheric S_2-tide by a factor 100 is untypically small; on the average it must be by a factor 1000, and sometimes by perhaps 10,000.

Holmberg himself does not take a position as to whether the S_2-tide is of gravitational or thermal origin. But in the case of thermal excitation the hypothesis loses much of its appeal. In order for a statistical balance to be achieved between the oceanic and atmospheric tidal couples, the former must be less or the latter larger, on the average, than their present values: this is to accomplished not by climbing up a sharp resonance curve, but by varying oceanic tides or thermal excitation. The resonance peak is broad and well removed from half the present l.o.d. A sharp atmospheric resonance within 2 min of half the l.o.d. requires explanation; a resonance within 2 hours of half the l.o.d. requires no more explanation than the resonance of wobble within 2 months of the length of year.

10. Interplanetary torque

It may turn out that some of the irregularities in the Earth's rotation have their source in the electromagnetic interaction of the Earth with interplanetary plasma. The study of this subject has just begun, and it is too early to arrive at any definite conclusions.

Cosmic rays interact with the magnetic field at great distances from the Earth and can be used to explore the external magnetic field. Simpson, Fenton, Katzman and Rose (1956) find that the geomagnetic equator effective for cosmic rays is simulated by a dipole having the same tilt as the dipole obtained from surface magnetic measurements but lagging behind it (i.e. displaced westward) by 40° to 45°. The dipole can be resolved into a component dipole coincident with the rotation axis plus a dipole fixed in the equatorial plane and rotating with angular velocity Ω. In terms of this description the external 'equatorial dipole' (derived from cosmic ray measurements) lags 45° behind the internal equatorial dipole (derived from surface magnetic measurements).

Beiser (1958) accounts for the lag of the external equatorial dipole in terms of an additional induced dipole resulting from the rotation of the Earth. Let M_i denote the magnetic moment of the internal equatorial dipole. Since this dipole is rotating it will give rise to an electric field in the interplanetary matter. If the matter is conducting, current loops will be set up in planes containing the equatorial dipole and the rotation-axis. These currents induce a magnetic field in the interplanetary fluid equivalent to a dipole of magnitude

$$M_{ind} = \tfrac{2}{3}\pi\kappa\Omega M(r^2 - a^2), \qquad (11.10.1)$$

where κ is the conductivity of the interplanetary matter, and r the distance from the Earth beyond which stray fields dominate over M. The induced dipole lies in the same plane as the internal dipole but is displaced 90° to the west. If we set

$$M = M_{ind}, \qquad (11.10.2)$$

then the total external equatorial dipole, being the vector sum of the internal and induced equatorial dipoles, will be 45° west of the internal dipole, as observed.

An equivalent interpretation of the westward displacement of the external magnetic field is also due to Beiser. The conducting interplanetary fluid exerts a drag on the lines of force of the internal rotating dipole. The work done against the drag comes from the energy of rotation. Angular momentum is lost by the Earth and gained by the interplanetary fluid. A rough estimate can be made of the rate at which the Earth is losing mechanical energy. The energy in the induced field equals

$$E = \frac{1}{8\pi}\int H_{ind}^2 \, dV = \frac{M_{ind}^2}{3a^2}. \qquad (11.10.3)$$

We have found $M_{ind} \approx M = 1\cdot59 \times 10^{25}$ g cm^3, the equatorial component of the geomagnetic dipole. Hence $E = 3\cdot26 \times 10^{23}$ ergs. The time constant for the free decay of a field in a body of linear dimension L is (Cowling, 1957)

$$4\pi\kappa L^2. \qquad (11.10.4)$$

From (11.10.1) and (11.10.2) we have

$$\frac{1}{\Omega} = \tfrac{2}{3}\,\pi\kappa(r^2 - a^2) \approx \tfrac{2}{3}\,\pi\kappa L^2 \qquad (11.10.5)$$

so that the energy decays in the order of a day:

$$-\frac{\mathrm{d}E}{\mathrm{d}t} \approx 2 \times 10^{19}\ \text{ergs sec}^{-1}, \quad \dot\Omega|_M = -\ 3{\cdot}3 \times 10^{-22}\ \text{rad sec}^{-2}$$
$$(11.10.6)$$

This corresponds to a torque of $L_M = 2{\cdot}7 \times 10^{23}$ dyne cm.

The calculation of the magnetic torque involves a number of uncertain points. The westward displacement of the external dipole depends to some extent on the initial geomagnetic coordinates used in the reduction of the data. In particular Beiser considers only the centered dipole components of the magnetic field. The total field and not just the dipole may account for the displacement of the cosmic ray magnetic equator. Storey *et al.* (1958) find difficulties with the dipole hypothesis in accounting for cosmic ray measurements in Australia. Quenby and Webber (1959) and Rothwell (1958) have questioned whether a distorted extra-terrestrial field is even required to account for the cosmic ray observations. There is also some question whether Beiser's step-wise analysis is an approximation to the appropriate hydromagnetic boundary value problem. Maeda (1958) has treated the problem starting with the equations for a fully ionized plasma. To a first approximation the field propagates into the surrounding plasma with the velocity of Alfvén waves. Maeda assumes that the plasma rotates with the Earth to a distance of ten Earth's radii and is motionless beyond. The phase lag of the external dipole can be interpreted in terms of this model. Maeda obtains 2×10^{16} ergs sec^{-1} for the loss of rotational energy by Alfvén radiation, 10^{-3} times Beiser's value.

Beiser's result (11.10.6) is of the same order as that obtained from ancient eclipses (see table 11.1) and he suggests that magnetic torque may play a part in the observed secular deceleration. But the slowing down of the Earth due to this torque would be accompanied by the speeding up of the interplanetary fluid, not of the orbital motion of the Moon. The effects of the magnetic drag should be compared with astronomical data only after removal of all solar and lunar effects implicit in the data. When the solar and lunar effects are removed, the angular velocity of the Earth appears to be increasing rather than decreasing (see table 11.1, fig. 11.19). There is then no astronomical evidence for a magnetic torque of the magnitude postulated, and if the torque exists, it must be overbalanced by other processes.

Interplanetary torques may play a part in the non-secular variation in the l.o.d. Fluctuations in the radiation of the Sun might be accompanied by variations in the conductivity κ and accordingly in torque. The time constants involved are quite short, and one might conceive that the 'turning points' in the twentieth century (§ 11.5) were produced by angular impulses accompanying solar outbursts. Values of $\dot{\Omega}$ at the turning points are of the order of 10^{-20} radians sec^{-2}, and the required torque 100 times that inferred by Beiser. Another difficulty is that the torque can act only in a direction of decelerating the Earth, whereas the astronomical evidence indicates equally sharp increases and decreases in the l.o.d. Beiser's model is, of course, greatly oversimplified. If, for example, the interplanetary fluid rotates more or less with the Earth, then conditions must be altogether different.

As a further example of interplanetary torque we may consider the effect of radiation pressure. Suppose there is a systematic difference in the albedo during the early morning hours as compared to the late afternoon hours. As an extreme case let the albedo be zero where local time is between 6^h and 12^h, and unity where it falls between 12^h and 18^h. The Sun's repulsive force at the Earth is $p = 4\cdot5 \times 10^{-5}$ dynes cm^{-2}. On the reflected side the radiation pressure is $2p$, on the absorbed side it is p, and the couple is then due to $\pm \frac{1}{2}p$. Let λ be the longitude east of local noon. Take an elementary area, $dS = a^2 \sin\theta\, d\theta\, d\lambda$. Its projection normal to the Sun's rays is $\sin\theta \cos\lambda\, dS$ (at equinox), and the differential torque is $\frac{1}{2}pa |\sin\lambda| \sin\theta \cos\lambda\, dS$. The couple equals

$$L_3 = 2 \times \tfrac{1}{2}pa^3 \int_0^{\pi/2} \int_0^{\pi/2} \sin^3\theta \sin\lambda \cos\lambda\, d\theta\, d\lambda$$

$$= \tfrac{1}{3}pa^3 \approx 4 \times 10^{21} \text{ dynes cm}$$

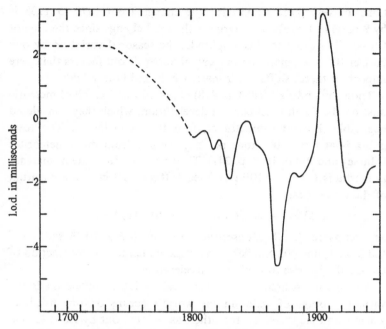

Fig. 11.19. Non-tidal variation in the l.o.d. Prior to 1800 a smoothed
curve is drawn.

compared to 4×10^{23} dynes cm for the lunar tidal couple. The effect
is negligible even under the present extreme assumption, but it is
perhaps surprising that it should turn out this large.

11. Inertia

In this section we shall consider the variation in the Earth's
moment of inertia on a historical time-scale. The problem is to
account for the non-tidal discrepancies in rotation. The fact that
there is no adequate geophysical theory for the tidal discrepancies
does not concern us here. The separation between these two discrepancies follows directly from the astronomical observations.

The astronomical evidence as presented in table 11.1, line 20 and
fig. 11.19 indicates a (non-tidal) increase in the angular velocity by
something like 10 parts per 10^8 in the last 2000 years (this figure is
uncertain by at least a factor of three) and by about 4 parts in 10^8
since 1700. Superimposed on these variations are 'decade fluctua-

tions' of very large amplitude. The change from 1870 to 1905 is by 9 parts in 10^8, about as large as the total change since the time of Christ. Compared to this amplitude, the seasonal variation by one part in 10^8 is so small that one can almost rule out factors that were important there: shifts in air mass, winds and bodily tides.

Thomson and Tait (1883, p. 418) observed that the fall of meteoric dust on the Earth must cause a deceleration, which they considered negligible. Present estimates confirm their conclusion. The total influx is at most 1000 tons per day, largely from micrometeorites (Massey and Boyd, 1958, p. 241). The corresponding rate of increase in inertia is $\dot{C} = 3 \times 10^{21}$ g cm^2 sec^{-1}. If we neglect the momentum of the meteorites,

$$\dot{\Omega} = - \Omega(\dot{C}/C) = - 3 \times 10^{-28} \text{ rad sec}^{-2}$$

as compared to the 'observed' rate of $+ 3 \times 10^{-22}$ rad sec^{-2} (table 11.1, line 20). Similarly any reasonable rates of contraction of the Earth give rise to negligible accelerations.

An obvious source for the historical variation in inertia is the variable storage of water in the world's oceans on one hand, in glaciers and ice sheets on the other, as pointed out by Lord Kelvin (Collected Papers, Vol. III, p. 337). The problem has recently been studied by Hosoyama (1952), Munk and Revelle (1952a, b), Young (1953) and Melchior (1954).

Consider the effect on rotation of a rise in global* sea-level due to melting of ice over Greenland and Antarctica, respectively. Let q be the variable load of water, in g cm^{-2} and q' that of ice. Now let $\mathscr{C}(\theta, \lambda) = 1$ where there are oceans and zero elsewhere; the associated coefficients a_n^m, b_n^m are tabulated in § A.1 under $\mathscr{C}$ (oceans). Set $c_n^m = a_n^m + ib_n^m$. Similarly let $\mathscr{C}'(\theta, \lambda) = 1$ over regions where there is melting (assumed uniform) and zero elsewhere. The associated coefficients that will be needed are

	$a_0'^0$	$a_2'^0$	$a_2'^1$	$b_2'^1$
$\mathscr{C}'$ (Greenland)	0·00414	0·0168	0·0088	− 0·0080
$\mathscr{C}'$ (Antarctica)	0·0282	0·129	− 0·0056	− 0·0160

* Melchior (1954) has treated the case of a rise in Atlantic sea-level only. It is not clear how the remaining oceans can be prevented from partaking in this rise.

The excitation function (9.5.1) can be written

$$\psi = -\frac{a^4}{C-A}\frac{2\pi}{5}[q c_2^1 + q' c_2'^1]$$

$$\psi_3 = -\frac{a^4}{C}\frac{8\pi}{3}[q(a_0^0 - \tfrac{1}{5}a_2^0) + q'(a_0'^0 - \tfrac{1}{5}a_2'^0)]$$

$$\left.\right\} . \quad (11.11.1)$$

But in order for water to be conserved,

$$q a_0^0 + q' a_0'^0 = 0. \qquad (11.11.2)$$

For frequencies low compared to the Chandler frequency, $m_i = \Psi_i$ according to (6.4.4); then allowing for load deformation and rotational deformation in accordance with table 6.1, we find that for a rise in sea-level by 1 cm ($q = 1$ g cm^{-2}):

Melting on	m_1	m_2	m_3	q'
	parts per 10^8			g cm^{-2}
Greenland	13·1	$-$ 9·9	$-$ 0·094	$-$ 178
Antarctica	$-$ 0·0	$-$ 1·9	$-$ 0·107	$-$ 25

$$\left.\right\} . \quad (11.11.3)$$

A melting of 178 g cm^{-2} on Greenland thus raises sea-level by 1 cm everywhere, and the pole is displaced towards Greenland by

$$[(13·1)^2 + (9·9)^2]^{\frac{1}{2}} \times 10^{-8} = 16·4 \times 10^{-8} \text{ rad} = 0''034 = 3·4 \text{ ft.}$$

Antarctic melting displaces the pole towards Chicago. The pole of rotation is a remarkably sensitive indicator of the source of melted water. Values in (11.11.3) lead to the ratios

Melting on	m_1/m_3	m_2/m_3
Greenland	$-$ 140	105
Antarctica	0	18

depending on the source of melted water. These ratios permit us to check the validity of the sea-level hypothesis. Between 1910 and 1940, m_3 increased by 6×10^{-8}. Had this been due to a change in sea-level, then the variation in the l.o.d. would have been accompanied by a polar displacement of $6 \times 10^{-8}[(140)^2 + (105)^2]^{\frac{1}{2}}$ $= 1·05 \times 10^{-5}$ rad $= 250$ ft away from Greenland, or 22 ft away from Chicago, again depending on the source of melted water. The astronomical evidence (§ 11.1) is for a displacement by roughly 10 ft *towards* Greenland, but this is uncertain. In all events displace-

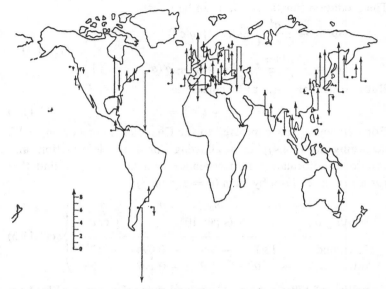

Fig. 11.20. Changes in recorded sea-level between the first and the second decades (*left arrow*) and between the second and the third decades (*right arrow*) of this century. The scale on the left is in cm. (From Munk and Revelle, 1952a.)

ments much larger than 10 ft can be ruled out, and this virtually disposes of the sea-level hypothesis, unless the source of melted water happens to be distributed just so as to cancel the asymmetrical distribution of land and sea. The present argument has the advantage of leading from one astronomical observation to another, making use of the Earth's sciences only with regard to the distribution of land and sea, and the location of Greenland.

A direct comparison with records of sea-level introduces many uncertainties. Fig. 11.20 shows the differences in the mean sea-level during the first and second, and during the second and third decades of this century, based on all available tide gauge records. At any one station the two values are usually quite different and in some instances of opposite sign. For any one decade the variation differs from place to place. Much of the scatter in time is due to meteorologic events, the ocean responding as an inverted barometer to variations in atmospheric pressure (§ 9.3). Changes in mean atmospheric pressure by 5 mb per decade have been documented over regions of the North Atlantic, and the expected sea-level response by 5 cm is of the same order as the observed variation. Part of the recorded oscillation may be due to crustal movements. To extract a global average is then a marginal undertaking. But in any event a drop in sea-level by 75 cm between 1910 and 19 10 as required by the observed variation in the l.o.d. can be ruled out. The decade variations during the last century are not due to changes in sea-level.

But this conclusion does not rule out the possibility that the mean drift over the last few hundred or few thousand years be associated with sea-level fluctuations. From an analysis of tide gauge records, Gutenberg (1941) suggests that there is at present a rise in sea-level by 10 cm per century; Kuenen (1950, pp. 533–4) obtains about twice this value and Disney (1955) 15 cm per century. All of these computations are based on more or less the same data, and the relatively good agreement may be misleading. There is also the difficulty that the observed changes in sea-level may be due to heating or cooling of the ocean waters. A heating by $1°$ C of the entire water column would raise the surface by 60 cm! This part of the changes in sea-level does not vary the load and can be neglected as far as the effect on the Earth's rotation is concerned.* A related difficulty involves the problem of isostatic adjustment. In the case of a steady increase in the ocean load we may eventually expect a compensating flow of mantle material from beneath the oceans to beneath the continents. The advantage of comparing directly the astronomically observed wobble with the astronomically observed l.o.d. is that these issues of baroclinicity in the ocean-crust system are side-tracked.

An alternate method is to consider the variation in the storage of ice and snow. Most of the storage is in Antarctica and Greenland. Seismic measurements conducted during IGY indicate an Antarctic ice cap of $2·1 \times 10^{22}$ g, as compared with a global total of $2·4 \times 10^{22}$ g, enough to raise sea-level by 67 m. Thorarinsson (1940) has extrapolated the average loss by 37 g cm^{-2} per year of five European and Arctic glaciers to all glaciated areas except Greenland and Antarctica. The resulting rise of global sea-level is 5 cm per century. Bauer (1955) has attempted a budget of the Greenland ice cap, based on measure-

* The problem is critical in the case of the annual variation (§ 9.5, *oceans*).

ments taken during the Expéditions Polaires Françaises (P.-E. Victor). He finds

Accumulation	0.45×10^{18} cm³/year
Ablation	$- 0.32$
Discharge from glaciers	$- 0.21$
Net	$- 0.08$

corresponding to a rise in sea-level by 3 cm per century. The values are, of course, extremely uncertain. At the moment nothing can be done about the Antarctic budget. Loewe (1956) reported

Accumulation	1.6×10^{18} cm³/year
Break-off of icebergs	$- 0.04$
Blow-off by winds	$- 0.3$

thus leading to a gain by 1.3×10^{18} cm³/year (36 cm per century drop in sea-level) which could have been ruled out on the basis of tide gauge observation, measurements in the l.o.d. or wobble. Sverdrup at the Antarctic base Maudheim discovered that warm water flowing *beneath* the ice leads to substantial melting. This process is capable of removing 10^{18} g of ice per year and thus balancing the budget (Wexler, 1958), but detailed calculations are out of the question.*

The estimates of melting outside of Antarctica thus produce a sea-level rise of 8 cm per century, about half the average of estimates based on tide gauge readings. Both methods are extremely uncertain, and the agreement is better than one has any right to expect. If we accept 10 g cm⁻² per century as representative of the increase in ocean load, we should expect a polar displacement of the order of 10–15 ft in fifty years, in the mean direction of the source of melted water. The reported displacement of 10 ft towards Greenland is not inconsistent with these considerations. For the last 200 years we should expect an increase in the l.o.d. by roughly $1\frac{1}{2}$ ms, whereas it appears to have *decreased* by 4 ms.†

* We have been informed by A. P. Crary that surveys conducted during the International Geophysical Year in the McMurdo area indicate little or no change in ice volume since the time of Scott's and Shackleton's visit. At Maudheim lichens were found growing down to snow-level in isolated nunataks, and this can only be explained by constant ice-level for at least a few tens of years.

† The apparent agreement reported by Young (1953) derives from the fact that a comparison is made with the total (tidal and non-tidal) variations in the l.o.d.

The earlier evidence is even more uncertain. Lawrence (1950) reports that the present rise in sea-level commenced at A.D. 1750. Prior to this time, and starting during the climatic optimum 3000 to 6000 years ago, the level was dropping. Daly (1920, 1927) and others (Kuenen, 1950, pp. 448, 537–9) estimate sea-level at the time of the optimum to have been 5 m above the present stand and 2 m above it at the time of Christ (Zeuner, 1946, pp. 93–4). On the basis of radiocarbon dating Shepard and Suess (1956) have extended the time-scale to 10,000 B.C. These data suggest a rise at the rate of 50–100 cm per century until roughly 1000 B.C.; the subsequent drop, as proposed above, is neither confirmed nor contradicted by the radiocarbon observations.

For a comparison with the ancient astronomical observations we may choose an uncompensated drop by 1 m. This gives $\delta m_3 = +8 \times 10^{-8}$. Astronomical estimates (table 11.1, line 21) range from $+ 10$ to $+ 33 \times 10^{-8}$. At least there is no contradiction here, so that a higher ancient sea-level may have been responsible for the non-tidal discrepancy in the ancient observations.

There are other possibilities. Urey (1952, pp. 80–87), who first noted that the ancient non-tidal discrepancies were consistent with an angular *acceleration*,* suggested that the acceleration was due to a rain of mantle iron into the core. According to this interpretation the core gains mass at the rate of 10^{11} g sec^{-1}, and at this rate the core would be formed in 0.6×10^9 years.

Munk and Revelle (1952a) have considered a 'turbulence of continents'; the Earth is cut into blocks each moving as a unit, with the assumption that there is no tendency for neighboring blocks to move together. Over the Earth as a whole as much mass must go up as goes down, but the tensor of inertia will fluctuate since the mean latitude and longitude of the rising blocks will not, in general, equal those of the sinking blocks. Let $\bar{z}$ designate the root-mean-square displacement of the blocks, and $\bar{q} = \rho_C \bar{z}$ the corresponding change in load. $\rho_C = 2.75$ g cm^{-3} is taken as the density of crustal rock. It can be shown that the r.m.s. values

---

* His method for separating the tidal from the non-tidal variation is different from ours, but the numerical results are compatible. He no longer considers the growth of the core as a possible cause of the secular acceleration.

of the rotation vector are

$$|\bar{\mathbf{m}}| = \frac{4\pi}{\sqrt{15}} \frac{a^4}{C - A} \bar{q} \, n^{-\frac{1}{2}}, \quad \bar{m}_3 = \frac{8\pi}{3\sqrt{5}} \frac{a^4}{C} \bar{q} \, n^{-\frac{1}{2}},$$

where n is the number of blocks. In these calculations $n \gg 1$.

Consider first the ratio of wobble to l.o.d., i.e.,

$$\frac{|\bar{\mathbf{m}}|}{\bar{m}_3} = \tfrac{1}{2}\sqrt{3} \, H^{-1} = 260.$$

From the wobble data we may set $|\bar{\mathbf{m}}|$ less than 5×10^{-7}. The twentieth-century wiggles suggest $\bar{m}_3 = 5 \times 10^{-8}$. The observed ratio $|\bar{\mathbf{m}}|/\bar{m}_3$ is less than 10, hence much smaller than the expected ratio for a random movement of crustal blocks. Again this speaks for a high axial symmetry of the processes responsible for the twentieth-century wiggle.

Set $n = 21$, corresponding to blocks of the size of North America. Then $|\bar{\mathbf{m}}| = 1 \cdot 2 \times 10^{-7} \bar{z}, \quad \bar{m}_3 = 4 \cdot 6 \times 10^{-10} \bar{z}$

with $\bar{z}$ in cm. Over the last fifty years we might consider 5 cm to be a possible value for the vertical motion of continents. Larger values are unlikely in view of the evidence from tide gauges (fig. 11.20); smaller values are possible. This gives $|\bar{\mathbf{m}}| = 6 \times 10^{-7}$ rad $= 0''\!.12 = 12$ ft for the polar displacement, and $\bar{m}_3 = 4 \cdot 6 \times 10^{-9}$ for the variation in rotation. The latter value is negligible compared to the observed rate; the former is in line with the upper limit set by the wobble observation. We may conclude that the indicated polar displacement toward Greenland might be due to continental unrest, but not the change in l.o.d. At the same time the wobble observations provide a meaningful restraint on how large such continental movements might be. Over the last two thousand years the 'observed' value of $\bar{m}_3 \approx 10^{-7}$ would call for $\bar{z} = 2$ m. This is not impossible; continental unrest may be a factor in the variation of the l.o.d. since ancient times.

Sedimentation has a small effect. 10^{15} g of sediments are deposited each year on the Mississippi Delta; this is 1 per cent of the mass assumed as lost each year from the Greenland ice cap.

In many regions of the world the ground water table has been lowered by tens of meters as a result of human activity and drought. In California a lowering by 60 m has taken place in an area 100 km $\times$ 300 km. The porosity of the ground is $0 \cdot 15$, and the mass of water lost equals $(100 \times 300 \times 10^{10})$ (6000) $(0.15) = 0.23 \times 10^{18}$ g. The resulting variation in sea-level is 3 mm, and the displacement of the pole is of the order of $0.''01$ towards the Pacific Southwest. This is at the limit of detectability, and if it were not for other, larger effects, the exhaustion of ground water would constitute the only case known to us where human activity might produce an observable motion of the pole.

In all of this discussion the question of isostatic compensation looms in the background. There can be no question that over sufficiently long periods any change in superficial loading is compensated by isostatic adjustment, and the effect of rotation is greatly reduced (but not annulled, see § 5.9). Over sufficiently short periods the adjustment is negligible. It has been traditional to refer to a time 'constant' of 10,000 years on the basis of the Fenoscandian adjustment, but the evidence is by no means clear cut. We prefer to leave to the reader the many possible alternatives that result when the present argument is combined with various degrees of isostatic compensation.

12. Core

We still have to find a source for the remarkably large decade fluctuations in the l.o.d. that have taken place during the last hundred years. The time-scale of vigorous events in sea and atmosphere is generally short compared with decades; the time-scale of geologic events is far longer. This leaves only the core. It is sufficiently large, sufficiently fluid and sufficiently remote so as not to be easily eliminated. Moreover, the surface magnetic field provides some evidence (if we assume a hydromagnetic origin) that motion in the core is complex and variable on a time-scale intermediate between that of atmospheric and geologic events. Bullard *et al.* (1950) first considered an 'electromagnetic coupling of the mantle to turbulent motions in the core' but found that the observed variations in the l.o.d. required a couple too large by a factor 100. With some effort this factor can now be reduced, as we shall see, but the difficulty remains.

A steady motion in the core, axially symmetric and driven by an internal energy source would not cause fluctuations in the l.o.d. or shift the pole. We shall examine the consequences if any of these conditions are not met.

Magnetic dissipation.—A secular deceleration of the Earth results if the energy of the magnetic field is derived by some means from the kinetic energy of rotation. The energy of the magnetic field is

$$E = (1/8\pi)\int(H_i)^2 \, dV, \qquad (11.12.1)$$

where the integration is over the whole Earth. Extrapolation of the surface field gives

$$E \approx 3 \times 10^{26} \text{ ergs.}$$

Elsasser (1950) estimates the decay time, t, for the field as 15,000 years. The resulting angular deceleration equals

$$\dot{\Omega} = \frac{1}{C\Omega} \frac{E}{t} \approx 10^{-26} \text{ rad sec}^{-1}.$$

which is tantalizingly small compared to the tidal deceleration of about 10^{-21} rad sec^{-1}. The conclusion is that this process, if operative, would not lead to an observable deceleration; at the same time the rotation of the Earth provides an adequate source of energy for the magnetic field.

Variable motion.—Halley in 1692 noted that many features of the Earth's magnetic field show a westward drift. Elsasser (1950) and Bullard (1950) interpret the drift in terms of material motion within the core. This suggests the possibility of comparing the observed westward drift with the rotation of the mantle. A correlation between the magnetic and astronomical observations has been suggested by Vestine (1952, 1953, 1954), Munk and Revelle (1952a), and Moore (Runcorn, 1954). The analysis of westward drift encounters two principal difficulties:

(1) It is difficult to assess the reality of apparent differences in the magnetic field at two epochs. The measurements are incomplete and different methods of reduction lead to widely differing results.

(2) The interpretation in terms of westward drift is not unique. The broad global magnetic features and localized 'storm' centers drift at different rates.

If the geomagnetic potential is expanded into spherical harmonics, then the following collection of terms are usually referred to in terms of a dipole which would produce an equivalent field:

axial dipole	p_1^0
equatorial dipole	p_1^1
centered dipole	p_1^0, p_1^1
eccentric dipole	$p_1^0, p_1^1, p_2^0, p_2^1, p_2^2$
non-dipole field	$p_2^0, p_2^1, p_2^2, p_3^0, \ldots$

The eccentric (or off-center) dipole is parallel to the centered dipole and at the present epoch displaced from the center of the Earth by about 300 km towards Indonesia. The position of the eccentric dipole as a function of time determines the drift of the quadrupole components of the field but not the motion of the higher harmonics.

Vestine (1953) has determined the position of the eccentric dipole from 1830 to 1945. For the epoch 1905 to 1945 the decade magnetic charts of Vestine et al. (1947) were used, and these are uniform with regard to the observational material and its reduction. Bullard et al. (1950) derived the mean westward drift of the non-dipole field for the epoch 1907·5 to 1945 using the data of Vestine et al. (1947). The calculated drift then averages the motion of all harmonics of degree 2 and higher in a least square sense. For the epoch 1907 to 1945 Bullard obtains $0°18 \pm 0°015$ per year for the non-dipole field, and Vestine $0°30$ per year for the eccentric dipole. This illustrates the uncertainty in interpretation.

Let x_i designate the coordinates of the position of the eccentric dipole. Table 11.4 lists the recalculated east longitude, $\lambda = \tan^{-1}(x_1/x_2)$, using Vestine's compilation. The decrease in λ corresponds to a westward drift.

Vestine notes that there is an apparent change in the rate of westward drift between 1905–1925 and 1925–1945. Using the values listed in table 11.4 we have

$$\dot\lambda = -0·85 \times 10^{-10} \text{ rad sec}^{-1} \text{ for } 1905–1925,$$
$$\dot\lambda = -1·72 \times 10^{-10} \text{ rad sec}^{-1} \text{ for } 1925–1945.$$

The marginal nature of the evidence for this change requires no comment (see fig. 11.21). Nonetheless, we wish to compare this

Table 11.4. The position of the eccentric dipole.

Source	Epoch	East Longitude
Gauss	1835	186°5
Erman-Petersen	1839	181°6
Adams	1845	181°1
Adams	1880	168°4
Fritsche	1885	168°5
Schmidt	1885	168°0
Neumayer-Petersen	1885	168°0
Vestine	1905	163°3
Vestine	1915	162°0
Dyson-Furner	1922	161°1
Vestine	1925	160°2
Vestine	1935	157°3
Afanasiera	1945	156°2
Vestine	1945	153°9

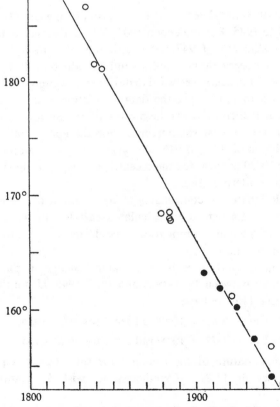

Fig. 11.21. East-west motion of the eccentric dipole. Closed circles are based on Vestine's analysis. The line corresponds to a westward drift by 0°27 per year.

apparent variation with astronomical evidence concerning the rotation of the mantle.

A simple model consists of a mantle and core rotating as two solid bodies about a common axis. Let $C_M = 7 \cdot 2 \times 10^{44} \, \text{g cm}^2$, $C_C = 0 \cdot 85 \times 10^{44} \, \text{g cm}^2$ designate their principal moments of inertia, and Ω_M, Ω_C their angular velocities. Conservation of momentum requires $C_M \Omega_M + C_C \Omega_C$ to remain constant, so that

$$\frac{\delta \Omega_M}{\Omega_M} = - \frac{C_C}{C_M} \frac{\delta \Omega_C}{\Omega_M}.$$

The rate of change of westward drift is given by

$$- \delta \dot{\lambda} = - (\delta \Omega_C - \delta \Omega_M)$$

and the equations can be combined into

$$\delta \Omega_M = - \frac{\delta \dot{\lambda}}{1 + C_M / C_C} = - 0 \cdot 105 \, \delta \dot{\lambda}. \qquad (11.12.2)$$

Equation (11.12.2) relates measurements at astronomic observatories with measurements at magnetic observatories under the present hypothesis. The reported increase in westward migration between the epochs 1905–25 and 1925–45 is $\delta \dot{\lambda} = - 0 \cdot 87 \times 10^{-10}$ rad sec^{-1}; (11.12.2) gives $\delta \Omega_M / \Omega_M = 12 \cdot 5 \times 10^{-8}$. Astronomic observations give $\delta \Omega_M / \Omega_M = 6 \times 10^{-8}$ for the period 1910 to 1930 (fig. 11.19), so that

$$\frac{(\delta \Omega_M) \text{ observed}}{(\delta \Omega_M) \text{ computed}} = 0 \cdot 48. \qquad (11.12.3)$$

The computed variation is based on the assumption that all of the core participates in the fluctuation and is therefore an upper limit. We conclude that the reported fluctuations in westward drift are consistent, in sign and magnitude, with the twentieth-century wiggle, provided a substantial part of the outer core participates in the fluctuation. Since all but the upper 100 km of the core are screened from magnetic view by electric induction, the present procedure is equivalent to estimating the angular momentum of the atmosphere by an extrapolation of surface winds. The resultant value of the annual change in the l.o.d. would have been poor indeed!

The present calculation implies that there exists an adequate couple to produce the observed change in angular velocity. The angular acceleration between 1910 and 1930 is roughly 0.7×10^{-20} rad sec^{-2}. Now if a_C designates the radius of the core, the mean lever arm is $\frac{1}{2}a_C$. Assuming a mean effective area of $2\pi a_C^2$, the mean stress that the core must exert on the mantle is

$$\frac{C_M \dot{\Omega}_M}{\pi a_C^3} = 0.03 \text{ dynes cm}^{-2}.$$

The electromagnetic stress resulting from a finite conductivity of the mantle is of the order (Elsasser and Takeuchi, 1955)

$$(2/5\pi)H_r H_\lambda,$$

where H_r is essentially the radial component of the main dipole field while H_λ represents the toroidal field which penetrates the mantle. H_r at the mantle boundary is 4 gauss, so that H_λ must be 0·1 gauss to account for the required stress of 0·03 dynes cm^{-2}. The toroidal field in the mantle is due to the leakage from the core. The core's toroidal field is of the same order of magnitude as its poloidal field, according to Elsasser, so that a value of 10 gauss might be appropriate though earlier workers had postulated much stronger toroidal fields in the core. The ratio of the toroidal fields in the core and mantle are in the inverse ratio of their conductivities. The calculation of the electromagnetic stress thus requires an estimate of the toroidal field in the core and the appropriate conductivities. Bullard (1949) estimates a core conductivity of $\kappa_C = 3 \times 10^{-6}$ e.m.u. on the basis of extrapolated laboratory data. This gives $4a_C^2\kappa_C/\pi = 14{,}000$ years for the time constant of the dipole field. The observed decrease by 5 per cent in a hundred years suggests a smaller conductivity, possibly $\kappa_C = 5 \times 10^{-7}$ e.m.u. The conductivity, κ_M, of the lower mantle must be less than 10^{-9} e.m.u. according to Runcorn's analysis of the high-frequency cut-off of the secular variation. McDonald's (1957) analysis gives about 10^{-9} e.m.u. Hughes (personal communication) suggests values as large as 10^{-8} e.m.u. may result from the effect on conductivity of a rise in temperature in the lower mantle. For a rough estimate we take

$$(H_\lambda)_{\text{mantle}} = \frac{\kappa_M}{\kappa_C}(H_\lambda)_{\text{core}} = \frac{10^{-9} \text{ e.m.u.}}{3 \times 10^{-6} \text{ e.m.u.}} \, 10 \text{ gauss} = 0.03 \text{ gauss}$$

for the toroidal field in the mantle, and this gives only one-third the amount required for a stress of 0·03 dyn cm^{-2}. The situation can be remedied by increasing κ_M to 3×10^{-9} e.m.u. or decreasing κ_C to 10^{-6} e.m.u., and these appear to be possible values. A toroidal field $(H_\lambda)_{\text{core}} = 30$ to 35 gauss would be adequate and might not lead to instability (Rikitake, 1955).

Some difficulties arise from consideration of the time constant of the relative motion of core and mantle. Bullard (1950) finds that the time needed for a disturbance in the relative rates of rotation to decay to $1/e$ is $2\rho_C/(\kappa_M H^2)$ where κ_M is the conductivity of the mantle and H the strength of the dipole field. Using a mantle conductivity of 10^{-9} e.m.u., the time constant is 30 years. The discussion of § 11.5 suggests that the rotation of the mantle undergoes sharp changes in a few years. This calls for $\kappa_M = 10^{-8}$ e.m.u., which is large but not impossible according to Hughes.

The stress could arise also from viscous coupling. The westward drift of about 0°2 per year indicates a differential velocity of

$$ u = a_C(\Omega_C - \Omega_M) = 0.04 \text{ cm sec}^{-1} $$

at the equator at the boundary. The boundary layer has a thickness

$$ s = (2\nu_C/\Omega)^{\frac{1}{2}} = 1.6 \times 10^2 \nu_C^{\frac{1}{2}} \text{ cm}, $$

where ν_C is the kinematic viscosity in the core. The viscous stress is

$$ \nu_C u/s = 2.5 \times 10^{-4} \nu_C^{\frac{1}{2}} \text{ dyn cm}^{-2}. $$

A viscosity of 10^4 cm^2 sec^{-1} is required, and the resulting thickness of the boundary-layer is 160 m.

The difficulty with viscous coupling lies in the very existence of the westward drift. Bullard (1950) points out that viscous coupling can lead only to an eastward drift. As the angular velocity of the mantle diminishes because of tidal friction, then at any moment the core rotates slightly faster than the mantle and eastward with respect to it. The picture is different in the case of electromagnetic coupling. If there is an interchange (convective or otherwise) of matter between the inner and outer portions of the core, then the inner portion rotates somewhat faster, and the outer portion somewhat slower than the core as a whole. The mantle is electromagnetically coupled to the entire core and the outer portion moves westward with respect to it.

In summary, we have found the faintest of hints in the magnetic data that motions in the core are associated with the twentieth-century wiggle. From angular momentum arguments, the suggested core motions are of the right sign and order of magnitude. The torque approach points out the difficulties. The electromagnetic torque might barely suffice provided the most favorable of the uncertain parameters are chosen.

So far we have dealt with the decade variations in the l.o.d., as observed during the last century. It is possible that the core may account for the variations on a longer time-scale, the eighteenth-century hump (Great Empirical Term of Newcomb, see fig. 11.19) and even the non-tidal variation since ancient days. The required torque is much less for these low-frequency terms, and the difficulty of finding an adequate couple does not arise.

Non-axial motion.—If the motion in the core is purely zonal, only changes in l.o.d. result. Any meridional component of the motion affects the position of the pole. Vestine (1953) finds a north-ward drift of the eccentric dipole of the same order as the westward drift, but with a much greater probable error. Using the northward motion as an indicator of the core motion, widely differing estimates of the resulting displacement of the pole have been published

Vestine (1953)	2 m
Elsasser and Takeuchi (1955)	0·002 cm
Munk and Revelle (1952a)	100 m

Elsasser and Munk (1958) purport to discuss these discrepancies (see § 3.3).

A simple model is that of a rigid mantle enclosing a spherical fluid core. For any frequencies low compared to the frequency of free nutation, $\mathbf{m} = \boldsymbol{\psi}$, $\dot{\mathbf{h}} \ll \Omega\mathbf{h}$; in accordance with (6.1.5) we then have

$$\mathbf{m} = \frac{\mathbf{h}}{\Omega(C - A)}.$$

Let $|\boldsymbol{\omega}_C|$ designate the non-axial component of the core's relative rotation. The northward drift of the eccentric dipole indicates

$|\boldsymbol{\omega}_C| = 0°3$ per year or 2.7×10^{-10} rad sec^{-1}. If the entire core partakes in this motion, $|\mathbf{h}_C| = 2.3 \times 10^{34}$ ergs sec^{-1}, and $|\mathbf{m}| = 1.2 \times 10^{-4}$ rad, corresponding to a polar displacement of 10^5 cm. The torque exerted by the core on the mantle maintaining this displacement is $\Omega|\mathbf{h}|$ or 1.7×10^{30} dynes cm^2, far greater than the maximum possible electromagnetic or viscous stresses. We are left with two possibilities: (1) the meridional component of the eccentric dipole drift is not real; (2) only a thin surface sheet of the core accounts for the northward drift. If C'_C is the moment of inertia of this sheet and C_C the total inertia of the core, then $C'_C/C_C < 10^{-5}$; for zonal motion this ratio is 0.48 (11.12.3). The northward drift must then involve a much smaller proportion of the outer core than the westward drift. The dynamics of such a situation are not obvious.

13. Summary

The astronomical material consists of the modern observations, 1680 to 1950, and the ancient observations, 1000 B.C. to 0. From the observed lunar and solar discrepancies relative to ephemeris positions it is possible to separate tidal effects from non-tidal effects. For the ancient observations the uncertainty is of the same order as the discrepancies themselves.

The problem of dissipation of energy in tides has been considered solved largely as a result of work by Jeffreys in 1920, who found that tidal dissipation in the Bering Sea alone could account for 80 per cent of the observed dissipation. We believe that his estimate of tidal friction in the Bering Sea is too large by a factor of ten, and his estimate of the astronomical discrepancy too small by a factor of three; the problem is open. One possible way out is the conversion of surface ocean tides into internal modes and the dissipation along interior layers. Dissipation by bodily tides may be larger by a factor of 100 than Jeffreys's estimate and accounts for an appreciable fraction (but apparently not all) of the observed discrepancies.

The rate, $\dot{r}_{\mathbb{C}}$, at which the Moon recedes from the Earth is proportional to the tidal torque. Accordingly, an increase by a factor of 3 in the torque, as compared to Jeffreys's estimate, would be associated with a corresponding increase in the present value of $\dot{r}_{\mathbb{C}}$. Without an understanding of the nature of contemporary tidal friction a discussion of the history of the Moon's orbit is premature.

The astronomers who have been concerned with the reduction of the observations all agree that the tidal effect during the last 200 years is half of what it was during the last 2000 years. Such a change is difficult to interpret from a geophysical point of view. Actually it turns out that the present rate of dissipation, which is rather well established, can be reconciled with the ancient observations without insuperable difficulties. On the other hand, minor fluctuation in tidal friction must be expected to take place at all times; and this is the reason why Ephemeris Time cannot be based on the Moon. The Moon must be used solely for smoothing and interpolating solar and planetary observations.

The non-tidal discrepancies indicate a *decrease* in the l.o.d. both over the last 2000 years and over the last 200 years. There are no insuperable difficulties of finding processes capable of explaining these secular changes. A drop in sea-level by 2 m since the time of Christ could explain the observed decrease in the l.o.d., and it is possible that a drop of this order has taken place. The decrease in the l.o.d. during the last 200 years is opposite in sign of the expected effect from the observed rise in sea-level, but a motion of continental blocks by reasonable amounts could be the important factor here.

There is some evidence of a ten-foot displacement of the pole towards Greenland during the twentieth century. This would be the expected displacement from a rise in sea-level by 10 cm, provided Greenland was the principal source of the melted water. Apparently this is the case. It should be emphasized that the geophysical and astronomical observations are indeed marginal, but as far as they go they are consistent.

So far the principal difficulty in arriving at a definitive solution is the multiplicity of geophysical processes capable of producing effects of the required order of magnitude. The principal application of the astronomical observations is that they offer a restraint with regard to the magnitude of various geophysical processes that may be postulated. But the situation is different with respect to the superimposed short-period (decade) fluctuations in the l.o.d. whose amplitudes are of the same order as the variation since Egyptian time. Concerning these, Newcomb (1909) wrote: 'I regard these fluctuations as the most enigmatical phenomenon presented by stellar motions, being so difficult to account for by the action of any known causes that we cannot but suspect them to arise from some action in nature hitherto unknown.' Sea-level variations, continental unrest, melting on Antarctica and other observable processes cannot possibly be the cause. The only known hope is with the core; we have arrived at this conclusion by what Sir Edward Bullard* has called the 'Sherlock Holmes procedure' of eliminating one possibility after another. Moreover, there is some faint evidence, based on the westward drift of the geomagnetic field, that fluctuations in the rotation of the core have taken place which are of the sign and order of magnitude required to explain the astronomical discrepancies. The fact that the discrepancies were not accompanied by a measurable wobble speaks for a process with a high symmetry relative to the rotation axis. A remarkable feature is that the astronomical observations prior to 1800 show no trace of these short-period fluctuations. Indeed, these constitute an enigmatic phenomenon.

* Symposium on Movements in the Earth's Core, UGGI, Rome, 1954.

CHAPTER 12

GEOLOGICAL VARIATIONS

The International Latitude Service provides information on wobble back to 1890. To go back further requires a new approach, depending on fossil records of various kinds. Compared to the astronomical observations the accuracy is down by a factor 10^6; accordingly wobble on a geologic time-scale must be a million times as large if there is to be any chance of detection. There is an additional uncertainty concerning the chronology of past evidence. No fossil evidence is known concerning the geological variations in the l.o.d. At the present rate of tidal deceleration, $5 \cdot 3 \times 10^{-22}$ rad sec^{-2} (table 11.1) the l.o.d. would have been 21 hr at the beginning of the Paleozoic.

1. Historical note

The possibility of large-scale polar wandering and of continental drift has, during the last hundred years, excited the imagination of many geologists. The occurrence of late Paleozoic glacial deposits near the present equator was largely responsible for the initial interest. Comte de Buffon founded the 'Catastrophic School' of polar wandering in the nineteenth century. Lubbock, De la Beche, Evans and others voiced their enthusiasm for the theory. Darwin (1877) published an analysis purporting to show that the pole 'may wander indefinitely from its primitive position' if the Earth were 'plastic', but a couple of degrees at most if it were 'sensibly rigid'. Darwin favored the latter view because 'Sir William Thomson has shown that the earth is sensibly rigid'. Some of Thomson's arguments are not pertinent (Munk, 1956). In any event Darwin's treatment did not in the least damp the fervor of the proponents of polar wandering (Barrell, 1914). In 1901 Reibisch proposed a theory of polar pendulation; during a geological period the pole performs a back-and-forth migration of some tens of degrees along some well-defined direction. Taylor in 1908 and the meteorologist Wegener (1912) are responsible for adding another degree of freedom in interpreting past climates. They pioneered the hypothesis of large-scale continental drift.

Wegener in particular marshalled an imposing collection of facts and opinions. Some of Wegener's evidence was cogent but many of his arguments were based on sheer speculation. For this latter reason neither continental drift nor polar wandering received wide acceptance, though a group of Southern Hemisphere geologists led by DuToit have continued the Wegener tradition. Furthermore physicists, unimpressed by geologic and biologic arguments, found convincing theoretical reasons for disposing of large-scale movements of the Earth's crust.

During the past ten years the subject has been reopened by developments in paleomagnetism. The orientation of magnetic minerals in certain ancient rocks presumably permits the location of the past magnetic pole. This evidence, when expressed in terms of the present latitude and longitude of the ancient pole, suggests large-scale movements of the rotation pole and of continents among themselves. Furthermore, Gold (1955) has demonstrated that polar wandering is not only physically feasible but is to be expected, provided the mantle has the anelastic properties proposed by Bondi and Gold from the damping of the Chandler wobble (§ 10.7). Physicists now have numbers to work with rather than faunal distributions, and their 'continental reconstructions' display an imagination at least equal to that of the nineteenth-century geologists. Munk summarized the situation in 1956: 'They (physicists) gave decisive reasons why polar wandering could not be true when it was weakly supported by paleoclimatic evidence; now that rather convincing paleomagnetic evidence has been discovered they find equally decisive reasons why it could not be otherwise.' Developments of the past few years lead us to question both the convincing nature of the evidence and the theoretical basis for polar wandering.

2. Paleomagnetic evidence

The natural remanent magnetization (NRM) of a rock specimen is the magnetization remaining after the geomagnetic field has been nulled. A wide range of rocks exhibit NRM; we will be principally interested in lavas and sediments for which the orientation of the specimen at the supposed time of magnetization can be determined. The iron oxide minerals (chiefly magnetite, Fe_3O_4, and hematite,

Fe_2O_3) are mainly responsible for NRM. A number of mechanisms by which the minerals acquire NRM have been proposed and studied. The reader is referred to the extensive reviews by Runcorn (1955a, 1955b, 1955c, 1956) and Nagata (1953).

There are two chief ways in which iron oxides can be magnetized without undergoing chemical changes. The magnetization obtained by cooling the material in a weak field from above its Curie point is termed thermo-remanent magnetization (TRM). Numerous laboratory studies confirm the existence of this mechanism. If the specimen is held in a field at a constant temperature, the magnetization remaining after removal of the field is termed isothermal remanent magnetization, IRM. TRM is much greater and much more stable than IRM. The magnetization of lavas is thought to be due to TRM with IRM superimposed.

The magnetization of sediments is not well understood. In some cases the magnetization is due to the mechanical orientation of grains settling through water in the Earth's magnetic field. Johnson, Murphy and Torreson (1948) carried out a detailed study of varved clays and showed that, on redeposition of clays in a weak magnetic field, the remanent magnetization was accurate in declination but not in inclination. This result indicates a tendency for the particles to settle horizontally regardless of the impressed field. Many sediments may become magnetized by the process of crystallization magnetization. The growth of a mineral such as hematite in a magnetic field results in the orientation of the hematite crystallites. The details of the process are not well understood.

Two methods of measuring the magnetization of specimens are in use. An electromagnetic method has been developed at the Department of Terrestrial Magnetism of the Carnegie Institution. A cylindrical specimen is rotated near the center of a pick-up coil; the intensity and phase of the (amplified) alternating voltage give intensity and direction of magnetization. Blackett in England has developed an astatic magnetometer. A cylindrical specimen is raised underneath the lower magnet of an astatic pair, and the deflexion noted. This is repeated for various orientations of the specimen. The use of the astatic pair (two oppositely directed magnets above one another) cancels the effect of the Earth's magnetic field to a first order;

Helmholtz coils surrounding the instrument cancel it to a second order. Both methods can *detect* fields of the order of 10^{-10} gauss and polarizations of the order of 10^{-8} gauss. To *measure* the direction of polarization, the intensity of the field should exceed 10^{-6} gauss. For lavas the intensity usually lies between 10^{-2} and 10^{-4} gauss; for sediments it is generally less than 10^{-4} gauss.

The use of NRM to determine the position of the ancient axis of rotation depends critically on two major questions:

(1) *Does the measured NRM record the magnetic field at the time of formation or have later changes in either the field or the rock altered or obliterated the initial recording?* One test for stability is to check whether the magnetic field in folded strata is distorted in the expected manner (Graham, 1949). However, most measurements are on horizontal strata to which the test cannot be applied. The effect of IRM can usually be eliminated by A.C. demagnetization, but there is no certain way of testing the stability of either TRM or crystallization magnetization. Magnetostriction may be a factor; laboratory experiments by Graham, Buddington and Balsley (1957) have shown that the direction of magnetization can be changed by directed stress. The effect of burial and later distortion of lavas and sediments has not been critically evaluated. However, Graham *et al.* are led by their studies to seriously question the validity of interpreting NRM in terms of past positions of the magnetic pole.

(2) *Does the magnetic pole provide meaningful evidence concerning the pole of rotation?* This involves, first of all, the question whether the past magnetic field is adequately portrayed by a dipole. The time constant of the magnetic field is 10^4 years. Creer, Irving and Runcorn (1957) have shown that over the past 50 million years there is little evidence for a quadrupole field (see below), and it is then reasonable to assume a dipole field in more remote times. If this dipole field is of hydromagnetic origin, then we may expect the magnetic pole to remain near the rotation pole provided the motion in the core is nearly symmetric with respect to the rotation axis. Such symmetry is a general consequence of the effect of rotation on fluid motion. But Cowling has demonstrated that a purely axisymmetric circulation cannot maintain any magnetic field, and there is then some uncertainty how far this argument for coincidence can

be carried. There is some evidence (see § 11.10) that the outer core rotates more slowly than the mantle, falling behind by one revolution every few thousand years. If this has held true during geologic times, then any non-axisymmetric component of the field would be smeared by the relative motion, and the average field would coincide with the rotation axis. Another argument is based on the evidence, to be presented below, that the poles remained close together during historical times and over the past few million years.

The last 50 million years.—Chevallier (1925) carried out a thorough study of the historical lava flows from Mt. Etna. He showed that the variations from 1600 onwards are in reasonable accord with the known secular variation of the geomagnetic field. Vases have furnished a record of magnetic inclination at the time they were fired (presumably in an upright position) as far back as the neolithic and bronze ages (Folgheraiter, 1899, Mercanton, 1906). Inclination and declination have been obtained from kiln walls. Thellier (1951) gives five values for a kiln abandoned in Carthage 146 B.C., and these are mutually consistent and agree with the present field.

The prehistoric field has been studied in New England and Swedish glacial varves. The indication is that the magnetic pole has remained near the geographic pole for at least 15,000 years (Johnson, Murphy and Torreson, 1948).

The late Tertiary or early Quarternary basalts of Iceland place the magnetic north-seeking pole either near the geographic north pole or the geographic south pole (fig. 12.1). This unexpected result indicates reversals in polarization along some preferred magnetic axis. Reversals have been found in rocks throughout the geologic column and represent an unsolved problem in paleomagnetism. Sigurgeirsson (1957) finds also occasional directions in the Icelandic basalts intermediate between the north and south pole orientation. He concludes that the dipole must have swung through 180° in times on the order of a few thousand years. Assami (1954) studied early Pleistocene basaltic lavas in Japan and found both reversed and normal directions. He attributes the reversals to solid state transformation in the ferrimagnetic material rather than to reversals in the field.

Campbell and Runcorn (1956) show that the Miocene basalts of the Columbia River have a NRM parallel to that of the present field with alternating zones of normal and reversed magnetization. They interpret the reversals in terms of changes in polarization of the field.

Numerous other studies of Tertiary rocks support the conclusion that for the past 50 million years the Earth's field has been mainly a

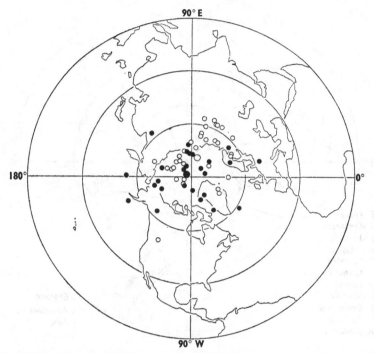

Fig. 12.1. Distribution of north-seeking poles (solid circles) and south-seeking poles (open circles) for Icelandic basalts. The larger circles show the average positions. From Sigurgeirsson (1957).

dipole coinciding in direction, but not always in polarization, with the present dipole. The only exceptions are to be found in the intermediate directions observed in the Icelandic basalts.

Eocene and older.—Creer, Irving and Runcorn (1957) have summarized the paleomagnetic measurements of American and British rocks. Irving and Green (1958) have investigated a wide range of Australian rocks. Both of these studies are summarized in fig. 12.2.

The first major deviation of the pole from its present position is found in rocks of Eocene and Cretaceous age. The American, Irish and Australian poles lie within about 30° of the present pole. However, Clegg, Deutsch, Everitt and Stubbs (1957) find that the lower Deccan traps, presumably of Cretaceous or Eocene age, give an

Fig. 12.2. Summary of pole positions during indicated geologic epochs, based on rock samples from England, Australia, America and India. The path AB through the present pole is drawn according to Milankovitch's theorem (12.6.11) to be discussed in § 12.6 and 12.7. The moment of inertia of the continent-ocean system is a minimum about an axis through A, and a maximum about an axis through B.

Indian pole near Florida (see fig. 12.2). Higher in the sequence but still in the Eocene, the Indian pole moves north towards its present position. In Jurassic time the Indian pole is north of the equator off the coast of Venezuela (Deutsch, Radakrishnamurty and Sahasrabudhe, 1958) while the Australian pole lies in the North Atlantic.

Both the American and English Triassic poles lie in southeast

Asia with the American pole displaced some 20° west in longitude relative to the English pole. Clegg *et al.* (1957) find that the Spanish Triassic pole coincides with the present pole while badly scattered (presumably only partly stable) French poles approximate the English

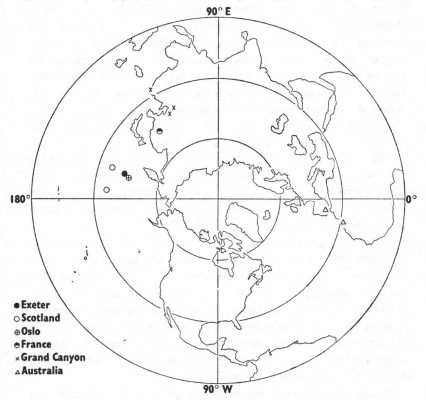

Fig. 12.3. Pole positions during the Permian, from European, American and Australian rock samples.

position. A rotation of Spain by some 35° is needed to bring her to the English position.

Studies on Permian rocks are summarized in fig. 12.3. Runcorn (Doell, 1955) and Graham (1955) have studied the Supai shale of the Grand Canyon and agree within 15° of latitude as to the position of the pole. The English, Scottish and Norwegian poles cluster about a point 40° east of the Grand Canyon results, whereas a

French Permian pole (Rutten, von Everdingen and Zijderveld, 1957) lies only slightly east of the American pole. Two Permian poles from different localities in Australia are in Spain and North Africa. Belshé (1957) has carried out extensive studies of British Carboniferous. His results are summarized in fig. 12.2. Clegg et al. (1957) studied additional Carboniferous formations in England and Scotland. The resulting poles differ from Belshé's by 50° and 80°, respectively, for these two areas. Clegg lists four possibilities for this discrepancy: (1) the rocks acquired magnetization at a time later than the Carboniferous; (2) by some unknown process the rocks acquired a magnetization oblique to the Earth's field; (3) a large change occurred in the position of the British land mass during the Carboniferous; (4) the field was not a dipole.

With the exception of Irving and Runcorn's (1957) work on the late pre-Cambrian, there have been no detailed studies of pre-Carboniferous rocks. The most striking feature of the middle Paleozoic results is the rapid shift of the Australian pole between Devonian and Carboniferous times (fig. 12.2).

Creer, Irving and Runcorn (1957) interpret the American and European data in terms of a polar wandering plus a relative drift of England and North America of some 20° since the Triassic. The indicated constant rate of polar wandering by 0°4 per million years as obtained from both continents is considered a remarkable demonstration for polar wandering. Clegg interprets the results from the Deccan traps as a rapid northward movement of India. Irving and Green (1958) postulate both a post- and pre-Carboniferous drift of Australia coupled with polar wandering.

Conclusions.—The evaluation of the paleomagnetic results is a difficult and perhaps dangerous task in such a youthful and rapidly developing subject. Several points are apparent from the previous discussion. The problem of the origin of NRM in many rocks remains to be solved. The stability of NRM under changing chemical and physical conditions is not understood. Graham's work on instability under directed stress hints at the complications. The phenomenon of magnetic reversals is still a mystery. The occurrence of reversals means that any one pole can be no more than 90° from any spot on the surface.

The consistency of results from a given locality for any geologic time and from different localities for the past 50 million years strongly argues for the method of paleomagnetism. The continuing increase in the degrees of freedom required to account for the paleomagnetic results argue against it. The early measurements of Runcorn in England and America could be explained by a northward motion of the pole through the Pacific. Further data from Great Britain and America now require, in addition to polar wandering, a relative movement of the two continents. The Australian and Indian results are inconsistent with each other and with American and European measurements. They can be brought into line only by further relative motions. Recent European results require that Spain be rotated relative to France, Scotland relative to England, and perhaps England with respect to itself. It is usually a bad omen for any method if the degrees of freedom required to interpret measurements grow at the same rate as the number of independent determinations.

3. Paleontological and paleoclimatic evidence

The classic case for (and against) polar wandering and continental drift has rested on geologic evidence. This evidence is based principally on the distributions of fossil plants, animals, beds of tillite and physical markings of glacial origin. The literature is vague, confusing and conflicting (see DuToit, 1937; Wegener, 1924). We can only give an indication of the problems.

Recent reviews of the paleontological evidence were given by J. W. Durham and W. J. Arkell during a symposium on various aspects of polar wandering (Day and Runcorn, 1955). According to Durham, early Tertiary marine fauna were of tropical character in the London, Paris and Volga basins, in northwestern India, southern Japan, South Africa, southwest Australia, southern New Zealand, and Patagonia, as well *as in the present* tropics. This evidence is consistent with the pole in its present position. If the Eocene pole relative to India were correctly indicated by the paleomagnetic data (fig. 12.2), then the Indian tropical fauna must have grown between 50° and 60° latitude.

Arkell states that in the Jurassic, the Arctic Ocean was a breeding ground of a rich marine molluscan fauna and could not have been covered by ice to the extent to which it is now. Arkell selects, possibly under pressure from the paleomagnetists, the only possible position of the Jurassic poles if they are assumed as cold as the present poles: the North Pole in the North Pacific and the South Pole in the South Atlantic. He goes on to say that this selection is based more on lack of evidence in the proposed polar regions than anything else.

Extensive glacial deposits of Permian or late Carboniferous age have been recognized in tropical India, South Australia, South Africa and Brazil. Associated with the Dwyka tillite of South Africa are striated bed rock, rôches moutonnées and interglacial deposits that indicate successive glaciation with the ice coming from the north. The paleomagnetic Permian pole in southern Asia would place South Africa near the equator, but India and Brazil would be nearer the poles. Polar wandering by itself cannot account for Permian glaciations.

Stehli (1957) in a detailed study of Permian brachiopods and fusulinids finds a latitudinal distribution (see fig. 12.4). The distributions are obviously inconsistent with a pole in southeast Asia. The advantage of Stehli's treatment is that the temperature dependence of the fauna need not be made explicit. Moreover, since the paleomagnetic rock samples were taken from the same formations, absolute time does not enter into the comparison of the two methods. The scarcity of Southern Hemisphere Permian outcrops does not allow a statement either for or against a northward drift of the southern continents.

There is abundant evidence for pre-Permian glaciation in Africa, perhaps as early as pre-Cambrian. There is also an indication of pre-Cambrian glaciations in Australia. This would suggest that the position of Australia and Africa, relative to the South Pole, remained relatively constant until some time after the Permian.

It appears that there is little positive evidence in the paleoclimatic and paleontological data for polar wandering of the kind suggested by paleomagnetic observations. There is a suggestion of continental drift in the paleoclimatic evidence, but this is not definitive.

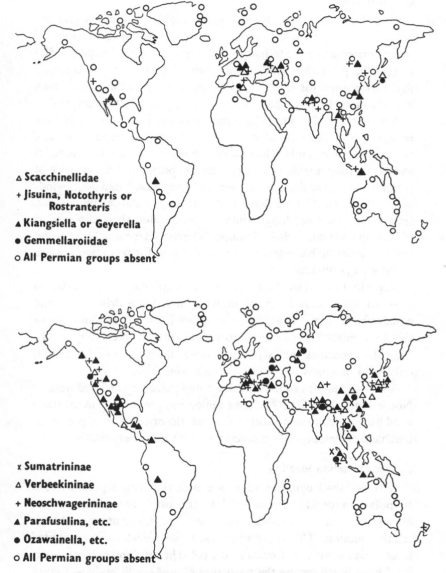

△ Scacchinellidae
+ Jisuina, Notothyris or
  Rostranteris
▲ Kiangsiella or Geyerella
● Gemmellaroiidae
○ All Permian groups absent

× Sumatrininae
△ Verbeekininae
+ Neoschwagerininae
▲ Parafusulina, etc.
● Ozawainella, etc.
○ All Permian groups absent

Fig. 12.4. Distribution of Permian brachiopods (*top*) and fusulinids (*bottom*)
according to Stehli (1957).

4. Paleo-winds

Sand dunes and ash falls may under certain conditions give the prevailing wind direction at the time of deposition. In regions of steady easterlies or westerlies, the pole lies in a direction perpendicular to the prevailing winds. It appears from a recent discussion* that this method for estimating the past position of the poles is even more tenuous than the paleomagnetic and paleontologic methods.

A major difficulty is the identification of wind-blown sand and the recognition of the type of dune. Many water-laid deposits closely resemble aeolian sands. The criteria for identifying aeolian material include frosting, excellent sorting, mineral purity and the absence of pebbles, but these features are not uniquely associated with wind-blown deposits. The barchans (crescent-shaped dunes) have a slip face on the lee side; longitudinal dunes develop their slip face at 90° to the average wind direction. Geologists generally interpret ancient dunes as barchans, though the reason for this identification is not always obvious.

Bagnold has shown that the rate of sand movement varies as $(u - u_t)^3$ where u_t is the threshold wind velocity for driven sand and equals about 10 m.p.h. at anemometer level. Thus the orientation of the dune might reflect severe storms rather than prevailing winds. Bagnold concludes that significant wind directions can be derived only from barchans in the steady trade-wind belts.

A Permian pole in southeast Asia would place England and western North America within 20° of the equator and presumably in the trade wind belt. Runcorn and Shotton find that the orientation of presumed barchans of Permian age is consistent with this interpretation.

5. The excitation function

Here we shall consider various causes that could possibly lead to polar wandering. These fall into the usual two classes: those associated with a distribution of matter and those associated with relative motion. The appropriate excitation function is computed in accordance with the formulae in § 6.8. The resulting polar wandering for an Earth having the properties of a Maxwell body is derived in § 12.6.

* Geophysical Discussion. *The Observatory*, 1958, **78**: 65–68.

Suppose the variation of density with depth is the same for all continents, to be designated by $\rho'(r)$, and the same for all oceans, designated by $\rho(r)$. Let

$$N = \int(\rho' - \rho)r^4 \, dr, \qquad (12.5.1)$$

and let c_2^1 be a complex coefficient of the ocean function (§ A.1). Then the excitation function is given by

$$\psi = \frac{2\pi}{5} \frac{N}{C - A} c_2^1. \qquad (12.5.2)$$

Recent seismic work has shown that the dividing line between continental and oceanic structure corresponds roughly to the 1000 fathom depth contour rather than the coast line. The corresponding values do not differ greatly:

$$
\begin{array}{llll}
\mathscr{C} \text{ (deep oceans),} & a_2^1 = -0.128 & b_2^1 = -0.173 \\
\mathscr{C} \text{ (oceans),} & a_2^1 = -0.101 & b_2^1 = -0.158
\end{array} \Bigg\} \quad (12.5.3)
$$

Using standard continental and oceanic sections proposed by Worzel and Shurbet (1955) the quantity N has been evaluated (Munk, 1958) at

$$N = 1.78 \times 10^{39} \text{ g cm}^2. \qquad (12.5.4)$$

In these standard sections, the mass per unit area is the same for continents and oceans, and the resultant effect depends on the fact that the continental mass is slightly further from the axis of rotation than the corresponding mass of the oceanic section.

Sir Harold Jeffreys has pointed out to us an intimate connexion between the foregoing discussion and the Eötvös force tending to make a floating body move towards the equator. This force plays a vital part in Wegener's theory concerning the drift of continents relative to one another. Here it enters in a more respectable way: only insofar as there is a torque on the crust as a whole. It turns out that the excitation function arising from the Eötvös torque on the continents is the same as that arising from the products of inertia of isostatically compensated continents. This is another example of the equivalence of the momentum approach with the torque approach.

Combining (12.5.1, 2, 3) gives

$$\psi_1 = -0.86 \times 10^{-4}, \quad \psi_2 = -1.16 \times 10^{-4} \qquad (12.5.5)$$

for $\mathscr{C}$ (deep oceans). Thus the distribution of continents and oceans is associated with an excitation pole displaced by 1 km towards the Eastern Pacific. This does not imply that the excitation pole for the whole Earth is displaced by one kilometer from the pole of rotation. If it were, there would be a one-kilometer wobble. It does imply that the continent-ocean system is balanced by density variations in the mantle or some compensating elastic distortion of the equatorial bulge (see § 12.7).

The Himalaya complex contains a mass of 3×10^{22} g above normal continental heights and the Andes about 1×10^{22} g. The Alps are negligible. The corresponding excitation functions have the magnitude

$$4 \cdot 9 \times 10^{-6}, \quad 1 \cdot 5 \times 10^{-6}$$

for the Himalayas and Andes, respectively, again assuming isostatic compensation. Compared to these values other effects appear to be minute. Assuming an Airy-type compensation, the Greenland icecap has an excitation function

$$\psi_1 = -0 \cdot 56 \times 10^{-8}, \quad \psi_2 = 0 \cdot 50 \times 10^{-8}. \qquad (12.5.6)$$

An upper limit to the effect of motion in the atmosphere can be found by assuming that the annual excitation function (table 9.10) is due entirely to winds (which it is not). The resultant magnitude is of the order 10^{-7}. The angular momentum of a single storm may produce at most an excitation function of 10^{-8}. For a severe change in atmospheric circulation the excitation is 10^{-7}. The effect of oceanic circulation is likely to be smaller by one order (§ 9·6).

Runcorn (1957) has suggested convection in the mantle as a cause for polar wandering. Consider a doughnut-like circulation around the Earth in a meridian plane, extending from the core to the surface. The mean distance from the Earth's center is $a' = 5000$ km, and the radius of the circular section equals $b = 1500$ km. Let $\rho_M = 4 \cdot 5$ g cm^{-3} designate the density of the mantle. The doughnut rotates relative to the Earth, with an angular velocity, ω, of one revolution in 100 million years. The linear relative velocity, $\omega a'$, equals 30 cm per year, considerably larger than the convection velocities of the order 1 cm per year estimated by Pekeris (Jeffreys,

1952, p. 325). The relative angular momentum of the doughnut equals

$$|\mathbf{h}| = (2\pi a')(\pi b^2)\rho_M a'^2\omega = 5 \times 10^{29} \text{ g cm}^2 \text{ sec}^{-1}$$

and the corresponding excitation function has a magnitude

$$|\psi| = \frac{|\mathbf{h}|}{\Omega(C - A)} = 2\cdot6 \times 10^{-9}$$

which is negligible compared to the other effects. But convection in the mantle may lead to variations in density whose effect is not negligible.

6. Polar wandering of a Maxwell Earth: an exercise

The Earth taken as a Maxwell body provides the simplest mathematical framework capable of exhibiting both free nutation and polar wandering. It is therefore a useful exercise provided the results are not taken too seriously. The shortcomings are discussed in § 10.3.

Polar wandering on a Maxwell Earth can be visualized as a slow wave-like propagation of the equatorial bulge with respect to the Earth (the geographic coordinates). From the point of view of an observer in space, the bulge remains fixed relative to the ecliptic plane, and the Earth slowly turns under it. This is not a slippage of the crust over the mantle! Rather there exists the prerequisite flow pattern in mantle and core to bring about the convergences and divergences associated with the propagation of the bulge. In this respect there is a resemblance to the daily wandering of the pole brought about by variations in the distribution of water by diurnal tidal currents.

The first attempt at a mathematical solution is contained in an extensive work by Sir George Darwin (1877). The excitation function is assumed to increase linearly with time starting at time 0. The (initial) polar wandering is at a constant *acceleration*, so that Darwin's excitation function is unnecessarily complicated for a discussion of polar wandering. At the time Darwin wrote, the theory of elasticity had not been systematically developed, and one of the dynamic laws is formulated incorrectly. There is also an algebraic error discovered by Lambert and listed by Jeffreys (1952, p. 343). Darwin's original solution led to polar wandering. The conclusion was reversed by Lambert's correction, and reversed once again by correcting the error in dynamics. Munk (1956) has related the history of these errors. It is not without interest to view, in the light of these developments, the extensive and at times heated controversy that has centered on Darwin's paper.

There appear to have been no further analytical treatments of this problem until a solution was published by Burgers (1955) and a discussion given by Inglis (1957). Burgers's solution confirms the conclusions arrived at by Gold (1955) from qualitative considerations.

The inertia of an elastic tidal bulge is given in (5.2.4) in terms of the Love number k. To treat the case of a Maxwell Earth, we need only to replace the Love number k by the operator (5.11.3, 4)

$$\hat{k} = \frac{k_f}{1 + \mu}\left(1 + \frac{\mu\alpha}{\hat{D} + \alpha}\right), \quad \alpha^{-1} = (1 + \mu)\tau = (1 + \mu)\frac{\bar{\eta}_M}{\bar{\mu}},$$

$$(12.6.1)$$

where μ is the dimensionless rigidity, $\hat{D}$ is the operator d/dt, and where τ and α are characteristic time constants. For events of very low frequency, $D \ll \alpha^{-1}$, and $\hat{k}$ approaches k_f, the appropriate fluid Love number; for high frequencies, $D \gg \alpha^{-1}$ and $\hat{k}$ approaches $k_f/(1 + \mu) = k$, the tidal-effective Love number. In accordance with the discussion in § 5·2 and 5·3 we set $k_f = k_s = 3GHCa^{-5}\Omega^{-2}$, $H = (C - A)/C$. Then from (5.2.4)

$$C_{ij} = \delta_{ij}I + \frac{\hat{k}}{k_f}(C - A)(m_im_j - \tfrac{1}{3}\delta_{ij}) + c_{ij},$$

where c_{ij} is some specified perturbation of inertia. Set $\omega_i\omega_i = \Omega^2$ = constant; then $H_i = C_{ij}\Omega m_j + h_i$, and

$$\frac{H_i}{\Omega} = (I + \tfrac{2}{3}J)\,m_i + J\frac{\mu\alpha}{\hat{D} + \alpha}(m_im_j - \tfrac{1}{3}\delta_{ij})\,m_j + c_{ij}m_j + \frac{h_i}{\Omega},$$

where

$$J = \frac{C - A}{1 + \mu}. \qquad (12.6.2)$$

In all previous problems we have been able to assume $m_1, m_2 \ll 1$; for the case of polar wandering this is no longer a permissible approximation. Rather we set

$$m_i = \bar{m}_i + m'_i,$$

where $\bar{m}_i$ designates the displacement of the mean drifting pole, and m_i' refers to the nutation about this position. For an initial-value problem the transient nutation decays in a time of order α^{-1} (§ 10.3) and thereafter we are concerned only with the mean drift. With this understanding we can write m_i for $\bar{m}_i$.

In the present scheme c_{ij}/I, $h_i/I\Omega$, $\hat{D}/\alpha$ (but not m_i) are small dimensionless numbers. Note that $I + \frac{2}{3}(1 + \mu)J = C$. Then to a first order

$$\frac{H_i}{\Omega} = Cm_i - \tfrac{2}{3}J\mu \frac{\hat{D}}{\alpha} m_i + c_{ij}m_j + \frac{h_i}{\Omega}.$$

The Eulerian equation (3.1.1) for $L_i = 0$ can be written in the operational form

$$\hat{D}(H_i/\Omega) + \Omega\varepsilon_{ijk}m_j(H_k/\Omega) = 0.$$

Substituting from above and neglecting the products of small quantities yields

$$C\dot{m}_i + \Omega\varepsilon_{ijk}m_j \left(c_{kl}m_l - \frac{J\mu}{\alpha} \dot{m}_k + \frac{h_k}{\Omega} \right) = 0, \qquad (12.6.3)$$

where $\dot{m}_i = \hat{D}m_i$. We now consider some special cases.

The Goguel-Fermi problem.—A simple example is that of a stepwise excitation function caused by relative motion starting at time 0 and maintained constant thereafter. This problem has been discussed by Goguel (1950) and the late Enrico Fermi (personal communication) in connexion with a secular effect of winds and ocean currents on the position of the pole, but we are not aware of any solutions.

We set $c_{kl} = 0$ in equation (12.6.3), and transform the equations to some new coordinate system so that the components, h_i, of relative momentum become 0, 0, $h = \sqrt{h_i h_i}$ in the new system (see, for example, Jeffreys and Jeffreys, 1950, 3·08). Thus if h_i is due to a single cyclone, the x_3-axis goes through the cyclone. The new coordinates are the principal axes of the excitation function, and not the principal axes of the planet Earth (except at the completion of polar wandering, $t = \infty$).

We shall require the following equations and definitions, all of which have been given previously. The frequency of the Chandler wobble equals

$$\sigma_0 = \frac{\mu}{1 + \mu} \sigma_r, \quad \sigma_r = \frac{C - A}{A} \Omega, \qquad (6.2.6, 6.1.4)$$

where σ_r is the wobble frequency for a rigid Earth. The damping time is α^{-1} and the Q of the system is

$$Q = \sigma_0/2\alpha. \tag{10.1.9}$$

Using the definition of J in (12.6.2) it follows that

$$\Omega J \mu/(C\alpha) = 2Q(A/C) \approx 2Q. \tag{12.6.4}$$

Set

$$\beta = \frac{h}{2QC}.$$

Then in the new coordinate system, (12.6.3) becomes

$$\left.\begin{aligned}
\left(\frac{1}{2Q}\right) \dot{m}_1 - (m_2\dot{m}_3 - m_3\dot{m}_2) + \beta m_2 &= 0, \\
\left(\frac{1}{2Q}\right) \dot{m}_2 - (m_3\dot{m}_1 - m_1\dot{m}_3) - \beta m_1 &= 0, \\
\left(\frac{1}{2Q}\right) \dot{m}_3 - (m_1\dot{m}_2 - m_2\dot{m}_1) &= 0
\end{aligned}\right\} \tag{12.6.5}$$

We can rotate the new coordinate system about the new x_3-axis so that m_2 is initially zero; i.e., the x_3-axis of the new system is drawn through the cyclone, and the present pole lies in the x_1, x_3-plane. The pole remains very nearly in this plane throughout the wandering (as will be shown). The perturbation scheme is as follows:

$$\dot{m}_i = \text{order } (\beta m_i), \quad m_2 = \text{order } (Q^{-1}), \quad m_1, m_3 = \text{order } (1).$$

In the first and third equations all terms are of the same order. The first term in the second equation is of the order Q^{-2} as compared to the remaining terms, so that, approximately

$$- m_3\dot{m}_1 + m_1\dot{m}_3 - \beta m_1 = 0.$$

Combining this with $m_i m_i = 1 \approx m_1^2 + m_3^2$

gives

$$\frac{dm_1}{dt} = \beta m_1\sqrt{1 - m_1^2}$$

with the solutions

$$m_1 = \text{sech } (\zeta + \beta t), \quad m_3 = \tanh (\zeta + \beta t), \tag{12.6.6}$$

where sech ζ and tanh ζ are the initial values of m_1 and m_3, respectively. As $t \to \infty$, $m_1 \to 0$ and $m_3 \to 1$, which are the expected asymptotic values. The colatitude of the pole in the new system is given by

$$\tan \theta = m_1/m_3 = \operatorname{csch} (\zeta + \beta t). \qquad (12.6.7)$$

The second or third of equations (12.6.5) can be used to obtain

$$m_2 = \left(\frac{1}{2Q}\right) (1 + \beta t) \operatorname{sech} (\zeta + \beta t)$$

which is always small compared to 1, as assumed.

The Milankovitch problem.—Milankovitch (1934) discussed polar wandering due to the distribution of continents. The coordinate system is transformed to the principal axis of the ocean-continent system (§ 12.7). In this transformed system $c_{kl} = 0$ for $k \neq l$. We write

$$\beta_1 = \frac{c_{11}\Omega}{2QC}, \quad \beta_2 = \frac{c_{22}\Omega}{2QC}, \quad \beta_3 = \frac{c_{33}\Omega}{2QC}, \qquad (12.6.8)$$

and select $\beta_3 \geqslant \beta_2 \geqslant \beta_1$. The equation (12.6.3) now becomes

$$\frac{1}{2Q} \frac{1}{m_2 m_3} \frac{dm_1}{dt} + \frac{d \ln m_2}{dt} - \frac{d \ln m_3}{dt} + \beta_3 - \beta_2 = 0 \quad (12.6.9)$$

plus two similar equations that follow from a cyclic rotation of suffixes. The present perturbation scheme differs from the preceding one insofar as all m_i's are now of order 1. Again setting $\dot{m}_i = $ order (βm_i) we find the first term of each equation to be small (by a factor Q^{-1}) compared to the remaining terms. (12.6.4) can then be integrated at once to give

$$\frac{m_1}{m_3} = \tan \theta_0 \cos \lambda_0 \, e^{-(\beta_3 - \beta_1)t}, \quad \frac{m_2}{m_3} = \tan \theta_0 \sin \lambda_0 \, e^{-(\beta_3 - \beta_2)t},$$

where θ_0, λ_0 are the initial coordinates of the pole in the transformed axes. By combining these expressions with $m_i m_i = 1$ we obtain

$$\left. \begin{array}{l} \tan \lambda = \dfrac{m_2}{m_1} = \tan \lambda_0 \, e^{(\beta_2 - \beta_1)t}, \\[2ex] \tan^2 \theta = \dfrac{m_1^2 + m_2^2}{m_3^2} \\[2ex] \qquad = \tan^2 \theta_0 \, e^{-2\beta_3 t} (\cos^2 \lambda_0 \, e^{2\beta_1 t} + \sin^2 \lambda_0 \, e^{2\beta_2 t}) \end{array} \right\} \qquad (12.6.10)$$

for the complete solution. It can be verified that

$$\tan \theta \cos \lambda \tan^\kappa \lambda = \text{const}, \quad \kappa = (\beta_1 - \beta_3)/(\beta_1 - \beta_2) \quad (12.6.11)$$

designates the path of the poles in the transformed system. This equation is referred to as the Milankovitch theorem (Scheidegger, 1958, 126–131). The path through the present pole is shown in fig. 12.2.

The incipient solution.—The preceding solutions place no restrictions on the total displacement of the pole. A treatment of the initial disturbance follows more conveniently from the perturbation scheme developed in § 6.1, based on $m_i \ll 1$. This is valid for polar displacements up to 1000 km, which is adequate for most purposes.

First take the case where the excitation function is due to a cyclone (no load deformation). Then according to (6.3.1, a) and (6.2.3)

$$\hat{\varphi} = \psi + \hat{\psi}_D = \psi + (\hat{k}/k_f)\mathbf{m}$$

so that (6.3.5) can be written in the form

$$i\frac{\dot{\mathbf{m}}}{\sigma_r} + \frac{k_f - \hat{k}}{k_f}\,\mathbf{m} = \psi.$$

The operational solution can be written down at once:

$$\mathbf{m} = -i\sigma_r \left[\frac{\hat{D} + \alpha}{\hat{D}(\hat{D} + \alpha - i\sigma_0)} \right] \psi. \qquad (12.6.12)$$

This is evaluated by following a few rules (for example, Jeffreys and Jeffreys, 1950, 7·051 – 7·08). The result is

$$\mathbf{m} = -\frac{i\sigma_r}{\alpha - i\sigma_0} \left[\alpha t - \frac{i\sigma_0}{\alpha - i\sigma_0}(1 - e^{-\alpha t}\,e^{i\sigma_0 t}) \right] \psi. \qquad (12.6.13)$$

For definiteness let $\psi_2 = 0$ and set $Q \gg 1$. Then, neglecting terms of order Q^{-1},

$$m_1 = \frac{1 + \mu}{\mu}\,\psi_1(1 + \alpha t - e^{-\alpha t}\cos\sigma_0 t), \quad m_2 = -\frac{1 + \mu}{\mu}\,\psi_1 e^{-\alpha t}\sin\sigma_0 t. \qquad (12.6.14)$$

We wish to compare the incipient solution with the exact solutions obtained previously. Differentiating (12.6.7) we obtain for the rate of polar wandering

$$-\dot{\theta} = \beta\cos^2\theta\,\operatorname{csch}(\zeta + \beta t)\operatorname{ctnh}(\zeta + \beta t)$$
$$= \beta\cos^2\theta\tan\theta\sec\theta = \beta\sin\theta$$
$$= \frac{1 + \mu}{\mu}\,\alpha\psi_1, \quad \psi_1 = \frac{h_1}{\Omega(C - A)}, \quad h_1 = h\sin\theta$$

which is consistent with the non-transient part of $\dot{m}_1$ from (12.6.13). A similar comparison can be made with (12.6.10).

A geometric interpretation is given in fig. 12.5, and this should be compared with fig. 6.1, center. (In both figures the excitation pole is on 19° E of x_1 rather than on x_1.) Initially the rotation pole, $\mathbf{m}$, revolves counter-clockwise about the mean position, $\mathbf{\Psi}$, of the excitation pole. So far this does not differ from the behavior for an elastic Earth. But there are two new features: the wobble is damped, and there is a continuous migration of the deformation pole (and

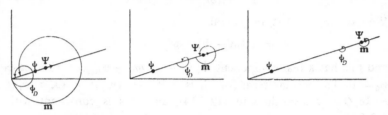

Fig. 12.5. Polar wandering of a 'Maxwell Earth'. The positions of the poles of rotation ($\mathbf{m}$), initial excitation ($\mathbf{\psi}$), deformation ($\mathbf{\psi}_D$), and modified excitation ($\mathbf{\Psi}$) are shown during the first wobble (0 to 1·2 years), the 11th wobble (12 to 13·2 years) and 21st wobble (24 to 25·2 years) following an abrupt excitation at time 0. The attenuation is based on a damping time, α^{-1}, of 10 years, hence a Q of 25. For a higher Q the damping is correspondingly smaller.

hence of $\mathbf{\phi}$, $\mathbf{\Psi}$, and $\mathbf{m}$) in the direction $\mathbf{\psi}$. The coordinates of the deformation pole are

$$\mathbf{\psi}_D = \frac{\hat{k}}{k_f}\,\mathbf{m} = \frac{1}{1+\mu}\,\frac{\hat{D}+\tau^{-1}}{\hat{D}+\alpha}\,\mathbf{m},$$

or, subject to the previous approximation,

$$\psi_{1(D)} = \mu^{-1}\psi_1[1 + \alpha(1+\mu)t - e^{-\alpha t}\cos\sigma_0 t],$$
$$\psi_{2(D)} = -\mu^{-1}\psi_1\,e^{-\alpha t}\sin\sigma_0 t.$$

Both the deformation pole and rotation pole move at a rate $\mu^{-1}(1+\mu)\alpha\psi_1$ (aside from the initial wobble), with the deformation pole lagging behind by a distance $\mathbf{m} - \mathbf{\psi}_D = \mathbf{\psi}$.

For simplicity terms of order Q^{-1} have been ignored. To this order the components of $\mathbf{m}$ and $\mathbf{\psi}_D$ normal to the initial excitation, $\mathbf{\psi}$, vanish with the damping of the nutation. But to trace the polar wandering, we shall require to expand (12.6.13) to the next order.

The non-transient components normal to ψ are

$$m_2 = -\frac{1+\mu}{\mu}\, Q^{-1}\psi_1(1 + \tfrac{1}{2}\,\alpha t),$$

$$\psi_{2(D)} = -\frac{1+\mu}{\mu}\, Q^{-1}\psi_1 \left(\frac{1+\tfrac{1}{2}\mu}{1+\mu} + \tfrac{1}{2}\alpha t\right).$$

Thus $\mathbf{m}$ and ψ_D move along parallel lines at an angle $\tfrac{1}{2}Q^{-1}$ to the right of the direction of ψ (fig. 12.6), and the two lines are separated by a distance

$$m_2 - \psi_{2(D)} = -\tfrac{1}{2}Q^{-1}\psi_1. \qquad (12.6.15)$$

We have shown that, in general,

$$\mathbf{m} = i\sigma_r(\mathbf{m} - \boldsymbol{\phi}), \quad \boldsymbol{\phi} = \psi + \psi_D$$

and this has a real component $\dot{m}_1 = -\sigma_r(m_2 - \psi_{2(D)})$ for the case $\psi_2 = 0$. The secular term for m_1 is $\mu^{-1}(1+\mu)\psi_1\alpha t = (\sigma_r/\sigma_r)\psi_1\alpha t = \tfrac{1}{2}\sigma_r Q^{-1}\psi_1 t$ according to (12.6.14), and this is consistent with (12.6.15).

The interpretation is as follows. The rotation pole is to the right (looking in the direction of wandering) of the excitation pole, $\boldsymbol{\phi}$, by an amount equal to $\tfrac{1}{2}Q^{-1}$ times the initial excitation,* ψ. In ordinary elastic problems $\mathbf{m}$ would then revolve about $\boldsymbol{\phi}$ in the usual fashion. But in the present case $\boldsymbol{\phi}$ can keep up with $\mathbf{m}$, and the two move parallel to one another. $\boldsymbol{\phi}$ moves because the bulge yields viscously (apart from the initial elastic yield), always tending to align itself normal to the rotation axis. Thus the deformation pole pursues the rotation pole, and the excitation pole, $\boldsymbol{\phi} = \psi + \psi_D$, is dragged along. It is, according to Gold (1955), 'as with the ass and the carrot hanging from a stick held by the rider'.

Eventually the rotation and deformation poles wander an appreciable distance from their initial position. The resulting decrease in the colatitude of the cyclone results in an exponential-like slowing down of the wandering, and the solution (12.6.6) gives the law of this deceleration. In the limit the pole reaches the cyclone, and the ass gets the carrot.

---

* For the continental excitation function (12.4.5), the angular distance between the rotation pole and the excitation pole (or pole of figure) is then 10^{-5} radians, or $2''$, at the limit of what might be observed. But the actual angle is likely to be 10^{-3} of this amount (§ 12.7).

The situation is nearly the same when the cyclone is replaced by a local *deficit* of mass (compensated for load deformation), except that the deceleration follows somewhat different laws (12.6.10 versus 12.6.6).

The Airy-Gold problem.—This is the problem of the wandering of the pole due to a sudden addition of superficial matter. In 1860 the Astronomer Royal discussed the possible effect of the elevation of a mountain mass by something like a gaseous explosion (Airy, 1860), and Gold (1955) considered a sudden upheaval of South America by 30 m.* Gold ignores load deformation but states explicitly that compensation needs to be taken into account.

We shall treat only the incipient case. To allow for load deformation (§ 5.9) all that needs to be done is to replace ψ in (12.6.12) by

$$(1 + \hat{k}')\psi = \frac{\iota + \hat{\mu}}{1 + \hat{\mu}}\,\psi = \frac{\iota + \mu}{1 + \mu}\,\frac{\hat{D} + \gamma}{\hat{D} + \alpha}\,\psi, \quad \gamma = \frac{\iota}{(1 + \mu)\tau}$$

so that
$$\mathbf{m} = -\,i\sigma_r\,\frac{\iota + \mu}{1 + \mu}\left[\frac{\hat{D} + \gamma}{\hat{D}(\hat{D} + \alpha - i\sigma_0)}\right]\psi. \quad (12.6.16)$$

Making use of $\iota \ll \mu$ and the approximations leading up to (12.6.16), we obtain

$$m_1 = \psi_1(1 + \gamma t - e^{-\alpha t}\cos\sigma_0 t), \quad m_2 = -\,\psi_1\,e^{-\alpha t}\sin\sigma_0 t,$$

which differs from the Goguel-Fermi solution in two ways: there is no amplification of the wobble over what it would be for a rigid Earth; and the rate of wandering is reduced by a factor

$$\frac{\gamma/\alpha}{\mu^{-1}(1 + \mu)} \approx \iota.$$

After the initial transient, polar wandering proceeds at a rate determined by the isostatically compensated load (as in the Milankovitch problem). This is not an unexpected result.

---

* The following is a quotation from Florian Cajori's 1934 translation of *Principia*: 'Suppose a uniform and exactly spherical globe to be first at rest in a free space; . . . let there be added anywhere between the pole and the equator a heap of new matter like a mountain, and this, by its continual endeavor to recede from the center of its motion, will disturb the motion of the globe and cause its poles to wander about its surface describing circles about themselves and the points opposite to them. Neither can this enormous deviation of the poles be corrected otherwise than by placing the mountain either in one of the poles . . .; or in the equatorial regions . . .'; There is enough resemblance between Newton's spherical Earth and a Maxwell Earth with equatorial bulge to make this quotation pertinent.

Kelvin–Voigt model.—It is instructive to compare the foregoing solution with a model that exhibits finite strength. A simple case is the Kelvin–Voigt (or firmo-viscous) model (§ 5.11). We have

$$\hat{k} = \frac{k_f}{1 + \hat{\mu}}, \quad \hat{\mu} = \mu(1 + \tau\hat{D}), \quad \tau = \frac{\bar{\eta}_{K-V}}{\tilde{\mu}}$$

$$\mathbf{m} = -i\sigma_r \frac{\hat{D} + \alpha''}{\hat{D}^2 + (\alpha'' - i\sigma_r)\hat{D} - i\sigma_r\tau^{-1}} \psi, \quad \alpha'' = \frac{1 + \mu}{\mu}\tau^{-1}.$$

Set
$$\alpha' = \frac{\sigma_0^2 \tau}{1 + \mu} \ll 1.$$

The incipient solution is

$$\mathbf{m} = \frac{1 + \mu}{\mu}\psi(1 - e^{-\alpha' t}e^{i\sigma_0 t}), \quad \psi_D = \frac{1}{\mu}\psi(1 - e^{-\alpha' t}e^{i\sigma_0 t}).$$

After the initial transient has died down, both $\mathbf{m}$ and $\boldsymbol{\phi} = \boldsymbol{\psi} + \boldsymbol{\psi}_D$ approach $(1 + \mu^{-1})\boldsymbol{\psi}$ and there is no polar wandering.

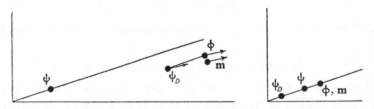

Fig. 12.6. The solution after the attenuation of the initial transient. This is drawn for an unrealistically small value of Q ($Q = 10$) to emphasize second-order terms which were ignored in fig. 12.5. The Maxwell model (*left*) leads to polar wandering, the Kelvin–Voigt model (*right*) does not.

The Darwin problem.—This is included for its historical interest. The excitation pole is specified by

$$\psi_1 = -Ut, \quad \psi_2 = 0 \text{ for } t \geqslant 0; \quad \psi = 0 \text{ for } t < 0,$$

corresponding to a constant rate of rise of the British Isles starting at time 0. The question concerning compensation does not arise in Darwin's paper. The simplest thing to do is to interpret $\boldsymbol{\psi}$ as a compensated excitation. Darwin does not allow for elastic yield, and accordingly we set $\mu = \infty$.

The procedure is to replace $\boldsymbol{\psi}$ (not a function of time in the previous solutions) by $- Ut$ in (12.6.12). Performing the indicated operations and writing $\sigma = \sigma_0 + i\alpha$ leads to

$$\mathbf{m} = \frac{\sigma_0}{\sigma} U \left[-\frac{1}{2} \alpha t^2 + \frac{\sigma_0}{\sigma} t - i \frac{\sigma_0}{\sigma^2} (1 - e^{i\sigma t}) \right].$$

The terms in t^2 give polar wandering at a constant *acceleration*, as compared to a constant velocity in the Goguel-Fermi problem. The difference arises because Darwin postulated an excitation pole of constant *velocity*, whereas in the previous cases the excitation pole had a constant displacement. The foregoing solution does not agree with Darwin's (Munk, 1956).

7. The present position of the pole

The usual starting point of any discussion on polar wandering (such as the one in § 12.6) is to presume the Earth to be in equilibrium until suddenly disturbed by some implausible rearrangement of matter. The ensuing motion of the pole is then computed for an Earth made of material that can be modeled by a combination of springs and dashpots. Finally, the computed polar path is found to be in agreement with a bewildering array of paleontological, and more recently, paleomagnetic evidence.

Francis Birch suggested to us that it may be appropriate to inquire first whether the *present* distribution of matter is consistent with the *present* position of the pole.

From an examination of various possible excitations of polar wandering (§ 12.5), it appears that the distribution of continents and oceans exceeds other possible causes by a factor 1000. Compared to the equatorial bulge, continents are small markings on the surface, still they might be the determining factor if the continental distribution is permanent and the bulge appropriately anelastic.* A rapidly spinning rubber ball would orient itself relative to the rotational axis in accordance with tiny surface markings. The final position of the pole is the one that places the continents as well as possible on top of the equatorial bulge. This puts the pole into the equatorial Pacific, as might be expected. The travel time depends on the anelasticity; it is less than 100,000 years according to the interpretation by Bondi and Gold (1955) of the damping of the Chandler

* See footnote, § 12.10.

wobble. The fact that the pole is not in the Pacific nor traveling towards it at this rate poses a dilemma.

The orientation of the principal 'continental axis' has been computed by Milankovitch (1934), Kuiper (1943) and Munk (1958). We start with the known components, c_{ij}, of the inertia tensor in the present x_i-system. The problem is to find the orientation of an x_i'-system for which the products of inertia,

$$c_{ij}' = - \int \rho x_i' x_j' \, dV, \quad i \neq j$$

vanish. Setting the third order determinant $|c_{ij} - \gamma \delta_{ij}|$ equal to zero yields a cubic equation in γ. It can be demonstrated that, in general, the three roots of this cubic equation are the principal moments of inertia i.e.,

$$\gamma_1 = c_{11}', \quad \gamma_2 = c_{22}', \quad \gamma_3 = c_{33}'.$$

Set $x_i' = l_{ij} x_j$, where l_{ij} is the transform matrix (not a tensor). We can select $\gamma_3 \geq \gamma_2 \geq \gamma_1$ so that the moment of inertia is a maximum about the x_3'-axis. The direction cosines of this axis relative to the present x_i-axes are l_{3j}, and these can be evaluated from the equations

$$(c_{ij} - \delta_{ij} \gamma_3) l_{3j} = 0, \quad l_{3j} l_{3j} = 1.$$

The colatitude, θ, and the east longitude, λ, of the principal pole of the ocean-continent system are given by

$$\cos \theta = l_{33}, \quad \cos \lambda = l_{32}/l_{31}.$$

Table 12.1 lists the positions of the pole and of the (dimensionless) principal moments of the ocean function.

The deep ocean values are preferable; in all events the results are not far different and agree roughly with Milankovitch's and Kuiper's

Table 12.1. Orientation and principal moments of the continent-ocean system (according to Munk, 1958).*

	θ	λ	γ_1/N	γ_2/N	γ_3/N
𝒞 (deep oceans)	76°	199°	2·69	3·14	3·47
𝒞 (oceans)	75°	176°	2·33	2·57	2·76

* The coefficients a_n^m, b_n^m on which these values are based differ somewhat from those given in § A.1.

values. The pole is in the vicinity of Hawaii, almost as far from the present pole as it can get. In fact, the pole corresponding to a minimum moment of inertia is near Archangel, at $\theta = 20°$, $\lambda = 66°$, and closest to the actual pole. Fig. 12.2 shows the polar path from the Archangel pole to the Hawaii pole through the present position in accordance with the Milankovitch theorem (12.6.11). It should be pointed out that these results do not depend on the continental and oceanic density distributions and the resulting integral N (12.5.4), provided that these distributions can be assumed as uniform over continents and uniform over oceans.

The computed direction of polar wandering is towards the principal pole near Hawaii. On the basis of "unzweideutig" paleoclimatic evidence, Milankovitch drew the arrow along his polar path pointing *away* from the Pacific, and Jardetzky (1949) seems to have gone along with this interpretation. This implies a negative value of the integral N: the inertia (per unit area) of continents is less than that of oceans. This interpretation contradicts any isostatic model and has been criticized by Gutenberg (1951, p. 203).

However, the suggestion of a reverse polarity requires further consideration. In the isostatic model, the inertia depends on second-order terms resulting from the slightly larger radial distance of continents than oceans. Suppose that isostatic balance does not hold precisely but that there is uncompensated erosion of continental matter and deposition on the sea floor. This represents a first-order effect which might reverse the sign of N. It is an intriguing possibility. The pole of maximum inertia is then at Archangel, relatively close to (but still 20° from) the actual location.

It turns out that 15 m of uncompensated sediments would reverse N. The resulting free air gravity anomaly over oceans is $+ 5$ mgals. Relatively large local gravity anomalies make it impossible to determine the sign of N on the basis of gravity observations. No compensation would lead to anomalies of $- 390$ mgals over oceans, and this is wildly wrong. But it is impossible to tell whether there is 100 per cent or 99 per cent compensation, and this is what we need to know. The problem might be resolved from the observed orbits of satellites.

Gold (1955) has suggested another possibility, namely that the pole is trapped in the Arctic Ocean. As the pole moves towards shore, additional snow is deposited along the coast which has the effect of driving the pole offshore. The difficulty here is that the effect of the snow is so minute. As an indication of magnitudes we note that the Greenland icecap has an excitation function 10^{-4} times that of the continent-ocean system (12.5.5, 6).

Next we consider the rate at which the pole moves towards the Hawaii pole. For a Maxwell Earth, the pole moves a distance $\mu^{-1}(1 + \mu)|\psi|$ during each damping interval, α^{-1} (12.6.13). According to (12.5.6) and § 9.7, $|\psi|$ equals $1\cdot41 \ [(0\cdot86)^2 + (1\cdot03)^2]^{\frac{1}{2}} \times 10^{-4} = 1\cdot89 \times 10^{-4}$ rad $= 1\cdot2$ km; following Bondi and Gold, α^{-1} is obtained from the damping of the Chandler wobble, and equals 10–20 years according to various interpretations of the latitude observations (table 10.1), or more than 50 years if the Q of the pole tide is pertinent (§ 10.4). The expected movement is then by 20 to 100 m per year towards Vancouver. This is from 300 to 1700 times the rate indicated by the latitude observations during the twentieth century (§ 11.1). The time required to travel most of the way to the Hawaii pole is of the order $(\beta_3 - \frac{1}{2}\beta_1 - \frac{1}{2}\beta_2)^{-1}$ according to (12.6.10), where $\beta_i = \gamma_i\Omega/(2QC)$ according to (12.6.8). Using the 1000 fathom values in table 12.1 together with (12.5.4) gives 20,000 and 80,000 years for $Q = 25$ and 100, respectively. Paleomagnetic evidence indicates that the pole has remained within $10°$ during the last 50 million years (§ 12.2).

There are ways out of this dilemma, as always. If we use the viscosity that has been associated with Fennoscandian uplift, 4×10^{22} g cm^{-1} sec^{-1}, then the damping time α^{-1} is *increased* by a factor of 400 and the rate of polar wandering reduced accordingly. At this rate a migration towards Hawaii cannot be ruled out. The observed damping of the Chandler wobble has then nothing to do with the Maxwellian viscosity of the mantle. The difficulty is that existing departures from isostasy in regions other than Fennoscandia require an increase in α^{-1}, not by a factor of 400, but by millions (§ 10.3); we then avoid a polar wandering that is too rapid at the expense of getting none at all.

Another possibility is that there are variations of density in the mantle which cancel the continent-ocean asymmetry, so that the combined excitation pole vanishes.* One such distribution leads to free air gravity anomalies with the extreme values of the second degree harmonic distributed as follows:

	130° W	50° E
Northern Hemisphere	+ 5 mgal	− 5 mgal
Southern Hemisphere	− 5 mgal	+ 5 mgal

Such a distribution cannot be ruled out on the basis of our present knowledge of gravity.

Finally there is the possibility that the Earth (or at least its shell) has sufficient finite strength to withstand the stresses imposed by the continent-ocean system. We can dispose of the dilemma this way, but then again we dispose of polar wandering.

8. Finite strength

The central issue in the dynamics of polar wandering is the problem of finite strength. On a Maxwell Earth there is no finite strength, and an infinitesimal stress (i.e., Gold's beetle) can turn the Earth around, given sufficient time. On a Kelvin–Voigt Earth there is finite strength and no polar wandering. In § 4.2 it was proven in general that an Earth without finite strength is secularly unstable.

There is abundant evidence from laboratory experiments that a wide variety of silicate materials resist either fracture or excessive deformation up to some finite stress difference. Once this finite strength is exceeded, the material fails, either by rupturing or undergoing a large deformation. But laboratory experiments are of short duration and may not be applicable on a geologic time scale. Evidence for a finite strength of the materials in the Earth is found in the existence of continents, oceans and mountains. Major gravity anomalies are associated with Paleozoic mountain chains so that the stress differences resulting from these distributions of mass must have persisted for very long times. Jeffreys (1952, p. 199) concludes that gravity anomalies indicate a strength of the order of 150 to 300 bars in the upper 600 km, the range in values depending upon whether the material beneath 600 km has any strength. The triaxial figure of the Moon indicates that the lunar material, which may be

* See footnote, § 12.10.

similar to mantle material, can support a stress difference of about 20 bars. Furthermore, the fact that the present pole is not moving towards the principal pole of the continent-ocean system can be interpreted in terms of finite strength (§ 12.7). The resultant stress is of the order $|\psi|\bar{\mu} = (1\cdot3 \times 10^{-4}$ radians$)$ $(5 \times 10^{11}$ dynes cm$^{-2})$ $= 65$ bars, and this corresponds to stress differences of the order of 10 bars. The minimum strength required to prevent polar wandering is then of the order of 10 bars.

Evidence against finite strength comes from two major sources: (1) The uplift of formerly glaciated areas of Fennoscandia and North America is thought to be evidence for finite viscosity (and zero strength). Jeffreys (1952, p. 327) criticizes this interpretation inasmuch as there are other regions with similar negative anomalies that are not rising. (2) The figure of the Earth closely approximates that of an equivalent rotating fluid. This is not necessarily an argument against finite strength, rather it imposes an upper limit. The question is whether this upper limit exceeds the ten bars required to prevent polar wandering in response to the continental excitation function. An estimate of the maximum allowable discrepancy between the observed eccentricity and the eccentricity of a fluid Earth is about one-third of one per cent (see § 5.4). Suppose that there does exist a discrepancy in eccentricity equal to the maximum allowed by observation, $\delta\varepsilon = 1 \times 10^{-5}$. If the Earth were a uniform, incompressible, elastic body, the maximum stress difference resulting from such a distortion occurs at the center and equals $(39/95)\rho ga\delta\varepsilon$ (Jeffreys, 1952, p. 367), or about 12 bars. Alternatively we can calculate the stress resulting from a change in angular velocity giving rise to a discrepancy in ellipticity of 10^{-5} (Stonely, 1924). Since the ellipticity varies as the square of the angular velocity

$$\frac{\delta\varepsilon}{\varepsilon} = 2\,\frac{\delta\Omega}{\Omega} = 3 \times 10^{-3}.$$

Assuming a uniform rate of tidal deceleration of 5×10^{-22} rad sec^{-2} (table 11.1), the discrepancy would develop in about 10 million years. The resulting stress is of the order $\psi_3\bar{\mu} = m_3\bar{\mu} = (1\cdot5 \times 10^{-3})$ $(5 \times 10^{11}) \approx 750$ bars, and the stress difference of the order of 100 bars. Considering the present uncertainty in the figure of the

Earth, a finite strength of 100 bars cannot be ruled out. On this hypothesis we should expect the bulge to be larger than the equilibrium values. According to Bullard (1948) the actual eccentricity is ($3\cdot3764 \pm 0\cdot0077$) × 10^{-3} from gravity observations, ($3\cdot3702 \pm 0\cdot0074$) × 10^{-3} from the motion of the Moon; and these values do, in fact, exceed the equilibrium value, ($3\cdot3632 \pm 0\cdot0006$) × 10^{-3}. Preliminary results from satellites indicate the opposite*, an actual eccentricity of ($3\cdot355$) × 10^{-3} (Merson and King-Hele, 1958) and ($3\cdot3521 \pm 0\cdot0006$) × 10^{-3} (Lecar, Sorenson and Eckels, 1959).

We note that the evidence from the pole position and the eccentricity are not inconsistent. The former indicates a minimum strength of the order of 10 bars, the latter a maximum of the order of 100 bars. Moreover, these values are not inconsistent with those obtained from gravity anomalies.

On the finite strength hypothesis, polar wandering is possible only if the excitation stress exceeds the 'strength' (see § 4.2). The theory of polar wandering for an inhomogeneous body having finite strength has not been developed. It is uncertain whether the Earth would respond principally by a plastic flow or by fracturing. The existence of earthquakes down to 600–700 km suggest that failure by fracturing is the principal mechanism. The inhomogeneity of the outer layers raises the problem of a local versus a global release of stresses.

Up to now polar wandering has been pictured as a wave-like motion involving the Earth as a whole. An alternative suggestion is that of a thin upper layer sliding over the interior. This requires a lubricating layer separating the 'shell' from the nucleus. If the lubricating layer is fluid and has a thickness greater than the difference of polar and equatorial radii, then the bulge of the nucleus does not contribute to the stability of the Earth. In this case the results of § 5.10 are applicable. Surface tension has the dimensions of rigidity times thickness. For a shell 100 km thick having a rigidity of 4×10^{11} dynes cm^{-2}, the equivalent surface tension is $\tilde{v} = 4 \times 10^{18}$ g sec^{-2}. The dimensionless surface tension is then $v = (15/2)\tilde{v}/(\rho g a^2) = 0\cdot02$. The amplification over a rigid body displacement is 50:1. For the

* According to the most recent (unpublished) values, the actual eccentricity is $3\cdot354 \times 10^{-3}$ and *exceeds* the equilibrium value, $3\cdot336 \times 10^{-3}$.

continent-ocean system, the pole would be displaced by 50 km. The resulting maximum stress difference in the shell is of the order of 20 bars. Kuiper (1943) uses a crust 35 km thick and derives a displacement of about 400 km and a corresponding maximum stress difference of 160 bars. The strength of the outer shell is of the order of 1000 bars. We conclude that the continent-ocean system could result in a displacement of the outer shell by a few degrees at most, and that the stress differences thus generated are too small to lead to failure. Seismic evidence rules out the possibility of amplifying these values by taking a much thinner shell.

9. Continental drift and polar wandering—rules of the game

We wish to discuss the rules to be followed if there is continental drift as well as polar wandering. The question concerning the reality of such a drift we can, fortunately, avoid.*

In the analysis of polar wandering, any movement of continents relative to one another enters in two ways: (1) the latitude and longitude of the observatories change over and above the changes resulting from the variable rotation, and (2) the variable distribution of the continent-ocean system leads to a variable excitation function. Under these circumstances it is impossible to solve uniquely for the position of the pole without some hypothesis concerning the behavior of the continents relative to the rest of the Earth.

A dated rock sample with remanent magnetization serves as our observatory. From this we obtain a record of the inclination, I, and declination, D, at a time t.† In principle this is no different than timed entries of latitude and meridian transits in the log of an astronomical observatory.

We consider n distinct 'continental' blocks. Assume there is no continental rotation and distortion. The hypothesis of no distortion

* There have been some reports of *measured* drift. For example, Jelstrup in 1932 found the longitude of Greenland by radio-link to exceed by 5″ the longitude determined by Börgen and Copeland in 1870 using lunar culminations and occultations. Markowitz (1945) has examined these and other reported drifts and ascribes them to experimental error. If these older determinations were to be accepted they would mean that '. . . large shifts occurred until the invention of the telegraph, moderate shifts until the invention of radio, and practically none since then'.

† The declination of the north-seeking pole is measured from the geographical meridian eastward, the inclination is measured from the horizontal plane, positive if the north-seeking pole is below the horizon. Coincidence of the axis of rotation with the magnetic dipole is assumed.

can be checked by comparing the results from many rock samples on one continent. We assume this experiment has been performed and that the results were reasonably consistent.

Let $\theta'(t)$, $\lambda'(t)$ designate the position of the pole of rotation, $\theta_n(t)$, $\lambda_n(t)$ the coordinates of a representative 'observatory' on the nth continent, and $I_n(t)$, $D_n(t)$ the magnetic readings, all referring to the time, t, the sample was magnetized. Let $\theta_n(0)$, $\lambda_n(0)$ designate the position where the sample was collected. Then for any one continent and any one time, the readings I, D, yield a relation between the positions of pole and continent (Creer, Irving and Runcorn, 1957):

$$\left.\begin{array}{l} \sin \theta' = \sin \theta \cos \phi + \cos \theta \sin \phi \cos D \\ \sin (\lambda' - \lambda) = \sec \theta \sin \phi \sin D \end{array}\right\}, \qquad (12.9.1)$$

where $\cot \phi = \frac{1}{2} \tan I$.

These are $2n$ equations in $2(n + 1)$ unknowns, θ_n, λ_n, θ', λ'. Without further assumptions there can be no solution.

We consider two possible assumptions:

(1) no continental drift, $\theta_n(t) = \theta_n(0)$, $\lambda_n(t) = \lambda_n(0)$;

(2) no polar wandering, $\theta'(t) = \theta'(0) = 0$.

In the first case we end up with $2n$ equations in 2 unknowns, θ', λ', and one can apply the usual statistical techniques to inquire whether the n values for the pole position are consistent in light of the (known) observational error. Under the second hypothesis we have $2n$ equations in $2n$ unknowns, and the only test is the one of reasonableness; for example, the overlap of two continents could be considered as unfavorable evidence. A further test for either hypothesis is afforded by a reasonable continuity in time of the motion of the pole or of the continents, respectively.

Suppose we do not wish to exclude either continental drift nor polar wandering. We then need some additional assumptions. As an example, suppose that each continent is conserved and that the rotation axis is at all times coincident with the principal axis of the ocean-continent system.* The resulting conditions

$$\theta' = \theta'[\theta_n(t),\ \lambda_n(t),\ \theta_n(0),\ \lambda_n(0)]$$
$$\lambda' = \lambda'[\theta_n(t),\ \lambda_n(t),\ \theta_n(0),\ \lambda_n(0)]$$

together with (12.9.1) form $2(n + 1)$ equations in $2(n + 1)$ unknowns.

* Thus ignoring the present situation in which the ocean-continent pole is at Hawaii.

These are some possible rules. There is no end of variation by which this game can be played.

10. Summary

The story of polar wandering is varied and complex. Our principal conclusion is that the problem is unsolved.

From the point of view of dynamic considerations and rheology (§ 12.5 to 12.8) the easiest way out is to assign sufficient strength to the Earth to prevent polar wandering, and the empirical evidence, in our view, does not yet compel us to think otherwise. If we must have polar wandering, a number of possibilities present themselves, all somewhat labored: (1) The continental excitation function is largely balanced by a density distribution in the mantle, and the finite strength of the Earth is so weak that minor readjustments can lead to stress differences exceeding this finite strength. (2) The continental excitation function is not balanced by a density distribution in the mantle, but there is sufficient finite strength to prevent polar wandering, but barely so. Minor adjustments can again initiate polar wandering. The intriguing aspect is that once the pole takes off for Hawaii, the continental excitation function *increases*, reaching a maximum of (sin 70°/sin 45°) = 1·3 times its present value at the time the Hawaii pole is 45° (as compared to the present value of 70°) from the pole of rotation. Thereupon the excitation function diminishes and the migration would stop presumably when Hawaii reached a colatitude of about 20°. (3) Inhomogeneities in the mantle may exist which give rise to an excitation function far larger than that of continents.* The previous remarks apply except that the finite strength required to maintain the present balance is higher.

All of these descriptions involve an 'all or none' aspect of the strength of the Earth which does not ring true. But once we allow

---

* *Note added in press.* From satellite orbits, O'Keefe, Eckels and Squires (1959) have obtained higher harmonics of the gravitational potential. The observed values of a_3^0 and a_4^0 are -62 and 34 times, respectively, those derived for the ocean-continent distribution (Munk and MacDonald, 1960). As yet there has been no determination of the longitude-dependent harmonics, and the orientation of the principal axis is undetermined. But it is simplest to assume that inhomogeneities in the mantle swamp those in the crust, and that the rotation axis does, in fact, coincide with the principal axis. The magnitude of the observed coefficients indicates that finite strength may be exceeded. If during some past geologic epoch the mantle inhomogeneities were differently distributed, then presumably the position of the pole was adjusted accordingly.

for inhomogeneities in the materials within the Earth, the problem becomes hopelessly tangled. The only case we have been able to consider is that of a shell on a fluid substratum. But the supposed strength of this shell exceeds by an order of magnitude the stress differences that are induced by the present continent-ocean system, and what evidence there is indicates that the present continentality is abnormally large. A crust gliding over the mantle does not appear to be a reasonable model for polar wandering.

APPENDIX

1. The Ocean Function

We 'shall follow the notation and conventions of Jeffreys and Jeffreys (1950, ch. 24) in treating spherical harmonics. The associated Legendre functions, $p_n^m(\cos \theta)$, are defined by

$$p_n^m(\cos \theta) = \frac{(n-m)!(n+m)!}{2^n}$$
$$\sin^m \theta \sum_{r=0}^{n-m} \frac{(\cos \theta - 1)^{n-m-r}(\cos \theta + 1)^r}{(m+r)!(n-r)!(n-m-r)!r!}. \quad (A.1.1)$$

In most mathematical texts, the definition of the Legendre function is

$$\frac{n!}{(n-m)!} p_n^m$$

rather than (A.1.1). In geophysical work Schmidt's version of the associated Legendre functions is often used (Chapman and Bartels, 1940), and these are defined by

$$\frac{n!}{(n-m)!} \left[\frac{2(n-m)!}{(n+m)!} \right]^{\frac{1}{2}} p_n^m.$$

A function $f(\theta, \lambda)$ of colatitude θ and east longitude λ can be expanded in a series of spherical harmonics

$$f(\theta, \lambda) = \sum_{n=0}^{\infty} \sum_{m=0}^{n} p_n^m \begin{pmatrix} a_n^m \cos m\lambda \\ b_n^m \sin m\lambda \end{pmatrix}$$
$$= a_0^0 + a_1^0 \cos \theta + \sin \theta \begin{pmatrix} a_1^1 \cos \lambda \\ b_1^1 \sin \lambda \end{pmatrix} + a_2^0 \left(\frac{3}{2} \cos^2 \theta - \frac{1}{2} \right)$$
$$+ \frac{3}{2} \cos \theta \sin \theta \begin{pmatrix} a_2^1 \cos \lambda \\ b_2^1 \sin \lambda \end{pmatrix} + \frac{3}{2} \sin^2 \theta \begin{pmatrix} a_2^2 \cos 2\lambda \\ b_2^2 \sin 2\lambda \end{pmatrix} + \cdots$$
$$(A.1.2)$$

provided f has a continuous second derivative. A case of particular interest is one in which f does not satisfy this condition but where f can be uniformly approximated over the sphere except at a set of arbitrarily small areas. In this case the series exists and approximates

f to within an arbitrarily small error except in the areas where f cannot be uniformly approximated.

Using the orthogonality conditions

$$\int_0^\pi \int_0^{2\pi} (p_n^m)^2 \binom{\cos^2 m\lambda}{\sin^2 m\lambda} \, \mathrm{d}s = \frac{2\pi}{2n+1} \frac{(n-m)!(n+m)!}{(n!)^2}$$

except for $m = 0$ when the right side is $4\pi/(2n+1)$, we obtain

$$a_n^0 = \frac{2n+1}{4\pi} \int_0^\pi \int_0^{2\pi} f(\theta, \lambda) p_n^0 \, \mathrm{d}s$$

$$\binom{a_n^m}{b_n^m} = \frac{(2n+1)(n!)^2}{2\pi(n-m)!(n+m)!} \int_0^\pi \int_0^{2\pi} f(\theta, \lambda) p_n^m \binom{\cos m\lambda}{\sin m\lambda} \, \mathrm{d}s$$

where $\mathrm{d}s = \sin\theta \, \mathrm{d}\theta \, \mathrm{d}\lambda$. The first few coefficients are then

$$\left.\begin{aligned}
a_0^0 &= \frac{1}{4\pi} \iint f \, \mathrm{d}s, \\[4pt]
a_1^0 &= \frac{3}{4\pi} \iint f \cos\theta \, \mathrm{d}s \\[4pt]
\binom{a_1^1}{b_1^1} &= \frac{3}{4\pi} \iint f \sin\theta \binom{\cos\lambda}{\sin\lambda} \, \mathrm{d}s, \\[4pt]
a_2^0 &= \frac{5}{4\pi} \iint f \left(\frac{3}{2}\cos^2\theta - \frac{1}{2}\right) \, \mathrm{d}s \\[4pt]
\binom{a_2^1}{b_2^1} &= \frac{5}{3\pi} \iint f \frac{3}{2}\cos\theta \sin\theta \binom{\cos\lambda}{\sin\lambda} \, \mathrm{d}s, \\[4pt]
\binom{a_2^2}{b_2^2} &= \frac{5}{12\pi} \iint f \frac{3}{2} \sin^2\theta \binom{\cos 2\lambda}{\sin 2\lambda} \, \mathrm{d}s
\end{aligned}\right\} \qquad . \quad (A.1.3)$$

We frequently require the coefficients associated with the 'ocean function', $\mathscr{C}$ (oceans), defined by

$$\left.\begin{aligned}
\mathscr{C}(\theta, \lambda) &= 1 \text{ where there are seas} \\
\mathscr{C}(\theta, \lambda) &= 0 \text{ where there is land}
\end{aligned}\right\} \qquad (A.1.4)$$

The spherical harmonic representation of the ocean function has been obtained by numerical integration;* the appropriate coefficients

---

* The oceanic boundaries were read off a world chart at every 5 degrees of latitude. Any body of ocean water or land having an east-west extent of greater than 2·5 degrees was taken into account. The actual numerical computations were carried out on an IBM 709 digital computer, with values rounded off to 5 places.

through degree eight are listed in Table A.1. An indication of the approximation can be obtained by comparing the $\mathscr{C} = 0.5$ contour with the actual coast line. Using only terms through the second degree, the Pacific Ocean and the Euroasian continent are

Table A.1. Coefficients for the Surface Spherical Harmonic Expansion of the Ocean Function.

	$\mathscr{C}$ (Oceans)	$\mathscr{C}$ (Deep Oceans)	Prey		$\mathscr{C}$ (Oceans)	$\mathscr{C}$ (Deep Oceans)	Prey
$a_0{}^0$	0·71436	0·62260	0·28658	$a_6{}^3$	− 0·00552	− 0·04416	− 2·71498
				$a_6{}^4$	− 0·15003	− 0·14788	− 10·18774
$a_1{}^0$	− 0·21348	− 0·27566	− 0·14748	$a_6{}^5$	0·06296	0·07095	6·45506
$a_1{}^1$	− 0·18781	− 0·17923	− 0·13184	$a_6{}^6$	− 0·00242	− 0·01909	− 0·62312
$b_1{}^1$	− 0·09486	− 0·14588	− 0·07753	$b_6{}^1$	0·09502	0·09567	0·36336
				$b_6{}^2$	− 0·02434	0·12460	1·44269
$a_2{}^0$	− 0·12969	− 0·19336	− 0·13231	$b_6{}^3$	− 0·18552	− 0·35941	− 6·59656
$a_2{}^1$	− 0·10117	− 0·12842	− 0·21101	$b_6{}^4$	0·12422	0·11974	9·25154
$a_2{}^2$	0·09994	0·15153	0·12203	$b_6{}^5$	0·06637	0·09447	11·47028
$b_2{}^1$	− 0·15800	− 0·17288	− 0·21168	$b_6{}^6$	− 0·01334	− 0·00806	− 0·08470
$b_2{}^2$	− 0·00691	0·08913	0·02987				
				$a_7{}^0$	0·19709	0·20904	0·08782
$a_3{}^0$	0·11587	0·08974	0·01435	$a_7{}^1$	− 0·03099	0·00482	0·14919
$a_3{}^1$	0·14883	0·13233	0·05068	$a_7{}^2$	− 0·20347	− 0·22773	− 3·03353
$a_3{}^2$	0·25675	0·26338	0·15893	$a_7{}^3$	0·12812	0·10865	0·36364
$a_3{}^3$	− 0·01426	− 0·07082	− 0·08006	$a_7{}^4$	0·19367	0·30853	42·49749
$b_3{}^1$	− 0·12782	− 0·03904	− 0·01966	$a_7{}^5$	− 0·01909	− 0·07751	− 3·11258
$b_3{}^2$	− 0·36678	− 0·40651	− 0·19122	$a_7{}^6$	0·01109	0·00079	11·11637
$b_3{}^3$	− 0·21081	− 0·25117	− 0·33518	$a_7{}^7$	0·00430	0·00403	1·54147
				$b_7{}^1$	− 0·17721	− 0·16175	− 0·56428
$a_4{}^0$	− 0·07762	− 0·19070	− 0·10848	$b_7{}^2$	− 0·00948	− 0·12384	0·23509
$a_4{}^1$	0·15258	0·16495	− 0·31341	$b_7{}^3$	− 0·09491	− 0·12980	− 1·38543
$a_4{}^2$	0·47062	0·36954	1·42560	$b_7{}^4$	− 0·04611	− 0·00541	− 0·54347
$a_4{}^3$	− 0·20737	− 0·24928	− 1·54335	$b_7{}^5$	0·10569	0·04868	− 0·07411
$a_4{}^4$	0·03489	0·07074	0·09142	$b_7{}^6$	0·05743	0·07635	27·92433
$b_4{}^1$	0·09491	0·08712	− 0·31847	$b_7{}^7$	0·02072	0·01537	5·36550
$b_4{}^2$	− 0·11532	− 0·14682	− 0·30369				
$b_4{}^3$	0·00987	− 0·00899	0·57627	$a_8{}^0$	0·03972	0·01835	− 0·01142
$b_4{}^4$	− 0·20560	− 0·25300	− 0·69706	$a_8{}^1$	− 0·00867	− 0·02559	− 0·04959
				$a_8{}^2$	− 0·09444	− 0·08060	− 3·95542
$a_5{}^0$	0·32449	0·32791	0·15340	$a_8{}^3$	− 0·15632	− 0·03652	− 6·21726
$a_5{}^1$	− 0·03384	0·03914	0·12886	$a_8{}^4$	− 0·00036	0·01978	0·59946
$a_5{}^2$	0·31448	0·13190	0·29492	$a_8{}^5$	0·14060	0·04098	45·21647
$a_5{}^3$	− 0·21237	− 0·08342	− 1·53596	$a_8{}^6$	− 0·04070	− 0·04115	− 93·91114
$a_5{}^4$	− 0·34021	− 0·38432	− 7·20707	$a_8{}^7$	− 0·04953	− 0·04538	− 161·45126
$a_5{}^5$	0·00010	0·00447	0·22350	$a_8{}^8$	0·00711	0·00672	16·94195
$b_5{}^1$	0·07910	0·03252	0·37119	$b_8{}^1$	0·14896	0·05192	0·19867
$b_5{}^2$	0·16681	0·17430	0·98238	$b_8{}^2$	0·11195	− 0·07223	− 0·66826
$b_5{}^3$	− 0·07088	− 0·19702	− 0·86419	$b_8{}^3$	− 0·09239	− 0·06708	− 3·64024
$b_5{}^4$	0·09870	0·07405	0·90863	$b_8{}^4$	− 0·11283	− 0·27402	− 38·53677
$b_5{}^5$	− 0·07607	− 0·04217	− 0·78982	$b_8{}^5$	− 0·00159	0·03261	24·45273
				$b_8{}^6$	− 0·06529	− 0·08038	− 110·51159
$a_6{}^0$	− 0·11756	− 0·11461	− 0·10085	$b_8{}^7$	− 0·02504	− 0·03325	− 48·37846
$a_6{}^1$	0·04501	0·00814	− 0·14776	$b_8{}^8$	0·00888	0·00870	− 29·29269
$a_6{}^2$	0·12352	0·12164	0·27402				

distinguished; but the regions occupied by South America, Antarctica, and a portion of Africa lie in the oceans, and the North Atlantic is on land. Using third-degree terms the major continental areas are delineated.

The function complementary to the ocean function is the 'continentality function', $\mathscr{C}$ (continents), defined by

$$\left.\begin{array}{l} \mathscr{C}\,(\theta,\lambda) = 1 \text{ where there is land} \\ \mathscr{C}\,(\theta,\lambda) = 0 \text{ where there is ocean} \end{array}\right\}. \qquad \text{(A.1.5)}$$

Thus $\mathscr{C}$ (oceans) $+ \mathscr{C}$ (continents) $= 1$, and a_n^m (oceans) $= -a_n^m$ (continents), b_n^m (oceans) $= -b_n^m$ (continents) except for the case $n = 0, m = 0$:

$$a_0^0 \text{ (oceans)} = 1 - a_0^0 \text{ (continents)}.$$

Table A.1 includes also the function $\mathscr{C}$ (deep oceans), defined by

$\mathscr{C}\,(\theta,\lambda) = 1$ for water deeper than 1000 fathoms,
$\mathscr{C}\,(\theta,\lambda) = 0$ for land and for water shallower than 1000 fathoms.

For comparison we have included Prey's function, defined as the distance from mean sea-level to the surface of land where there is land and the bottom of the sea where there is sea. The harmonics pertaining to this 'solid boundary' were computed by Love (1908) up to degree 3 and extended by Prey* (1922) to degree 16. The tabulated values refer to Prey's coefficients multiplied by $[(n-m)!/n!]^2$ to conform to Jeffreys's notation, and divided by -8.597×10^3 to 'normalize' the topography (originally given in meters). The normalization is such as to make

$$a_0^0 \text{ (oceans)} + a_0^0 \text{ (Prey)} = 1.00,$$

and all other coefficients comparable in magnitude and sign. In the language of communication engineering, $\mathscr{C}$ (oceans) is a 'clipped' version of Prey's solid boundary. At small wave numbers the two spectra should be comparable, and they are. At large wave numbers the effect of topography, subaerial and submarine, is to raise the harmonics of the solid boundary relative to $\mathscr{C}$ (oceans). For still higher harmonics the process of clipping must raise the harmonics

---

* Prey computed harmonics for two other surfaces as well: (1) the *visible* surface, that is to the surface of land where there is land and zero elsewhere; and (2) to the bottom of the sea and zero elsewhere. The solid boundary is a sum of these.

of $\mathscr{C}$ (oceans). But the known steepness of the continental slope introduces a somewhat clipped topography in fact, so that this effect is relatively unimportant. Some of Prey's high harmonics look suspiciously large.

Integrals of products of three Legendre functions.—The systematic development of Love numbers (§ 5.12) requires products of three associated Legendre functions. Using the recurrence relations

$$\sin \theta \, p_n^m = \frac{1}{(2n + 1)n} [n(n + 1)p_{n+1}^{m+1} - (n - m)(n - m - 1)p_{n-1}^{m+1}]$$

and $\quad \cos \theta \, p_n^m = \frac{1}{(2n + 1)n} [n(n + 1)p_{n+1}^m + (n + m)(n - m)p_{n-1}^m]$

we obtain the integrals of triple products required to deal with second-degree disturbing potentials:

$$\left.\begin{aligned}
\int_0^\pi p_k^u p_k^u p_0^0 \sin \theta \, d\theta &= \frac{2(k - u)! \, (k + u)!}{(2k + 1)(k!)^2} \\[2mm]
\int_0^\pi p_{k+1}^u p_k^u p_1^0 \sin \theta \, d\theta &= \frac{2(k + u + 1)! \, (k - u + 1)!}{(2k + 3)(2k + 1)(k + 1)! \, k!} \\[2mm]
\int_0^\pi p_{k+1}^{u+1} p_k^u p_1^1 \sin \theta \, d\theta &= \frac{2(k + u + 2)! \, (k - u)!}{(2k + 3)(2k + 1)(k + 1)! \, k!} \\[2mm]
\int_0^\pi p_{k-1}^{u+1} p_k^u p_1^1 \sin \theta \, d\theta &= \frac{- 2(k + u)! \, (k - u)!}{(2k + 1)(2k - 1)(k + 1)! \, k!} \\[2mm]
\int_0^\pi p_{k+2}^u p_k^u p_2^0 \sin \theta \, d\theta & \\[2mm]
&= \frac{3(k + 2 - u)! \, (k + 2 + u)!}{(2k + 5)(2k + 3)(2k + 1)(k + 2)(k + 1)! \, k!} \\[2mm]
\int_0^\pi p_{k+2}^{u+1} p_k^u p_2^1 \sin \theta \, d\theta & \\[2mm]
&= \frac{3(k + u + 3)! \, (k - u + 1)!}{(2k + 5)^2(2k + 1)(k + 2)(k + 1)! \, k!}
\end{aligned}\right\} \quad \text{(A.1.6)}$$

2. Power spectra*

We are concerned with time series whose statistical properties are assumed not to vary with a translation of the time axis. Let $u(t)$

---

* The reader is referred to Bartlett (1956), Davenport and Root (1958), and to Blackman and Tukey (1958) for a treatment of this subject.

designate such a *stationary time series*; furthermore let the time average, $\langle u(t) \rangle$, be zero. The autocorrelation (un-normalized)

$$R(\tau) = \langle u(t)u(t - \tau) \rangle \qquad \text{(A.2.1)}$$

is related to the power spectrum $S(f)$ by the transforms

$$S(f) = 4 \int_0^\infty R(\tau) \cos 2\pi f\tau \, d\tau, \quad R(\tau) = \int_0^\infty S(f) \cos 2\pi f\tau \, df,$$
$$\text{(A.2.2)}$$

where $f = \sigma/2\pi$ is the frequency. Note that

$$R(0) = \langle u^2(t) \rangle = \int_0^\infty S(f) \, df \qquad \text{(A.2.3)}$$

is the mean-square value, or the total 'power' of the record. Accordingly $S(f) \, df$ is the contribution towards this total power from a frequency band $f - \frac{1}{2} \, df$ to $f + \frac{1}{2} \, df$. We call $S(f)$ the 'power density', or spectral density. If $S(f)$ is constant over some range of frequencies,

$$S(f) = S \text{ for } f_1 < f < f_2,$$

then all frequencies contribute equally to the power $(f_2 - f_1)S$, and the spectrum is said to be 'white' over this frequency range.

It can be proven that the expression (A.2.2) for the power spectrum in terms of the autocorrelation is at least formally equivalent to

$$S(f) = \lim_{T \to \infty} \frac{1}{T} \left| \int_{-\frac{1}{2}T}^{\frac{1}{2}T} u(t) \, e^{-2\pi i f t} \, dt \right|^2$$

expressed directly in terms of the record, $u(t)$. For a record of finite length T, the power spectrum can be obtained either from the transform of the 'complete' autocorrelation (using lags up to $\tau = T$) or directly from the record by squaring and summing the cosine and sine amplitudes for each harmonic. From either method we obtain estimates of the continuous power spectrum at the frequencies 0, T^{-1}, $2T^{-1}$, $3T^{-1}$, . . . The resultant values will be extremely ragged, and some smoothing over neighboring harmonics is desirable. Two methods of smoothing have been used:

(1) *Faded autocorrelation.* The autocorrelation is multiplied by a 'lag window' $r(\tau)$, which vanishes for all values τ larger than τ_0;

generally τ_0 is chosen between $0 \cdot 01T$ and $0 \cdot 1T$. The transform (A.2.2) of the 'faded' autocorrelation $r(\tau)R(\tau)$ gives

$$\bar{S}(f) = \int_{-\infty}^{\infty} s(\phi)S(f + \phi)\, \mathrm{d}\phi \qquad (A.2.4)$$

in place of $S(f)$. We may regard $s(\phi)$ as the 'spectral window' of the method. The lag window and spectral window are Fourier pairs. We have used a pair proposed by Tukey;

$$r(\tau) = \tfrac{1}{2}[1 + \cos{(\pi\tau/\tau_0)}] \text{ for } \tau \leqslant \tau_0; \; r(\tau) = 0 \text{ for } \tau > \tau_0,$$
$$(A.2.5)$$

$$s(u) = 2\tau_0 \frac{\sin \pi u}{2\pi u(1 - u^2)}, \quad u = 2\phi\tau_0. \qquad (A.2.6)$$

The spectral window is roughly triangular with $s(0) = \tau_0$, $s(\pm 1) = \tfrac{1}{2}\tau_0$, $s(\pm 2) = 0$; beyond $|u| = 2$ there are sidebands of the order u^{-3}.

The principal problem is to choose the proper compromise between resolution and statistical reliability. Making τ_0 smaller broadens the spectral window (diminishes the resolution) but improves reliability. Any computed value, $\bar{S}(f)$, is an estimate of the true spectrum having a chi-square distribution with degrees of freedom equal to twice the width of the spectral window measured in units of T^{-1}.

(2) *Method of maximum likelihood.* [*] Let $S_n T^{-1}$ designate the (unsmoothed) power of the nth harmonic of a record of length T ($f_n = nT^{-1}$). Suppose that harmonics in the neighborhood of a frequency f_0 are peaked, and we wish to fit these to some preconceived analytical function. Consider the resonance response of a linear system (§ 10.1). Each S_n is then assumed to be an independent variable with a mean value

$$\bar{S}_n = \frac{A}{B + (f_n - f_0)^2} \qquad (A.2.7)$$

and a (Rayleigh) probability density $(\bar{S}_n)^{-1} \exp{(- S_n/\bar{S}_n)}$. The joint probability for all S_n is

$$\varphi = \prod_n (\bar{S}_n)^{-1} \exp{(- S_n/\bar{S}_n)}$$

---

[*] See Cramer, 1946, chs. 33 and 34.

II

and this is maximized by choosing A, B, f to satisfy

$$\frac{d}{dA}(\log \varphi) = 0, \quad \frac{d}{dB}(\log \varphi) = 0, \quad \frac{d}{df_0}(\log \varphi) = 0,$$

or

$$\sum[(S_n/\bar{S}_n) - 1] = 0, \quad \sum(\bar{S}_n - S_n) = 0, \quad \sum(f_n - f_0)(\bar{S}_n - S_n) = 0.$$

Degrees of freedom. These are given by the ratio

$$\nu = 2\Delta f/(\Delta f)_0, \qquad (A.2.8)$$

where Δf is the actual resolution of the smoothed record, and $(\Delta f)_0 = T^{-1}$ the ultimate resolution for a record of duration T. Thus ν equals twice the number of harmonics that have been combined into an estimate of spectral density. The factor 2 arises from the combination of sine and cosine terms, phase having been sacrificed. It can be shown that $\nu = 2T/\tau_0$, where τ_0 is the duration of the faded autocorrelation.

For the unsmoothed spectrum, the individual spectral line has a standard deviation equal to its mean value; for the smoothed spectrum, this is reduced by a factor $f(\nu)$. For large ν, $f(\nu)$ approaches $\nu^{-\frac{1}{2}}$. The conflict between the desirability of high resolution (small Δf) and high reliability (large ν) is made explicit by (A.2.8).

Cross spectra. The foregoing discussion can be generalized to a crosscorrelation (not a tensor)

$$R_{uv}(\tau) = \langle u(t)v(t - \tau) \rangle \qquad (A.2.9)$$

between two records, $u(t)$ and $v(t)$. For $u = v$ we have $R_{uu}(\tau) = R(\tau)$ as in (A.2.1); for $u \neq v$ we can define the co-spectrum and quadrature-spectrum according to the transforms

$$\left. \begin{array}{l} S_{uv}(f) = 2 \displaystyle\int_{-\infty}^{\infty} R_{uv}(\tau) \cos 2\pi f\tau \, d\tau, \\[2mm] S'_{uv}(f) = 2 \displaystyle\int_{-\infty}^{\infty} R_{uv}(\tau) \sin 2\pi f\tau \, d\tau \end{array} \right\} . \qquad (A.2.10)$$

The coherence, $Co(f)$ and phase difference, $\phi(f)$, are defined by

$$Co^2 = \frac{S_{uv}^2 + S_{uv}'^2}{S_{uu}S_{vv}}, \quad \tan \phi = \frac{S'_{uv}}{S_{uv}}.$$

A value of $Co(f)$ above $2\nu^{-\frac{1}{2}}$ indicates a meaningful phase relation, ϕ, between the two records for frequencies in the band $f \pm \frac{1}{2}\Delta f$.

Aliasing. The use of digital (rather than continuous) records introduces a troublesome ambiguity. Harmonic oscillations with frequencies

$$f, \quad (\Delta t)^{-1} - f, \quad (\Delta t)^{-1} + f, \quad 2(\Delta t)^{-1} - f, \ldots \quad \text{(A.2.11)}$$

all look alike. If, for example, the frequency f is slightly lower than the sampling frequency $(\Delta t)^{-1}$, then it cannot be distinguished from a low-frequency $(\Delta t)^{-1} - f$; it is like viewing a pendulum with a stroboscope which is not precisely synchronized. Thus the high frequency f appears under the *alias** of the low frequency $(\Delta t)^{-1} - f$.

To avoid aliasing one must make certain that frequencies higher than the "Nyquist frequency"

$$f_N = (2\Delta t)^{-1} \quad \text{(A.2.12)}$$

do not contain any appreciable power. This is achieved either by smoothing the record before taking digital values, or by sampling at sufficiently small time intervals Δt. The requirement may be bothersome, but it cannot be avoided; it is an inherent, irrevocable characteristic of digital analysis.

If the spectrum is contained entirely in the frequency range 0 to f_N, all the aliased frequencies in (A.2.11) are absent; it can be proven that the record is then uniquely determined by values taken at intervals $\Delta t = (2 f_N)^{-1}$.

Differentiation and differences. Let $\bar{S}(f)$ designate the smoothed power spectrum of $u(t)$. Then the smoothed power spectra of $\mathrm{d}u/\mathrm{d}t$, $\mathrm{d}^2 u/\mathrm{d}t^2, \ldots$, are given by

$$(2\pi f)^2 \bar{S}(f), \quad (2\pi f)^4 \bar{S}(f), \ldots, \quad \text{(A.2.13)}$$

respectively.

Suppose u_n is the value of u at time $n\Delta t$. The first differences, $u_2 - u_1, u_3 - u_2, \ldots$, can be written

$$u(t + \tfrac{1}{2}\Delta t) - u(t - \tfrac{1}{2}\Delta t) = \int_{-\infty}^{\infty} u(t + \tau)\kappa(\tau)\,\mathrm{d}\tau, \quad \text{(A.2.14)}$$

where $\qquad \kappa(\tau) = \delta(\tau - \tfrac{1}{2}\Delta t) - \delta(\tau + \tfrac{1}{2}\Delta t).$

---

* This descriptive term has been coined by Tukey. The phenomenon has long been recognized in diffraction spectroscopy, where the periodic repetition of the main peak yields the 'overlapping' (or aliased) orders. There is also a close connexion to the *synodic* tides.

The Fourier transform of $\kappa(\tau)$ is

$$\int_{-\infty}^{\infty} \kappa(\tau) \cos 2\pi f\tau \, d\tau = 0, \quad \int_{-\infty}^{\infty} \kappa(\tau) \sin 2\pi f \, d\tau = 2 \sin \left(\tfrac{1}{2}\pi f/f_N\right),$$

where f_N is the Nyquist frequency (A.2.12). Forming Fourier transforms of both sides of (A.2.14) we find that the power spectrum of the first, second, . . ., differences equals the power spectrum of $u(t)$ multiplied by

$$[2 \sin \left(\tfrac{1}{2}\pi f/f_N\right)]^2, \quad [2 \sin \left(\tfrac{1}{2}\pi f/f_N\right)]^4, \ldots \qquad \text{(A.2.15)}$$

The spectrum of the time derivative, du/dt, is obtained from (A.2.15) by dividing by $(\Delta t)^2$ and then letting Δt approach 0. The result is

$$(\Delta t)^{-2}[2(\tfrac{1}{2}\pi f/f_N)]^2 = (2\pi f)^2$$

as in (A.2.13).

BIBLIOGRAPHY

AIRY, G. (1860). Change of climate. *Athenaeum*, **7171**, 384.

ANDERSSON, F. (1937). Berechnung der Variation der Tageslänge infolge der Deformation der Erde durch fluterzeugende Kräfte. *Arkiv för Matematik, Astronomi och Fysik*, **26A**, 1.

ASSAMI, E. (1954). *Proc. Japan Acad.*, **30**, 102.

BAKHUYZEN, H. (1913). Über die Änderung der Meereshöhe und ihre Beziehung zur Polhöhenschwankung. *Vierteljahrschrift der Astronomischen Gesellschaft, Leipzig*, **47**, 218.

BANNON, J. and STEELE, L. (1957). Average water-vapour content of the air. *Meteorological Research Committee, London*, **1075**.

BARRELL, J. (1914). The status of hypotheses of polar wanderings. *Science*, **40**, 333.

BARTLETT, M. S. (1956). *An Introduction to Stochastic Processes*. Cambridge University Press.

BAUER, A. (1955). The balance of the Greenland ice sheet. *Journal of Glaciology*, **2**, 456.

BAUSSAN, J. (1951). La composante de Chandler dans la variation des niveaux marins. *Annales de Géophysique*, **7**, 59.

BEISER, A. (1958). The external magnetic field of the Earth. *Il Nuovo Cimento, Serie X*, **8**, 160.

BELSHÉ, J. (1957). Palaeomagnetic investigations of carboniferous rocks in England and Wales. *Advances in Physics*, **6**, 187.

BENTON, G. and ESTOQUE, M. (1954). Water-vapor transfer over the North American continent. *Journal of Meteorology*, **11**, 462.

BIRCH, F. and BANCROFT, D. (1942). The elasticity of glass at high temperatures, and the vitreous basaltic substratum. *American Journal of Science*, **240**, 457.

BLACKMAN, R. and TUKEY, J. (1958). The measurement of power spectra from the point of view of the communications engineer. *Bell System Technical Journal*, **37**, 185.

BLASER, J. and BONANOMI, J. (1958). Comparison of an ammonia maser with cesium atomic frequency standard. *Nature*, **182**, 859.

BLASER, J. and DE PRIUS, J. (1958). Comparison of astronomical time measurements with atomic frequency standards. *Nature*, **182**, 859.

BONDI, H. and GOLD, T. (1955). On the damping of the free nutation of the Earth. *Monthly Notices, Royal Astronomical Society*, **115**, 41.

BONDI, H. and LYTTLETON, R. (1948). On the dynamical theory of the rotation of the Earth. *Proceedings of the Cambridge Philosophical Society*, **44**, 345.

BOWDEN, K. (1953). Note on wind drift in a channel in the presence of tidal currents. *Proceedings of the Royal Society, A*, **219**, 426.

BOWDEN, K. and FAIRBAIRN, L. (1956). Measurement of turbulent fluctuations and Reynolds stresses in a tidal current. *Proceedings of the Royal Society, A*, **237**, 422.

BRIDGMAN, P. (1949). *The Physics of High Pressures*. London: G. Bell & Sons, Ltd.

BROUWER, D. (1952a). A new discussion of the changes in the Earth's rate of rotation. *Proceedings of the National Academy of Sciences*, **38**, 1.

BROUWER, D. (1952b). A study of the changes in the rate of rotation of the Earth. *Astronomical Journal*, **57**, 125.

BROUWER, D. and WATTS, C. (1942). A comparison of the results of occultations and meridian observations of the moon. *Astronomical Journal*, **52**, 169.

BROWN, E. (1926). The evidence for changes in the rate of rotation of the Earth and their geophysical consequences, with a summary and discussion of the deviations of the moon and sun from their gravitational orbits. *Transactions of the Astronomical Observatory of Yale University*, **3**, 207.

BULLARD, E. (1948). The secular change in the earth's magnetic field. *Monthly Notices, Royal Astronomical Society Geophysical Supplement*, **5**, 248.

BULLARD, E. (1949). The magnetic field within the Earth. *Proceedings of the Royal Society, A*, **197**, 433.

BULLARD, E. (1950). The origin of the Earth's magnetic field. *The Observatory*, **70**, 139.

BULLARD, E., FREEDMAN, C., GELLMAN, H., and NIXON, J. (1950). The westward drift of the Earth's magnetic field. *Phil. Trans. Am.*, **243**, 67.

BURGERS, J. (1955). Rotational motion of a sphere subject to visco-elastic deformation. I, II, III: *Nederland. Akad. Wetensch. Proc.*, **58**, 219.

CAMPBELL, C. and RUNCORN, S. (1956). Magnetization of the Columbia River basalts in Washington and northern Oregon. *Journal Geophysical Research*, **61**, 449.

CECCHINI, G. (1928). Il problema della variazione delle latitudini. *Pubblicazioni del Reale Osservatorio Astronomico di Brera in Milano*, **61**, 7.

CECCHINI, G. (1950). Le variazioni di latitudine e il movimento del polo di rotazione terrestre. *Bulletin Géodésique*, **17**, 325.

CECCHINI, G. (1952). Le variazioni di latitudine e il movimento del polo di rotazione terrestre in base alle osservazioni fatte nelle Stazioni Internazionali di Latitudine nel biennio 1949–50. *Bulletin Géodésique*, **26**, 423.

CHANDLER, S. (1891a). On the variation of latitude. *Astronomical Journal*, **11**, 83.

CHANDLER, S. (1891b). On the supposed secular variation of latitude. *Astronomical Journal*, **11**, 109.

CHANDLER, S. (1892). On the variation in latitude. *Astronomical Journal*, **12**, 17.

CHAPMAN, S. and BARTELS, J. (1940). *Geomagnetism.* Oxford University Press.

CHARNEY, J. (1955). The generation of oceanic currents by wind. *Journal of Marine Research*, **14**, 477.

CHEVALLIER, R. (1925). L'aimantation des laves de l'Etna et l'orientation du champ terrestre en Sicile du XIIe au XVIIe siècle. *Ann. de Phys.*, **4**, 5.

CHREE, C. (1889). The equations of an isotropic elastic solid in polar and cylindrical co-ordination, their solution and application. *Transactions of the Cambridge Philosophical Society*, **14**, 250.

CHRISTIE, A. (1900). The latitude variation tide. *Bull. Phil. Soc. of Washington*, **13**, 103.

CLEGG, J., DEUTSCH, E., EVERITT, C., and STUBBS, P. (1957). Some recent paleomagnetic measurements made at Imperial College, London. *Advances in Physics*, **6**, 216.

CLEMENCE, G. First-order theory of Mars. *Astr. Pap. Am. Eph. and Nau. Alm.*, **11**, part I.

COLBORNE, D. (1931). The diurnal tide in an ocean bounded by two meridians. *Proceedings of the Royal Society*, **131**, 38.

COMSTOCK, G. C. (1954). The secular variation of latitude. *Astr. J.*, **12**, 24.

COWAN, R. (1950). Polar wanderings and the shifting of the atmospheric mass. *M.S. Thesis, Massachusetts Institute of Technology.*

COWELL, P. (1905). Lunar theory from observation. *Nature*, **73**, 80.

COWLING, T. (1957). *Magnetohydrodynamics.* New York: Interscience Publishers, Inc.

CRAMER, H. (1946). *Mathematical Methods of Statistics.* Princeton University Press.

CREER, K., IRVING, E., and RUNCORN, S. (1957). Geophysical interpretations of palaeomagnetic directions from Great Britain. *Philosophical Transactions of the Royal Society*, *A*, **250**, 144.

DALY, R. (1920). Sea level. *Proceedings of the National Academy of Sciences*, **6**, 246.

DALY, R. (1927). The geology of St. Helena Island. *Proceedings of the American Academy of Arts and Sciences*, **62**, 31.

DANJON, A. (1958). The George Darwin lecture. *Royal Astronomical Society, The Observatory*, **78**, 107.

DARWIN, G. (1887). On the influence of geological changes on the Earth's axis of rotation. *Philosophical Transactions of the Royal Society, A,* **167,** 271.

DAVENPORT, W., Jr. and ROOT, W. (1958). *An Introduction to the Theory of Random Signals and Noise.* New York: McGraw-Hill.

DAY, A. and RUNCORN, S. (1955). Polar wandering; some geological, dynamical and paleomagnetic aspects. *Nature,* **176,** 422.

DEFANT, A. (1950). On the origin of internal tide waves in the open sea. *Journal of Marine Research,* **9,** 111.

DE SITTER, W. (1927). On the secular accelerations and the fluctuations of the longitudes of the moon, the sun, Mercury and Venus. *Bulletin of the Astronomical Institutes of the Netherlands,* **4,** 21.

DEUTSCH, E., RADAKRISHNAMURTY, C., and SAHASRABUDHE, P. (1958). Remanent magnetism of some lavas in the Deccan Traps. *Philosophical Magazine,* **3,** 170.

DICKE, R. (1957). Principle of equivalence and the weak interaction. *Reviews of Modern Physics,* **29,** 355.

DICKE, R. (1958). Gravitation, an enigma. *Journal of the Washington Academy of Science,* **48,** 213.

DIETRICH, G. (1944). Die Gezeiten des Weltmeeres als geographische Erscheinung. *Z. Ges. Erdkunde.*

DISNEY, L. (1955). Tide heights along the coasts of the United States. *Proceedings of the American Society of Civil Engineers,* **81,** 666.

DOELL, R. (1955). Palaeomagnetic study of rocks from Grand Canyon of the Colorado River. *Nature,* **176,** 1167.

DOODSON, A. T. and WARBURG, H. D. (1941). *Admiralty Manual of Tides.* London: H.M. Stationery Office.

DUPREE, A. (1957). *Science in the Federal Government.* Harvard University Press.

DuTorr, A. (1937) *Our Wandering Continents.* Edinburgh: Oliver & Boyd.

ECKART, C. (1948). The thermodynamics of irreversible processes. IV. The theory of elasticity and anelasticity. *The Physical Review,* **73,** 373.

ELSASSER, W. (1950). The Earth's interior and geomagnetism. *Reviews of Modern Physics,* **22,** 1.

ELSASSER, W. and MUNK, W. (1958). Geomagnetic drift and the rotation of the Earth. *Contributions in Geophysics: In Honor of Beno Gutenberg.* Pergamon Press.

ELSASSER, W. and TAKEUCHI, H. (1955). Nonuniform rotation of the Earth and geomagnetic drift. *Transactions of the American Geophysical Union,* **36,** 584.

ENGELS, F. (1954). *Dialectics of Nature.* Moscow: Foreign Languages Publishing House.

ESSEN, L. and PARRY, J. (1955). An atomic standard of frequency and time interval. *Nature*, **176**, 280.

ESSEN, L., PARRY, J., MARKOWITZ, W., and HALL, R. (1958). Variation in the speed of rotation of the Earth since June 1955. *Nature*, **181**, 1054.

FEDOROV, E. (1949). On the influence of changes of sea level induced by motion of the Earth's poles, on their motion. *Doklady Akad. Nauk. U.S.S.R.*, **67**, 647.

FINCH, H. (1950). On a periodic fluctuation in the length of the day. *Monthly Notices, Royal Astronomical Society*, **110**, 3.

FOLGHERAITER, M. (1899). Sur les variations séculaires de l'inclinaison magnétique dans l'antiquité. *Journ. de Phys.*, **8**, 660.

FOTHERINGHAM, J. (1909). The eclipse of Hipparchus. *Monthly Notices, Royal Astronomical Society*, **69**, 204.

FOTHERINGHAM, J. (1918). The secular acceleration of the Sun as determined from Hipparchus's equinox observations; with a note on Ptolemy's false equinox. *Monthly Notices, Royal Astronomical Society*, **78**, 406.

FOTHERINGHAM, J. (1920). Secular accelerations of sun and moon as determined from ancient lunar and solar eclipses, occultations, and equinox observations. *Monthly Notices, Royal Astronomical Society*, **80**, 578.

FOTHERINGHAM, J., and LONGBOTTOM, G. (1915). The secular acceleration of the moon's mean motion, as determined from occultations in the Almagest. *Monthly Notices, Royal Astronomical Society*, **75**, 377.

GEOLOGIC DIVING CONSULTANTS. (1956). *Oceanographic Studies for Sewage Outfall, City of San Diego, California*.

GINZEL, F. K. (1906). *Handbuch der Mathematischen und Technischen Chronologie*. Leipzig: J. C. Henrichs'sche Buchhandlung.

GLAUERT, H. (1915). Rotation of the Earth. *Monthly Notices, Royal Astronomical Society*, **75**, 489.

GOGUEL, J. (1950). Les déplacements séculaires du pôle. *Ann. de Geoph.*, **6**, 139.

GOLD, T. (1955). Instability of the earth's axis of rotation. *Nature*, **175**, 526.

GOLDSBROUGH, G. (1913). The dynamical theory of the tides in a polar ocean. *Proceedings of the London Mathematical Society*, **14**, 31.

GOLDSBROUGH, G. (1914). The dynamical theory of the tides in a zonal ocean. *Proceedings of the London Mathematical Society*, **14**, 207.

GONDOLATSCH, F. (1953). Erdrotation, Mondbewegung und das Zeitproblem der Astronomie. *Veröffentlichungen des Astronomischen Rechen-Institus zu Heidelberg*, No. 5.

GRAHAM, J. (1949). The stability and significance of magnetism in sedimentary rocks. *Journal of Geophysical Research*, **54**, 131.

GRAHAM, J. (1955). Evidence of polar shift since Triassic time. *Journal of Geophysical Research*, **60**, 329.

GRAHAM, J., BUDDINGTON, A., and BALSLEY, J. (1957) Stress-induced magnetizations of some rocks with analyzed magnetic minerals. *Journal of Geophysical Research*, **62**, 465.

GREAVES, W. and SYMMS, L. (1943). The short-period erratics of free pendulum and quartz clocks. *Monthly Notices, Royal Astronomical Society*, **103**, 196.

GROVES, G. (1957). Day to day variation of sea level. *Meteorological Monographs*, **2**, 32.

GROVES, G. and MUNK, W. (1959). A note on tidal friction. *Journal of Marine Research* (in press).

GUTENBERG, B. (1941). Changes in sea level, postglacial uplift, and mobility of the Earth's interior. *Bulletin of the Geological Society of America*, **52**, 721.

GUTENBERG, B. (1951). *Internal Constitution of the Earth*. New York: Dover.

GUTENBERG, B. (1956). Damping of the Earth's free nutation. *Nature*, **177**, 887.

HASSAN, E. M. (1960). Fluctuations in the atmospheric inertia: 1873–1950. *Meteorological Monographs* (in press).

HAUBRICH, R. and MUNK, W. (1959). The pole tide. *Journal of Geophysical Research*, **64**, 2373.

HAURWITZ, B. (1956). The geographical distribution of the solar semidiurnal pressure oscillation. *Meteorol. Pap.* **2**, *No. 5, New York University, College of Engineering*.

HEISKANEN, W. (1921). Über den Einfluss der Gezeiten auf die säkuläre Acceleration des Mondes. *Ann. Acad. Scient. Fennicae A*, **18**, 1.

HEISKANEN, W. (1957). The Columbus geoid. *Transactions of the American Geophysical Union*, **38**, 841.

HEISKANEN, W. and VENING MEINESZ, F. (1958). *The Earth and Its Gravity Field*. New York: McGraw-Hill.

HELMERT, F. (1915). Neue Formeln für den Verlauf der Schwerkraft im Meeresniveau beim Festlande. *Sitzungsberichte der Königlichen Preussischen Akademie der Wissenschaften*, **676**.

HESS, G. (1958). The annual variation of the length of the day as evidence relating to a theory of gravity. *Senior Thesis, Department of Physics, Princeton University*.

HOLMBERG, E. (1952). A suggested explanation of the present value of the velocity of rotation of the Earth. *Monthly Notices, Royal Astronomical Society, Geophysical Supplement*, **6**, 325.

HOSOYAMA, K. (1952). On secular change of latitude. *Transactions of the American Geophysical Union*, **33**, 345.

HOYLE, F. (1955). *Frontiers of Astronomy.* New York: Harper Brothers.

HUAUX, A. (1951). Sur des déplacements concomitants de l'axe instantané de rotation et de l'axe principal instantané d'inertie de la terre. *Bulletin de l'Academie Royale de Belgique, Classe des Sciences,* 37, 53.

HYLCKAMA, T. (1956). The water balance of the earth. *Drexel Institute of Technology, Laboratory of Climatology, Publications in Climatology,* 9, 57.

INGLIS, D. (1957). Shifting of the Earth's axis of rotation. *Reviews of Modern Physics,* 29, 9.

IRVING, E. and GREEN, R. (1958). Polar movement relative to Australia. *Royal Astronomical Society, Geophysical Journal,* 1, 64.

IRVING, E. and RUNCORN, S. (1957). Analysis of the palaeomagnetism of the torridonian sandstone series of northwest Scotland. *Philosophical Transactions of the Royal Society, A,* 250, 83.

JARDETZKY, W. (1949). On the rotation of the Earth during its evolution. *Transactions of the American Geophysical Union,* 30, 797.

JEFFREYS, H. (1915). The viscosity of Earth. *Monthly Notices, Royal Astronomical Society,* 75, 648.

JEFFREYS, H. (1916). Causes contributory to the annual variation of latitude. *Monthly Notices, Royal Astronomical Society,* 76, 499.

JEFFREYS, H. (1917). The viscosity of the Earth. *Monthly Notices, Royal Astronomical Society,* 78, 116.

JEFFREYS, H. (1920). Tidal friction in shallow seas. *Philosophical Transactions of the Royal Society, A,* 221, 239.

JEFFREYS, H. (1926). On the dynamics of geostrophic winds. *Quarterly Journal of the Royal Meteorological Society,* 52, 85.

JEFFREYS, H. (1928). Possible tidal effects on accurate timekeeping. *Monthly Notices, Royal Astronomical Society, Geophysical Supplement,* 2, 56.

JEFFREYS, H. (1933). The formation of cyclones in the general circulation. *Proceedings of the International Meteorological Assoc., Lisbon,* 219.

JEFFREYS, H. (1940). The variation of latitude. *Monthly Notices, Royal Astronomical Society,* 100, 139.

JEFFREYS, H. (1949). Dynamic effects of a liquid core. *Monthly Notices, Royal Astronomical Society,* 109, 670.

JEFFREYS, H. (1952). *The Earth.* Cambridge University Press.

JEFFREYS, H. (1956). The damping of the variation of latitude. *Monthly Notices, Royal Astronomical Society,* 116, 26.

JEFFREYS, H. (1958a). A modification of Lomnitz's law of creep in rocks. *Geophysical Journal of the Royal Astronomical Society,* 1, 92.

JEFFREYS, H. (1958b). Review. *Journal of Fluid Mechanics,* 4, 335.

JEFFREYS, H. and JEFFREYS, B. (1950). *Methods of Mathematical Physics.* Cambridge University Press.

JEFFREYS, H. and VICENTE, R. (1957a). The theory of nutation and the variation of latitude. *Monthly Notices, Royal Astronomical Society,* **117,** 142.

JEFFREYS, H. and VICENTE, R. (1957b). The theory of nutation and the variation of latitude: The Roche model core. *Monthly Notices, Royal Astronomical Society,* **117,** 162.

JENKINSON, A. (1955). Average vector wind distribution of the upper air in temperate and tropical latitudes. *The Meteorological Magazine,* **84.** 140.

JOHNSON, E., MURPHY, T., and TORRESON, O. (1948). Pre-history of the Earth's magnetic field. *J. Terr. Mag.,* **53,** 349.

JONES, SIR H. SPENCER (1932). *Annals of the Cape Observatory,* **13,** Part 3.

JONES, SIR H. SPENCER (1939a). The tidal effect on the variation of latitude at Greenwich. *Monthly Notices, Royal Astronomical Society,* **99,** 196.

JONES, SIR H. SPENCER (1939b). The rotation of the earth, and the secular accelerations of the sun, moon and planets. *Monthly Notices, Royal Astronomical Society,* **99,** 541.

JONES, SIR H. SPENCER (1956). The rotation of the earth. *Handbuch der Physik,* **47,** 1.

KELVIN, LORD. (*See* SIR W. THOMSON.)

KLEIN, F. and SOMMERFELD, A. (1903). *Theorie des Kreisels.* Heft III. Leipzig: Teubner.

KNEISSL, M. (1955). Nachweis systematischer Fehler beim Feinnivellement. *Abh. d. Bayer. Akad. d. Wiss., Mat.-Naturw. Klasse, Neue Folge, H.* **68.**

KNOPOFF, L. and MACDONALD, G. (1958). Attenuation of small amplitude stress waves in solids. *Reviews of Modern Physics,* **30,** 1178.

KUENEN, P. (1950). *Marine Geology.* New York: John Wiley & Sons, Inc.

KUIPER, G. (editor). (1954). *The Earth as a Planet.* University of Chicago Press.

KUIPER, H. (1943). Poolbewegingen tengevolge van poolvluchtkracht. *Doctoral Thesis, University of Utrecht,* N. Hollandsche Uitg. Mij, Amsterdam.

KULIKOV, K. A. (1950). The motion of the poles of the Earth and the variation of latitude. *Ouspekhi Astr. Nauk,* **5,** 111.

LAMB, H. (1932). *Hydrodynamics.* Cambridge University Press.

LAMBERT, W. (1922). The interpretation of apparent changes in mean latitude. *The Astronomical Journal,* **34,** 103.

LAMBERT, W. (1928). The importance from a geophysical point of view of a knowledge of the tides in the open sea. *Bulletin* 11, *Section d' Océanographie, Conseil International de Recherches, Union Géodésique et Géophysique Internationale*, **52**.

LAMBERT, W. (1931). Chapter 5, Earth tides; Chapter 6, Tidal friction. *Bulletin* 78 *of the National Research Council.*

LAMBERT, W., SCHLESINGER, F., and BROWN, E. (1931). The variation of latitude. *Bulletin* 78 *of the National Research Council*, **16**, 245.

LAMP, J. (1891). Über Niveauschwankungen der Ozeane als eine mögliche Ursache der Veränderlichkeit der Polhohe. *Astron. Nachrichten*, **126**, 223.

LANDSBERG, H. (1933). Über Zusammenhänge von Tieferdbeben mit anderen geophysikalischen Erscheinungen. *Gerlands Beiträge zur Geophysik V. Conrad-Wien*, **40**, 238.

LANDSBERG, H. (1948). Note on deep-focus earthquakes, pressure changes, and pole motion. *Geofisica Pura e Applicata*, **12**, 177.

LARMOR, J. (1896). On the period of the earth's free Eulerian precession. *Proceedings of the Cambridge Philosophical Society*, **9**, 183.

LARMOR, J. (1909). The relation of the Earth's free precessional nutation to its resistance against tidal deformation. *Proceedings of the Royal Society, A*, **82**, 89.

LAWFORD, A. and VELEY, V. (1951). Variations in the length of day. *Nature*, **167**, 684.

LAWRENCE, D. (1950). Glacier fluctuation for six centuries in southeastern Alaska and its relation to solar activity. *Geogr. Rev.*, **40**, 191.

LECAR, M., SORENSON, J., and ECKELS, A. (1959). A determination of the coefficient *J* of the second harmonic in the Earth's gravitational potential from the orbit of satellite 1958 β_2. *Journal of Geophysical Research*, **64**, 209.

LEDERSTEGER, K. (1949). Numerische Untersuchungen über die Perioden der Polbewegung. *Österreichischen Zeitschrift für Vermessungswesen*, Sonderheft **7**, 1.

LISITZIN, E. and PATTULLO, J. (1960). The principal factors influencing the seasonal variation in sea level. *Jour. of Geophysical Research* (in press).

LOEWE, F. (1956). *Études de Glaciologie en Terre Adélie*, 1951–52, *Act. Scient. et Industr.* 1247, *Exp. Pol. Fran.* Paris: Hermann & Cie.

LOMNITZ, C. (1956). Creep measurements in igneous rocks. *Journal of Geology*, **64**, 473.

LOMNITZ, C. (1957). Linear dissipation in solids. *Journal of Applied Physics*, **28**, 201.

LONGUET-HIGGINS, M. (1958). On the intervals between successive zeros of a random function. *Proceedings of the Royal Society, A*, **246**, 99.

LOVE, A. (1908). Note on the representation of the Earth's surface by means of spherical harmonics of the first 3 degrees. *Proceedings of the Royal Society, A*, **80**, 553.

LOVE, A. (1927). *A Treatise on the Mathematical Theory of Elasticity*. New York: Dover Publications.

LYONS, H. (1952). Spectral lines as frequency standards. *Annals of the New York Academy of Sciences*, **55**, 831.

LYTTLETON, R. A. (1953). *The Stability of Rotating Liquid Masses*. Cambridge University Press.

MAEDA, K. (1958). Distortion of the magnetic field in the outer atmosphere due to the rotation of the Earth. *Annales de Géophysique*, **14**, 154.

MARKOWITZ, W. (1942). The free polar motion, 1916·0–1940·0. *The Astronomical Journal*, **50**, 17.

MARKOWITZ, W. (1945). Redeterminations of Latitude and Longitude. *Transactions of the American Geophysical Union*, **26**, 197.

MARKOWITZ, W. (1954). Photographic determination of the moon's position, and application to the measure. *The Astronomical Journal*, **59**, 69.

MARKOWITZ, W. (1955). The annual variation in the rotation of the Earth, 1951–4. *Astronomical Journal*, **60**, 171.

MARKOWITZ, W. (1959). Variations in rotation of the Earth, results obtained with the dual-rate moon camera and photographic zenith tubes. *Astronomical Journal*, **64**, 106.

MASON, W. (1958). *Physical Acoustics and the Properties of Solids*. New York: Van Nostrand.

MASSEY, H. and BOYD, R. (1958). *The Upper Atmosphere*. London: Hutchinson.

MAXIMOV, I. (1956). The polar tide in the Earth's ocean. *Doklady Akademii Nauk. U.S.S.R.*, **108**, 799.

McDONALD, K. (1957). Penetration of the geomagnetic secular field through a mantle with variable conductivity. *Journal of Geophysical Research*, **62**, 117.

McLELLAN, H. (1958). Energy considerations in the Bay of Fundy system. *Journal of Fisheries Research Board of Canada*, **15**, 115.

MEINARDUS, W. (1942). Die bathygraphische Kurve des Tiefseebodens und die hyposographische Kurve der Erdkruste. *Ann. d. Hydr. u. Marit. Met.*, **70**, 225.

MELCHIOR, P. (1954). On the motions of the instantaneous axis of rotation relative to the earth (Contribution à l'étude des mouvements de l'axe instantané de rotation par rapport au Globe terrestre). *Observatoire Royale de Belgique, Monographie No. 3*.

MELCHIOR, P. (1955). Déplacements séculaires du pôle moyen et catalogues d'étoiles. *Observatoire Royal de Belgique, Communications* **79**.

MELCHIOR, P. (1957). Latitude variation. *Progress in Physics and Chemistry of the Earth*, **2**. Pergamon Press.

MERCANTON, P. (1906). Magnetic inclination in prehistoric times. *Comptes Rendus*, **143**, 139.

MERSON, R. and KING-HELE, D. (1958). Use of artificial satellites to explore the earth's gravitational field: results from Sputnik 2 (1957). *Nature*, **182**, 640.

MILANKOVITCH, M. (1934). Der Mechanismus der Polverlagerungen und die daraus sich ergebenden Polbahnkurven. *Gerlands Beitr. z. Geoph.*, **42**, 70.

MINTZ, Y. (1951). The geostrophic poleward flux of angular momentum. *Tellus*, **3**, 195.

MINTZ, Y. (1954). The observed zonal circulation of the atmosphere. *Bulletin of the American Meteorological Society*, **35**, 208.

MINTZ, Y. and MUNK, W. (1951). The effect of winds and tides on the length of day. *Tellus*, **3**, 117.

MINTZ, Y. and MUNK, W. (1954). The effect of winds and bodily tides on the annual variation in the length of day. *Monthly Notices, Royal Astronomical Society, Geophysical Supplement*, **6**, 566.

MUNK, W. (1950). On the wind-driven ocean circulation. *Journal of Meteorology*, **7**, 79.

MUNK, W. (1956). Geophysical discussion. *Royal Astronomical Society, The Observatory*, **76**, 56.

MUNK, W. (1958). Remarks concerning the present position of the pole. *Geophysica*, **6**, 335.

MUNK, W. and GROVES, G. (1952). The effect of winds and ocean currents on the annual variation in latitude. *Journal of Meteorology*, **9**, 385.

MUNK, W. and HASSAN, E. M. (1961). Atmospheric excitation of the Earth's wobble. *Geophysical Journal*, **4**, Jeffreys Jubilee number (in press).

MUNK, W. and HAUBRICH, R. (1958). The annual pole tide. *Nature*, **182**, 42.

MUNK, W. and MACDONALD, G. (1960). Continentality and the gravitational field of the Earth. *Journal of Geophysical Research* (in press).

MUNK, W. and MILLER, R. (1950). Variation in the Earth's angular velocity resulting from fluctuations in atmospheric and oceanic circulation. *Tellus*, **2**, 93.

MUNK, W. and REVELLE, R. (1952a). On the geophysical interpretation of irregularities in the rotation of the earth. *Monthly Notices, Royal Astronomical Society, Geophysical Supplement*, **6**, 331.

MUNK, W. and REVELLE, R. (1952b). Sea level and the rotation of the earth. *American Journal of Science*, **250**, 829.

MURRAY, C. (1957). The secular acceleration of the moon, and the lunar tidal couple. *Monthly Notices, Royal Astronomical Society*, 117, 478.

NADAI, A. (1950). *Theory of Flow and Fracture of Solids*, 2nd ed. New York: McGraw-Hill.

NAGATA, T. (1953). *Rock Magnetism*. Tokyo: Maruzen.

NEUGEBAUER, O. (1957). *The Exact Sciences in Antiquity*, 2nd ed. Brown University Press.

NEWCOMB, S. (1892). Remarks on Mr. Chandler's law of variation of terrestrial latitudes. *Astronomical Journal*, 12, 49.

NEWCOMB, S. (1909). Fluctuations in the Moon's motion. *Monthly Notices, Royal Astronomical Society*, 69, 164.

NISHIMURA, E. (1950). On earth tides. *Transactions of the American Geophysical Union*, 31, 357.

NOMITSU, T. (1932). Note on Dr. Rosenhead's papers, 'The annual variation of latitude' and 'Tides on a two-layer Earth.' *Memoirs of the College of Science, Kyoto Imperial University*, A-15, 123.

O'KEEFE, J., ECKELS, A., and SQUIRES, K. (1959). The gravitational field of the Earth. *Astronomical Journal*, 64, 245.

PARIYSKI, N. and BERLYAND, O. (1953). The effect of seasonal changes in atmospheric circulation on the rotational velocity of the Earth. *Trans. of Geophysical Institute, Academy of Sciences of U.S.S.R.*, 19, 103.

PATTULLO, J., MUNK, W., REVELLE, R., and STRONG, E. (1955). The seasonal oscillation in sea level. *Journal of Marine Research*, 14, 88.

PEKERIS, C. (1937). Atmospheric oscillations. *Proceedings of the Royal Society*, A, 158, 650.

PEKERIS, C. (1939). The propagation of a pulse in the atmosphere. *Proceedings of the Royal Society*, A, 171, 434.

POINCARÉ, H. (1910). Sur la precession des corps deformables. *Bulletin Astronomique*, 27, 321.

POLLAK, L. (1927). Das Periodogramm der Polbewegung. *Gerlands Beiträge zur Geophysik*, 16, 108.

PREY, A. (1922). Darstellung der Höhen- und Tiefenverhältnisse der Erde. *Abhandlungen der Königlichen Gesellschaft der Wissenschaften zu Göttingen, Mathematisch-Physikalische Klasse, neue Folge*, 11.

PROUDMAN, J. (1913). Limiting forms of long period tides. *Proceedings of the London Mathematical Society*, 13, 273.

PROUDMAN, J. (1916). On the dynamical equations of the tides. *Proceedings of the London Mathematical Society*, 18, 1.

PRZBYLLOK, E. (1919). Veröff. des Preuss. Geodäsischen Inst., N.F., 80.

PRZBYLLOK, E. (1927). Über die Ursachen des nichtperiodischen Teiles der Polhöhenschwankungen. *Schriften der Königsberger Gelehrten Gesellschaft*, 3, 43.

QUENBY, J. and WEBBER, W. (1959). Cosmic ray cut-off rigidities and the Earth's magnetic field. *Philosophical Magazine*, **8**, 90.

REICHEL, E. (1949). Zum Dampfgehalt und Wasserkbreislauf der Atmosphäre. *Met. Rundschau*, **2**, 206.

REID, J. (1956). Observations of internal tides in October 1950. *Transactions of the American Geophysical Union*, **37**, 278.

RIKITAKE, T. (1955). Growth of the magnetic field of the self-exciting dynamo in the earth's core. *Tokyo University Earthquake Research Institute Bulletin*, **33**, 571.

RILEY, G. (1944). The carbon metabolism and photosynthetic efficiency of the earth as a whole. *Amer. Sci.*, **32**, 129.

ROSENHEAD, L. (1929). The annual variation of latitude. *Monthly Notices, Royal Astronomical Society, Geophysical Supplement*, **2**, 140.

ROTHWELL, P. (1958). Cosmic rays in the Earth's magnetic field. *The Philosophical Magazine*, **3**, 961.

ROUTH, E. (1905). *Dynamics of a System of Rigid Bodies, Part II*. New York: Macmillan. Republished by Dover Publications, 1955.

RUDNICK, P. (1953). The detection of weak signals by correlation methods. *Journal of Applied Physics*, **24**, 128.

RUDNICK, P. (1956). The spectrum of the variation in latitude. *Transactions of the American Geophysical Union*, **37**, 137.

RUNCORN, S. (1954). The Earth's core. *Transactions of the American Geophysical Union*, **35**, 49.

RUNCORN, S. (1955*a*). Rock magnetism, geophysical aspects. *Phil. Mag., Adv. Phys.*, **4**, 244.

RUNCORN, S. (1955*b*). The permanent magnetization of rocks. *Endeavour*, **14**, 152.

RUNCORN, S. (1955*c*). The Earth's magnetism. *Scientific American*, **193**, 152.

RUNCORN, S. (1956). The present status of theories of the main geomagnetic field. *Geologie en Mijnbouw*, **18**, 347.

RUNCORN, S. (1957). Convection currents in the mantle and recent developments in geophysics. *Verhandelingen Koninklijk Nederlandsch Geologisch-Mijnbouwkundig Genootschap, Geologische Serie*, **18** (Gedenkboek F. A. Vening Meinesz).

RUTTEN, M., VON EVERDINGEN, R., and ZIJDERVELD, J. (1957). Paleomagnetism in the Permian of the Oslo Graven (Norway) and of the Esterel (France). *Geologie en Mijnbouw*, **19**, 193.

SCHEIBE, A. and ADELSBERGER, U. (1950). Die Gangleistungen der PTR-Quarzuhren und die jährliche Schwankung der astronomischen Tageslänge. *Zeitschrift für Physik*, **127**, 416.

SCHEIDEGGER, A. (1958). *Principles of Geodynamics*. Berlin: Springer.

310   THE ROTATION OF THE EARTH

SCHWERDTFEGER, W. (1960). The seasonal variation of the strength of the
southern circumpolar vortex (in press).

SCHWERDTFEGER, W. and PROHASKA, F. (1956). Der Jahresgang des
Luftdruckes auf der Erde und seine halbjährige Komponente.
Meteorologische Rundschau, **9**, 33.

SCHWEYDAR, W. (1916). Die Bewegung der Erdachse der elastischen Erde
in Erdkörper und im Raume. *Astronomische Nachrichten*, **203**, 103.

SCHWEYDAR, W. (1919). Zur Erklärung der Bewegung der Rotationspole
der Erde. *Preuss. Akad. Wiss. Ber.*, **20**, 357.

SEKIGUCHI, N. (1954). On a character about the secular motion of the pole
of the earth. *Publ. Astr. Soc. Japan*, **5**, 109.

SEVARLIC, B. (1957). *Sur le problème de la variation des latitudes et du
mouvement du pôle instantané de rotation à la surface de la Terre.*
Beograd.

SHEPARD, F., REVELLE, R., and DIETZ, R. (1939). Ocean-bottom currents
off the California coast. *Science*, **89**, 488.

SHEPARD, F. and SUESS, H. (1956). Rate of postglacial rise of sea level.
Science, **123**, 1082.

SIEBERT, M. (1954). Zur Theorie der thermischen Erregung gezeitenartiger
Schwingungen der Erdatmosphäre. *Naturwissenschaften*, **41**, 446.

SIGURGEIRSSON, T. (1957). On the direction of magnetization in Icelandic
basalts. *Advances in Physics*, **6**, 240.

SIMPSON, G. (1918). The twelve-hourly barometer oscillation. *Quarterly
Journal of the Royal Meteorological Society*, **44**, 1.

SIMPSON, J., FENTON, K., KATZMAN, J., and ROSE, D. (1956). Effective
geomagnetic equator for cosmic radiation. *Physical Review*, **102**,
1648.

SMITH, H. (1953). Quartz clocks of the Greenwich time service. *Monthly
Notices, Royal Astronomical Society*, **113**, 67.

SMITH, H. and TUCKER, R. (1953). The annual fluctuation in the rate of
rotation of the Earth. *Monthly Notices, Royal Astronomical Society*,
113, 251.

SPITALER, R. (1901). Die periodischen Luftmassenverschiebungen und ihr
Einfluss auf die Langenänderungen der Erdachse (Breitenschwan-
kungen). *Petermanns Mitteilungen, Ergänzungsband*, **29**, 137.

STARR, V. (1948). An essay on the general circulation of the Earth's
atmosphere. *Journal of Meteorology*, **5**, 39.

STARR, V. and PEIXOTO, J. (1958). On the global balance of water vapor
and the hydrology of deserts. *Tellus*, **10**, 188.

STARR, V., PEIXOTO, J., and LIVADAS, G. (1958). On the meridional transport of water vapor in the northern hemisphere. *Geofisica Pura e
Applicata*, **39**, 174.

STEHLI, F. (1957). Possible Permian climatic zonation and its implications. *American Journal of Science*, **255**, 607.

STEVENSON, R., TIBBY, R., and GORSLINE, D. (1956). The oceanography of Santa Monica Bay, California. Report to Hyperion Engineers, Inc. by Geology Department, University of California.

STOMMEL, H. (1957). A survey of ocean current theory. *Deep-Sea Research*, **4**, 149.

STONELEY, R. (1924). The shrinkage of the Earth's crust through diminishing rotation. *Monthly Notices, Royal Astronomical Society, Geophysical Supplement*, **1**, 149.

STOREY, J., FENTON, K., and MCCRACKEN, K. (1958). Effective magnetic meridian for cosmic rays. *Nature*, **181**, 34.

STOYKO, N. (1936). Sur l'irrégularité de la rotation de la terre. *Comptes Rendus des Séances de l'Académie des Sciences*, **203**, 29.

STOYKO, N. (1937). Sur la périodicité dans l'irrégularité de la rotation de la terre. *Comptes Rendus des Séances de l'Académie des Sciences*, **205**, 79.

STOYKO, N. (1950). Sur la variation saisonnière de la rotation de la terre. *Comptes Rendus des Séances de l'Académie des Sciences*, **230**, 514.

STOYKO, N. (1951). La variation de la vitesse de rotation de la terre. *Bulletin Astronomique*, **15**, 16.

STREET, R. (1917). Dissipation of energy in tides. *Proceedings of the Royal Society, A*, **93**, 349.

STRUVE, O. (1952). The variation of latitude. *Sky and Telescope*, **11**, 109, 142.

SVERDRUP, H., JOHNSON, M., and FLEMING, R. (1946). *The Oceans*. New York: Prentice-Hall.

SWALLOW, J. (1955). A neutral-buoyancy float for measuring deep currents. *Deep-Sea Research*, **3**, 74.

TAKEUCHI, H. (1950). On the Earth tide of the compressible Earth of variable density and elasticity. *Transactions of the American Geophysical Union*, **31**, 651.

TAKEUCHI, H. (1951). On the Earth tide. *Journal of the Faculty of Science, University of Tokyo*, Sect. II, **7**.

TAYLOR, G. I. (1919). Tidal friction in the Irish Sea. *Philosophical Transactions of the Royal Society, A*, **220**, 1.

TAYLOR, G. I. (1929). Waves and tides in the atmosphere. *Proceedings of the Royal Society, A*, **126**, 169.

THELLIER, E. (1951). Sur la direction du champ magnétique terrestre, retrouvée sur des parois des fours des époques punic et romaine, à Carthage. *Comptes Rendus*, **233**, 1476.

312   THE ROTATION OF THE EARTH

THOMSON, SIR W. (1876). *Presidential Address, British Association.* Reprinted in Mathematical and Physical Papers, 3. Cambridge University Press.

THOMSON, SIR W. (1882). On the thermodynamic acceleration of the Earth's rotation. *Proc. Roy. Soc. Edinb.*, 11, 396. See also *Collected Works III*, 341.

THOMSON, SIR W. and TAIT, P. (1879). *Treatise on Natural Philosophy.* Cambridge, Vol. 1, part 1.

THOMSON, SIR W. and TAIT, P. (1883). *Treatise on Natural Philosophy.* Cambridge, Vol. 1, part 2.

THORARINSSON, S. (1940). Present glacier shrinkage, and eustatic changes of sea level. *Geog. Annaler*, 22, 131.

THORNTHWAITE, C. (1948). An approach toward a rational classification of climate. *The Geographical Review*, 38, 55.

TISSERAND, F. (1891). *Traité de Mécanique Céleste.* Vol. II. Paris: Gauthier Villars.

TOWNES, C. (1951). Atomic clocks and frequency stabilization. *Journal of Applied Physics*, 22, 1365.

UREY, H. (1952). *The Planets, Their Origin and Development.* New Haven: Yale University Press.

VAN DEN DUNGEN, F., COX, J., and VAN MIEGHEM, J. (1949). Sur les fluctuations de périod annuelle de la rotation de la terre. *Bulletin de l'Académie Royale de Belgique, Classe des Sciences*, 35, 642.

VAN DEN DUNGEN, F., COX, J. and VAN MIEGHEM, J. (1950). Sur les fluctuations saisonnières de la rotation du globe terrestre. *Bulletin de l'Académie Royale de Belgique, Classe des Sciences*, 36, 388.

VERNE, J. (1889). *Les Voyages Extraordinaires; Sans Desus Dessous.* Paris: J. Hetzel et Cⁱᵉ. (We understand an English translation entitled *Topsy-Turvy* was published by Ogilvie and Company, New York, in 1890).

VERONIS, G. and STOMMEL, H. (1956). The action of variable wind stresses on a stratified ocean. *Journal of Marine Research*, 15, 43.

VESTINE, E. (1952). On variations of the geomagnetic field, fluid motions, and the rate of the Earth's rotation. *Proceedings of the National Academy of Sciences*, 38, 1030.

VESTINE, E. (1953). On variations of the geomagnetic field, fluid motions, and the rate of the Earth's rotation. *Journal of Geophysical Research*, 58, 127.

VESTINE, E. (1954). Discussion in symposium on the Earth's core. *Transactions of the American Geophysical Union*, 35, 63.

VESTINE, E., LAPORTE, L., LANGE, I., COOPER, C., and HENDRIX, W. (1947). Description of the Earth's main magnetic field and its secular change, 1905-1945. Washington, D.C. *Carnegie Inst. Pub. 785.*

VOLTERRA, V. (1895). Sulla teoria dei movimenti del polo-terrestre. *Astron. Nachr.*, **138**, 33.

VON OPPOLZER, T. (1886). *Traité de la determination des orbites des comètes et des planètes.* Vol. I. Paris: Gauthier Villars.

WALKER, A. and YOUNG, A. (1955). The analysis of the observations of the variation of latitude. *Monthly Notices, Royal Astronomical Society,* **115**, 443.

WALKER, A. and YOUNG, A. (1957). Further results on the analysis of the variation of latitude. *Monthly Notices, Royal Astronomical Society,* **117**, 119.

WANACH, B. (1919). Die Chandlersche und die Newcombsche Periode der Polbewegung. *Zentralbureau der Internationalen Erdmessung,* **34**, 23.

WEGENER, A. (1912). Die Entstehung der Kontinente und Ozeane. *Petermanns Mitteilungen, Erganzungsband* **58**, 185, 253, 305.

WEGENER, A. (1924). *The Origin of Continents and Oceans.* New York: E. P. Dutton & Co.

WEXLER, H. (1958). Some aspects of Antarctic geophysics. *Tellus,* **10**, 76.

WILKES, M. (1949). *Oscillations of the Earth's Atmosphere.* Cambridge University Press.

WOOLARD, E. (1953). Theory of the rotation of the earth around its center of mass. *Astr. Pap., Amer. Ephem. and Nau. Alm.,* **15**, part I.

WOOLARD, E. (1959). Inequalities in mean solar time from tidal variations in the rotation of the earth. *Astronomical Journal,* **64**, 140.

WORZEL, J. and SHURBET, G. (1955). Gravity interpretations from standard oceanic and continental section. *Geological Society of America Spec. Paper* 62, 87.

YOUNG, A. (1953). The effect of the movement of surface masses on the rotation of the earth. *Monthly Notices, Royal Astronomical Society, Geophysical Supplement,* **6**, 482.

ZACHARIAS, J., YATES, J. and HAUN, R. (1955). An atomic frequency standard. *Proceedings of the Institute of Radio Engineers,* **43**, 364.

ZEUNER, F. (1946). *Dating the past; an introduction to geochronology.* London: Methuen.

INDEX OF AUTHORS AND SUBJECTS